Tribocatalysis, Tribochemistry, and Tribocorrosion

Tribocatalysis, Tribochemistry, and Tribocorrosion

edited by

Czesław Kajdas
Ken'ichi Hiratsuka

PAN STANFORD PUBLISHING

Published by

Pan Stanford Publishing Pte. Ltd.
Penthouse Level, Suntec Tower 3
8 Temasek Boulevard
Singapore 038988

Email: editorial@panstanford.com
Web: www.panstanford.com

British Library Cataloguing-in-Publication Data
A catalogue record for this book is available from the British Library.

Tribocatalysis, Tribochemistry, and Tribocorrosion

ISBN 978-981-4316-95-8 (Hardcover)
ISBN 978-1-315-36498-8 (eBook)

Contents

Preface

This 3T book project started when Pan Stanford Publishing invited Prof. Czesław Kajdas to write or edit a book on tribocorrosion. Prof. Kajdas then asked me to be a co-editor and to extend the book's theme to tribochemistry and tribocatalysis, as we had been collaborating in research on these subjects. Our plan for the book was to demonstrate tribochemistry, tribocatalysis, and tribocorrosion from the same perspective, i.e., the promotion of electron behavior through tribological processes. The emission of low-energy electrons from friction surfaces is a trigger for subsequent chemical reactions. They make friction surfaces protective or deteriorative, which results in lubrication or wear. With electron emission, the negative-ion-radical action mechanism (NIRAM) approach takes a key role in tribochemistry. Although the NIRAM approach has not been fully accepted, a number of tribochemical phenomena can be explained by applying this approach.

This book consists of three main chapters. After the introduction, tribochemistry is discussed. Tribochemistry is a general term dealing with chemical reactions taking place at a sliding interface. In this chapter, tribochemical phenomena in the lubrication of hard disk drive as well as fundamental mechanisms are described. Then it is followed by the physical bases of tribochemistry: exo-electron emission and charge/discharge in oil flow. This chapter ends with the materials science aspects of tribochemical reactions/wear of silicon nitride.

The next chapter on tribocatalysis introduces its definition, mechanisms, and examples. The central theme is the activation energy, as it characterizes catalytic reactions. The combination of three models—NIRAM, α_i, and thermodynamics of irreversible processes—illustrates how the activation energy is decreased in tribochemical and tribocatalytic reactions.

The last main chapter is devoted to tribocorrosion. At this point, tribochemistry is closely related to electrochemistry. The relationship leads to triboelectrochemistry, which is one of the central issues of chemical-mechanical polishing. The case study of planarization of tantalum is reviewed.

The topics in this book are rather specific and may appear to be diverse from chapter to chapter. However, the underlying mechanism, which is low-energy electron emission/transfer at the interface to initiate reactions, is the same. This mechanism is repeatedly developed in all the chapters.

I am profoundly grateful to all the authors for their hard work and intensive contribution to this book. Their insightful chapters will benefit the readers who seek an understanding of complicated tribological problems. My appreciation also goes to the authors' families for helping them accomplish the manuscripts, especially to the lead editor Prof. Kajdas' wife, Maria; their daughter, Hania; together with her family for constant support to Prof. Kajdas and me. I am indebted to Pan Stanford Publishing, in particular, Mr Arvind Kanswal, Ms Jenny Rompas and Mr Stanford Chong (Publishers), for their guidance and patience throughout this project.

To conclude, I express my sincere gratitude and deepest respect to Prof. Czesław Kajdas. Without his enthusiasm and dedication, this book would never have been realized. I am confident that all the authors share the same thoughts. In May last year, Czesław unexpectedly left us. This book is dedicated to him.

Ken'ichi Hiratsuka
March 2018

Chapter 1

Introduction

Czesław Kajdas[a] **and Ken'ichi Hiratsuka**[b]

[a]*Automotive Industry Institute in Warsaw, Poland*
[b]*Department of Innovative Mechanical and Electronic Engineering, Chiba Institute of Technology, Japan*

khiratsuka@it-chiba.ac.jp

1.1 General Information and Aim of the Book

Tribology as the science and technology of interacting mating elements in relative motion is well presented in many books. The present book aims at describing the recent trends in tribology concerning scientifically sophisticated topics, particularly tribocatalysis, selected fragments of tribochemistry and tribocorrosion. Both tribocorrosion and tribochemistry are very complex processes mostly related to specific wear-accelerating or wear-reducing phenomena of interacting surfaces being in relative motion. They clearly concern the effect of contact pressure and chemical reactions of the interacting solids with the environment, which encompasses either a corrosive medium or a wear-reducing lubricant.

The book has been written according to the level of senior and graduate students and is intended for everyone who wants to solve

Tribocatalysis, Tribochemistry, and Tribocorrosion
Edited by Czesław Kajdas and Ken'ichi Hiratsuka

ISBN 978-981-4316-95-8 (Hardcover), 978-1-315-36498-8 (eBook)
www.panstanford.com

special problems and get a better understanding of new findings and/or tribology-related discoveries. It also serves as a highly specialized text book.

Tribophysics encompasses a wide range of phenomena such as tribocharging/triboelectrification, triboemission, UV-triboemission, acoustic emission in friction, flash temperature, and triboplasma originating from magma-plasma model presented by P. A. Thiessen over 50 years ago. Tribomechanics concerns processes under the influence of mechanical energy on solids and relates to mechanical activation of solids. Tribochemistry interconnects tribocatalysis and tribocorrosion. Typical phenomena in each field of tribology are shown in Fig. 1.1. This chapter presents a brief description of typical physical tribological phenomena and their significance for tribochemistry and mechanochemistry.

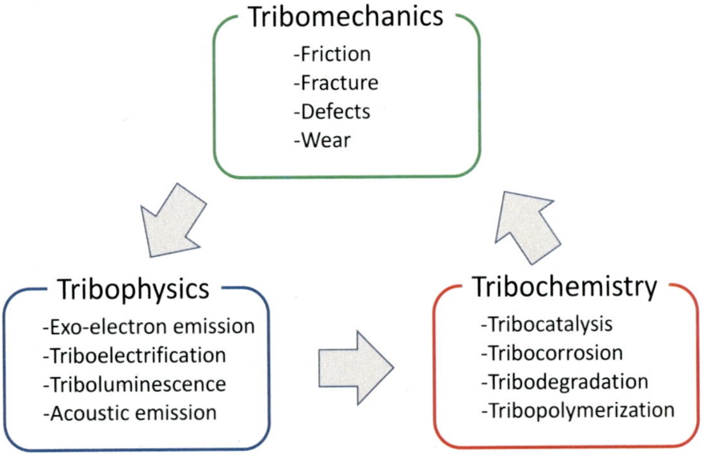

Figure 1.1 Tribomechanical, tribophysical, and tribochemical interactions.

1.2 Mechanochemistry and Tribochemistry

1.2.1 General Information Concerning Mechanochemistry

Detailed description of tribochemistry is in Chapter 2 of this book. By and large, we consider tribochemistry as a subfield of mechanochemistry. The same is due to mechanocatalysis/

mechanosynthesis and tribocatalysis. A common denominator for all these specific, mechano-activated processes is the fact that they are governed by lower activation energies than those regular thermochemical or catalytic processes.

Accordingly, mechanochemical reactions are clearly distinct from those of thermochemical ones. Processes initiated by mechanical action are also widely known as tribochemical reactions. The book [1] provides a good evidence for that.

Research on very specific mechanochemical reactions (mechanochemistry) is growing rapidly and, the number of published papers/books is regularly increasing. However, it was also mentioned that the universally agreed definition of the mechanochemistry term is not yet widely and explicitly used. A recent paper [2] provides thoughtful and authoritative discussion of the use of the term mechanochemistry. In particular, it demonstrates the distinction between the mechanical breaking of covalent bonds, where mechanical energy is used for bond breaking as the first step in chemical transformations.

Recently, a good historical review on the development of mechanochemistry, with the emphasis on the mechanochemistry of solids was described in [3]. Another paper reviewing this topic considers the mechanical activation of covalent bonds and describes/discusses 399 references [4]. It demonstrates that today's chemist is used to consider the activation of chemical reactions in terms of thermochemistry, electrochemistry, and photochemistry, which reflects the standard physical chemistry textbooks. The next way of chemical activation mechanochemistry/tribochemistry, is not widely known.

Mechanochemistry is a coupling of specific mechanical and chemical phenomena on a molecular scale leading to initiation of heterogeneous chemical reactions due to specific mechanical bond breakage of the molecules. Actually, the activation energies of such reactions are lower than those of regular thermochemical processes. Thus these reaction types might also be considered catalyzed ones. In other words, mechanical interaction of a solid-solid-chemical environment system generates a "*catalyst.*" At this point, catalyst can be defined as a powerful substance/agent that enables a chemical reaction to be initiated by lower activation energy and/or proceed at a faster rate than thermochemical reactions.

1.2.2 Brief Historical Background of Mechanochemistry

The first document of a mechanochemical reaction concerns a book by Theophrastus of Ephesus (371–286 B.C.), a student of Aristotle, *De Lapidibus* (*On Stones*). Most probably the first considered mechanochemical reaction was the cinnabar reduction process. Since the cinnabar was rubbed in a brass mortar with a brass pestle in the presence of vinegar, leading to metallic mercury generation, the reaction

$$HgS + Cu \rightarrow Hg + CuS$$

can also be considered a first tribochemical process clearly demonstrating a common point between mechanochemistry and tribochemistry. Action mechanism of this reaction is not known yet. However, in Chapter 2, we present a detailed reaction approach, allowing one to comprehend how this specific reaction process might proceed.

The introduction of the term mechanochemistry usually is attributed to Ostwald. The authors of work [4] note that while in Ostwald's textbook on general chemistry, mechanical work is discussed in the context of the mechanical heat equivalent discovered by Mayer and Joule, they were unable to locate the alleged section on mechanochemistry in the three book editions [5–7]. Furthermore, it has been emphasized that the origin of the term mechanochemistry remains obscure [4]. Presently both mechanochemistry and tribochemistry are well established fields in materials science and solid-state chemistry [8, 9]. According to [9] mechanochemistry is a field that deals with athermal or ultra-fast chemical reactions between solids or between solids and surrounding gaseous, or liquid molecules under mechanical forces. The subordinate subject tribochemistry is one specific domain of mechanochemistry that concerns with chemical reactions between tribomaterial surfaces and gaseous/liquid lubricant molecules under shearing action and thermal influences [2].

1.2.3 Initiation of Mechanochemical/Tribochemical Reactions

Recently, tribochemical principles of in situ reconditioning of rubbing metal surfaces have been presented in detail [10]. It was emphasized that during frictional processes, generation and dissipation of mechanical energy make the tribopairs' contact surfaces subjected to very high flash temperatures and high local stresses and strains. The atoms of the tribomaterials and the third body components on and in the vicinity of contact points become highly excited to trigger complex physical and chemical reactions [11].

Mechanical forces produce strain fields in solids, which make the atoms leave their equilibrium positions due to lattice vibration, alter bond lengths and angles of their atomic arrangements leading to electronically excited states. Generally, the energy field incurred by the tribophysical effects, for example, the exo-electron emission (EEE) process, triboluminescence, triboplasma, crystal defects, etc., trigger specific tribochemical (mechanochemical) reactions.

All types of chemical reactions/processes are accompanied by a change in energy. Some of them release energy to their surroundings, mostly in the form of heat and thus are called exothermic. Conversely, some reactions need to consume heat from their surroundings to proceed.

Figure 1.2 demonstrates an approach to the triggering process of mechanochemical/tribochemical reactions of the water molecule and various lubricant molecules. We assume, however, that electronically initiated reaction by both thermionic electrons and exoelectrons are also of importance, because both electron types are low-energy species. The detail will be discussed in Section 1.3.

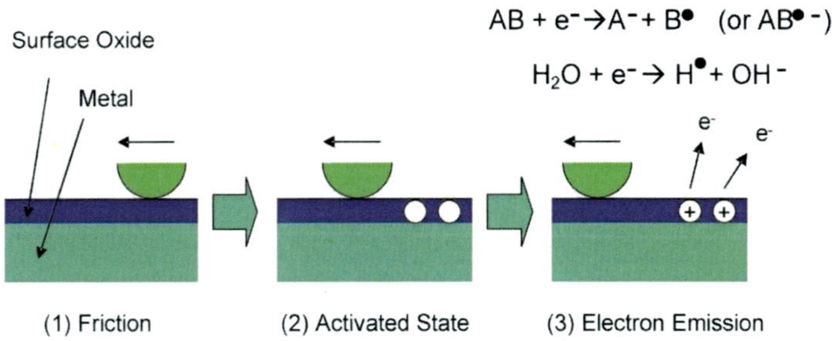

Figure 1.2 An initiation process of tribochemical reactions activated by the electron emission associated with friction.

Works [4, 12] present and discuss a kinetic theory of the breakdown of solids, which is based on an experimentally established relationship between the life time of the material, the tensile stress acting on the material, and temperature. The modified Arrhenius equation accounts for the lowering of the activation energy by mechanical work

$$K = K_0 e^{\left(-\frac{E_a - \alpha\sigma}{RT}\right)}, \qquad (1.1)$$

where σ is the tensile stress measured in Nm^{-2}, which gives the coefficient α the dimension of m^3mol^{-1}, and their product is the mechanical work lowering the activation energy E_a. K_0 is the Arrhenius frequency factor, and K is the rate of bond-rupture events occurring in the polymer solid. More detailed information on this equation is described in paper [4].

It is of note at this point that most recently a general theoretical model of E_a lowered by mechanical action was proposed [13]; it is so-called α_ι model. Another theoretical work based on thermodynamics of irreversible processes (TIP) evidences α_ι model and provides its better understanding [14]. Detailed information is included in Section 3.3 on "Theoretical advances in kinetics of tribochemistry based on TIP." Paper [15] provides additional confirmation of α_ι model. Detailed description of the model is in the book chapter on activation energy (E_a).

Work [16] points out that measuring the action of energy on matter is a complex problem, e.g., in the case of granular materials. It is due to the fact that particle size reduction by grinding generally demonstrates poor overall energetic yields and a significant challenge in this area is to accurately estimate the energy consumed, including that stored in the particles. Additionally, it demonstrates that fine/ultra-fine milling/grinding processes require a lot of energy, a part of which is stored, i.e., becomes internal energy and can initiate mechanochemical reactions. This type of internal energy might relate to the lowering process of activation energy. Interestingly, the studied appearance of free radicals during the grinding of α-lactose monohydrate using ESR technique, similar radicals as those induced by γ-radiation had been found [16]. It provides with the well convincing novel concept of mechanical energy action on matter in fine grinding by using molecular probes and somehow confirms α_ι model [13, 15].

1.3 Flash Temperature

Flash temperature is very short and holding in esteem the mechanically initiated chemistry and thermally triggered chemistry at the contact of asperities caused by the flash temperature effect, the following question can be asked. Might this reaction initiation process be considered in terms of overcoming the activation energy by heat or by another form of energy? The common denominator of these reactions is that they might be triggered by low-energy electrons. This hypothesis is also relevant to the next authors' hypothesis saying that the intermediate reactive species of both tribochemical/mechanochemical reactions and some heterogeneous catalytic/tribocatalytic and/or mechanocatalytic reactions are produced by the same mechanism.

Thermionic emission is the flow of electrons from a metal or metal oxide surface, caused by thermal vibrational energy overcoming the electrostatic forces holding electrons to the surface. The effect increases significantly at higher temperatures (over 300°C) but is always present at much lower temperatures. The science dealing with this phenomenon is thermionics and the charged particles are called thermions (thermal electrons). Both thermions and exo-electrons are low-energy species.

In any metal, there are electrons that are free to move from atom to atom. This is sometimes referred to as a "*sea of electrons*." Their velocities follow a statistical distribution, rather than being uniform, and occasionally an electron will have enough velocity to exit the metal without being pulled back in. The minimum amount of energy needed for an electron to leave the surface is the work function. It varies from metal to metal. A thin oxide coating is often applied to metal surfaces in vacuum tubes to give a lower work function, as it is easier for electrons to leave the surface of the oxide.

Richardson's law, also called the Richardson–Dushman equation, states that the emitted current density J (A/m^2) is related to temperature T by the following equation:

$$J = AT^2 e^{\frac{-W}{kT}}, \tag{1.2}$$

where T is the metal temperature in Kelvin. W is the work function of the metal, k is the Boltzmann constant. The proportionality constant A, known as Richardson's constant, is given by

$$A = \frac{4\pi mk^2 e}{h^3} = 1.20173 \times 10^6 \left[\frac{A}{m^2K^2}\right], \qquad (1.3)$$

where m and e are the mass and charge of an electron, and h is Plank's constant. Because of the exponential function, the current increases rapidly with temperature.

It is of note at this point that the Richardson–Dushman equation must be corrected for the Schottky effect; the current emitted from the metal cathode into the vacuum depends on the metal's thermionic work function, and that this function is lowered from its normal value by the presence of image forces and by the electric field at this cathode. This enhancement is given by the field-enhanced thermionic emission (FEE) equation:

$$J(E_s, T, W) = AT^2 e^{\frac{-(W - \Delta W)}{kT}}$$

$$\Delta W = \left[\frac{eE_c}{(4\pi\varepsilon_0)}\right]^{1/2} \qquad (1.4)$$

where E_c is the electric field strength at the cathode spot, ε_0 is the vacuum permittivity.

This equation is relatively accurate for electric field strengths lower than about 108 $\mathrm{V\,m^{-1}}$. For electric field strengths higher than 108 $\mathrm{V\,m^{-1}}$ the use of the Murphy and Good equation for thermo-field (T–F) emission is more appropriate.

Physically, temperature is a measure related to the average kinetic energy of a substance. To be more precise, it is a measure of the random agitation of matter under the effect of thermal fluctuations. It is a fundamental parameter in thermodynamics and it is conjugate to entropy. More quantitatively, the order of magnitude of the fluctuations of the energy associated with an atom, molecule, or another elementary constituent of a physical system is $k_B T$, where k_B is the Boltzmann constant and T is temperature expressed in Kelvin. Temperature in respect to matter is a property only of macroscopic amounts and serves to gauge the average intensity of the random actual motions of the individually mobile particulate constituents. Temperature plays an important role in determining the rate and extent to which chemical and/or mechanochemical/tribochemical reactions occur. Temperature also controls the type and quantity of thermal radiation emitted from a surface.

Flash temperature expressed as the maximum computed friction temperature, can also be expressed in the form of thermionic emission [17]. Accordingly, the thermionic emission is assumed to bridge "*hot spots*" with rubbing surface spots activated by triboemission of electrons [18]. Figure 1.3 demonstrates a thermionic spike of the emitted current density J (A/m^2). Therefore, flash temperatures can also be expressed in the form of electronic energy.

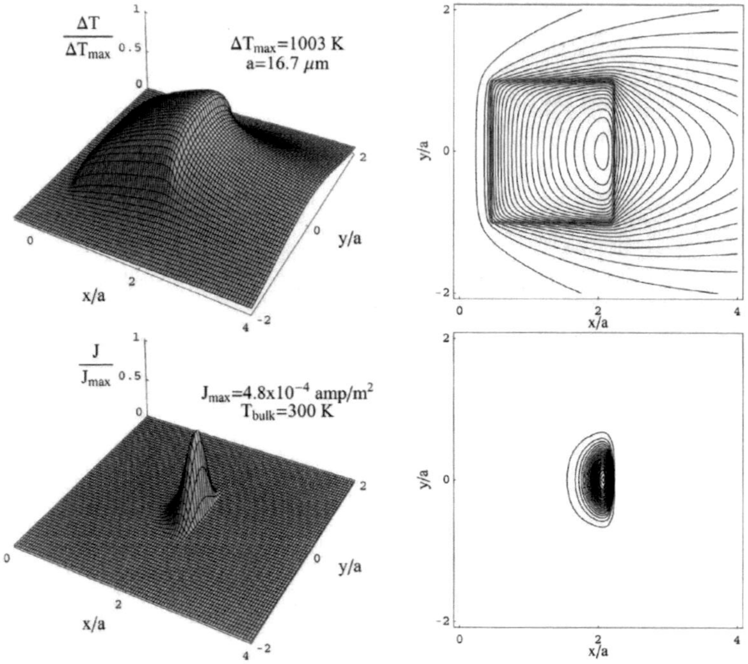

Figure 1.3 Spike of thermionic emission J (A/m^2). *Tribol. Lett.*, An examination of thermionic emission due to frictionally generated temperatures, 13, 2002, 147–153, B. Vick, M. J. Furey, C. Kajdas, with permission of Springer.

1.4 Milling/Grinding as an Example of Typical Mechanochemical Processes

The name mechanochemistry was applied to the field of reactions caused by mechanical energy [19]. A narrower field, tribochemistry, is used for reactions initiated by friction, e.g., during the milling of solid reagents [8, 12]. The grinding of two solid substances produces a complex series of transformations, the mechanical energy braking the order of the crystalline structure, generating cracks, and new surfaces. At the point of collision of the edges, the solids deform and even melt, forming hot points where the molecules can reach very high vibrational excitation leading to bond breaking. Such stochastic processes appear in a period of 10^{-7} s. In which thermal equilibrium does not exist [8]; this period, noted as the plasma phase, is followed by a post-plasma period of 10^{-6} s or more in which relaxation processes dissipate the energy reaching the Maxwell–Boltzmann distribution.

Plasma is a mixture of electrons, negatively and positively charged particles along with neutral atoms and molecules. Thus, it is similar to as in which a significant portion of the particles are ionized. Often it is called the fourth state of materials. The basic premise is that heating a gas dissociates its molecular bonds, rendering it into its constituent atoms. Then, further heating leads to a loss of electrons (ionization), turning it into a plasma: containing charged particles and electrons. Considering the plasma generation during milling processes, it is possible to understand specific mechanochemical reactions.

The above-mentioned post-plasmatic reactions are responsible for many of the products generated. A wide range of processes take place on milling/grinding or solids, e.g., they include [19] (i) comminution of the particles to a very small size, (ii) formation of dislocations and point defects in the crystalline structure, (iii) generation of large new surfaces and phase transformations in polymorphic materials, and (iv) activation of different chemical reactions, etc. Milling/grinding can be carried out in a variety of ways; however, the simplest is historic laboratory mortar and pestle used by Carry-Lea over 120 years ago [20] who is commonly recognized as a pioneer of mechanochemistry/tribochemistry owing to his works on the silver halides. Notably, he clearly differentiated the discovered reactions from thermally activated processes.

Ball mills are used when higher energy is required and when the milling time involves many hours. Very high energy vibrators are used for prolonged high energy milling as in mechanical alloying (MA). Milling and MA continues to be a fascinating field of investigation for over five decades. MA is a powder processing technique that allows production of homogeneous material starting from blended elemental powder mixture. The technique originally developed for the production of oxide dispersion strengthened nickel-base superalloys has now branched out to the synthesis of equilibrium and non-equilibrium alloy phases, including solid solutions, intermetallics, quasicrystals, amorphous alloys, and bulk metallic glasses [21]. Book [21] surveys the vast field of MA and the related techniques from a scientific and technological point of view.

The MA process was developed in 1970 [22] as a new technology approach to make new powder materials or new alloys that cannot be produced by regular melting, casting, and solidification processes. It is the process of alloying in the solid state by mechanical treatment of the materials in a ball mill; initially aimed at producing superalloys dispersion-strengthened with oxides. Presently, MA along with mechanochemical synthesis is of particular importance due to the unique microstructures and properties of materials made by this technique. For example, it is powerful mode for the production of light metal systems (Al, Mg, Ti), which are used in aerospace and marine industries [21].

In MA processes there are two temperature effects considered. The first one is the local temperature pulse due to ball collisions. These local temperature pulses have a short duration, about 10^{-5} s, approximately the same as the collision time between balls [22]. Since it is very difficult to experimentally measure such temperature pulses, they can only be estimated theoretically based on appropriate models. Such models should encompass the extent of plastic deformation the powder particles undergo and the efficiency of heat transfer [23]. These temperatures are estimated to be a few hundred degrees Celsius.

Fine and ultrafine milling activation is an effective procedure where an improvement in technological processes can be attained via a combination of several effects which influence the properties of applied solids [24]. Mechanochemistry applying this type of milling demonstrates the differences in comparison with traditional technological procedures. The environmental aspects of these processes are particularly attractive. The combination of mechanosynthesis and spark plasma sintering has been recently used for the first time to process dense fine-grained ceramics of the $(1-x)NaNbO_{3-x}SrTiO_3$ system [25].

High-speed vibration milling (HSVM) method can also be applied for mechanochemica/tribochemical reactions of fullerenes. It was found that by the use of a HSVM the complexation of fullerenes and γ-cyclodextrin proceeds most effectively [26]. Therefore, when the regular fullerene (C_{60}) was intensively shaken with 4 molar equivalents of γ-cyclodextrin for only 600 s by HSVM, a magenta-colored solution of C_{60} was obtained by filtration with a membrane filter. The concentration of C_{60} was 1.4×10^{-3} mol L^{-1}, which is the highest value for C_{60} dissolved in water; after this aqueous solution was let stand for two weeks, the C_{60}-γ-cyclodextrin 1:2 complex was obtained as purple crystals [27].

Other mechanochemical/tribochemical reactions of that type include C_{60} dimerization and trimerization along with cycloaddition processes. Figure 1.4 demonstrates a possible reaction mechanism for the C_{60} dimerization in the solid state and Fig. 1.5 shows the C_{60} equilibrated reaction with anthracene [26].

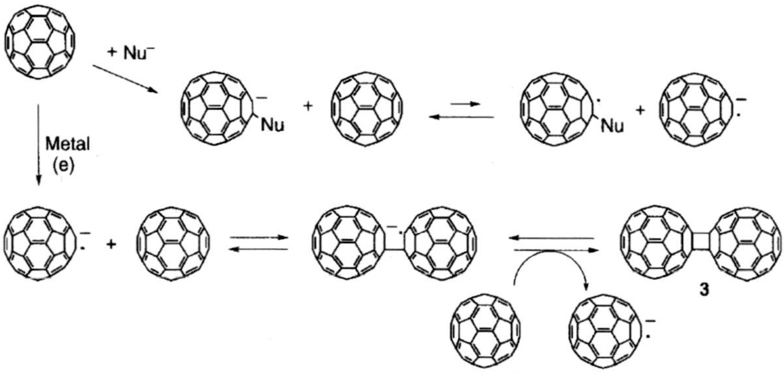

Figure 1.4 Possible mechanism for the dimerization of C_{60} in the solid state. *Organic Solid State Reactions,* Topics in Current Chemistry (ed. F. Toda), The mechanochemical solid-solid state reaction of fullerenes, 254, 2005, 185–206, K. Komatsu, with permission of Springer.

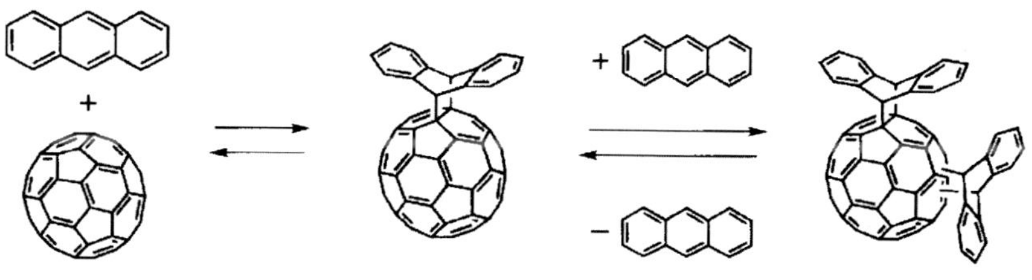

Figure 1.5 Equilibrated reaction of C_{60} with anthracene. *Organic Solid State Reactions, Topics in Current Chemistry* (ed. F. Toda), The mechanochemical solid-solid state reaction of fullerenes, 254, 2005, 185–206, K. Komatsu, with permission of Springer.

High-speed vibration milling method can also be applied for mechanochemica/tribochemical reactions of fullerenes. It was found that by the use of a HSVM the complexation of fullerenes and γ-cyclodextrin proceeds effectively, shown in Figs. 1.4 and 1.5.

Most recent paper [28] demonstrates a standard procedure for TG (thermogravimetric analysis) of carbonaceous materials including single-walled carbon nanotubes (SWCN). A good example for SWCN application is storage of hydrogen.

1.5 Mechanocatalysis and Tribocatalysis

According to [8], mechanocatalysis and/or tribocatalysis is the subsection of tribochemistry dealing with the change of catalytic properties of solids due to the effect of mechanical energy. Most recent works concern mechanocatalysis [29] and application of mechanochemistry in waste management [30]. It is accepted that heterogeneous catalysis cannot be easily applied to cellulose. On the other hand, by mechanically grinding the adequate catalyst and reactant material, it is possible to induce solid-solid mechanocatalysis. For instance, applying the best catalysts that possess high surface acidities and layered structures, up to 84% of the available cellulose can be converted to water soluble compounds in a single pass [29]. Mechanocatalytic processing of materials provides significant advantages in comparison with currently used technologies. Last but not least the mechanochemical technology is also important for waste management. Mechanochemical processes need only simple apparatus. They are fast, economic and environmentally friendly. Review paper [30] summarizes the present state of the art.

First regular basic research on friction/tribocatalysis was started as early as 1990 by Hiratsuka et al. [31–33]. The important starting point was the fact that gases were consumed when two materials were slid together [31]. This research has been focused on clarifying the common and different points of tribocatalytic reactions as compared to thermal, photo- and electrocatalytic reactions. The research was started with a simple sliding system where the catalyst was slid against itself [31, 32]. Soon it was hypothesized that low wear would lead to high reaction rate on the surface if friction enhances the surface diffusion of adsorbed species [32]. For example, it appeared clear that the wear of palladium against aluminum oxide is far less than that between palladium and itself. It was found that the synthesis of carbon dioxide from carbon monoxide and oxygen proceeds more easily from the former than from the latter. Thus, the hypothesis has been proved and this research demonstrated that catalytic reactions are greatly increased when catalysts are in sliding contact.

Further research aimed at finding the role of the aluminum oxide catalyst support on catalytic/tribocatalytic oxidation of carbon monoxide and other reaction [33, 34].

A good example of mechanocatalysis relates to hydrogen generation from water. Production of hydrogen from water is an attractive way to establish a clean energy system of the future [35]. The hydrogen reaction generation was accomplished when some simple oxide powders such as NiO, Co_3O_4, Cu_2O, and Fe_3O_4 were suspended in distilled water and rubbed against the bottom wall of the reaction vessel by magnetic stirring. The reaction proceeded steadily without any supply of electric or photon energy as long as the suspension was stirred by a rotating stirring rod. This implies that the mechanical energy given by the rotating stirring rod is essential for the reaction to take place. The reaction rate depends strongly on the materials of both the catalyst and the bottom of the vessel. Typical efficiency of the mechanical-to-chemical energy conversion was reported to be 4.3% (for NiO). A more recent paper [36] focuses on water splitting on tungsten-based photocatalysts.

1.6 Tribocorrosion

Tribocorrosion is the surface damage due to the combined effect of a chemical reaction which weakens the surface material and a mechanical action which consequently removes it. The removal of surface material causes the production of active fresh surface, which would be subjected to subsequent chemical reactions with the environment. The synergy effect of mechanical and chemical interactions is also found in adhesive wear processes where oxygen adsorption influences wear. Compared with this gas atmosphere, tribocorrosion is characterized by its liquid environment. Especially when the rubbing surface is exposed to a corrosive media such as acid, the surface material is transformed into a weak compound. In that case, wear depth is much more increased as compared to that without any corrosive media, resulting in enhanced wear.

The interaction of metal with the environment, on the other hand, induces the production of surface protective film such as oxide. In that case, wear is significantly reduced. How can we understand the mechanisms of the two environmental effects which are in contradiction to each other?

Metal surface can be either protective or self-destructive depending on the type of interaction with the environmental molecules. In general, while some surface oxide films on metal protect the inside from wear, some oxides induce the destruction of the soft metal underneath. On the other hand, surface hydroxide, which is generally soft, does not protect the metal, leading to tribocorrosion. The same is true for the cases of chloride and fluoride. The mechanism of these detrimental effects is due to the successive formation of mechanically weak film which can be easily worn out by the shear stress.

The book by E. McCafferty, *Introduction to Corrosion Science*, Springer (2010), [37] should be considered one of the referenced books. The same is due to the book edited by D. Landolt and S. Mischler [38]. The latter book reviews recent research on the tribocorrosion of passive metals and coating, with particular reference to the design of tribocorrosion test equipment, data evaluation, and the optimization of materials' properties for tribocorrosion systems. It also includes a comprehensive selection of tribocorrosion problems in engineering and medicine, such as the tribocorrosion of medical implants, and tribocorrosion issues in nuclear power plants, marine environments, automotive cooling circuits, and elevated-temperature metal working [38]. Part one of that book reviews basic notions of corrosion and tribology and discusses tribocorrosion mechanisms under fretting, sliding, and erosion conditions. Its second part focuses on methods for measuring and preventing tribocorrosion. It also discusses tribocorrosion mechanisms under fretting. The third part demonstrates a selection of tribocorrosion problems in engineering and medicine and also concerns chemical-mechanical polishing. Thus, it also makes a good bridge between Sections 4.1 and 4.2 of this book.

Since the most recent book references [37, 38] include all the major scientific and practical information concerning the generally accepted tribocorrosion knowledge, it is not necessary to include this basic data in the present book's Section 4.1. Our approach to tribocorrosion is based on what is known and unknown in tribocorrosion and tribochemistry.

1.7 Concluding Remarks

As presented in the section on tribochemistry (Section 2.1), tribochemistry is a subset of mechanochemistry. Although both terms are well known and their specific reaction effects, particularly mechanochemical ones [39, 40], have been widely discussed for over one century, but even in the past decade, it has been stated that the precise nature of these reactions is still not well understood [41].

The most recent detailed review work [42] presents the main concepts and features of mechanochemistry and tribochemistry disciplines. Both relate to chemistry induced by mechanical action combined with chemical phenomena leading to initiation of heterogeneous chemical reactions. Since there is a kind of mechanochemistry/tribochemistry overlap, it reviews adequately selected issues of such reactions. It has been proposed to consider the overlap of these disciplines only in terms of inorganic and organic chemistry. Applying these terms, *mechanochemistry is the inorganic mechanochemistry* and *tribochemistry is the organic mechanochemistry.*

References

1. Z. V. Todres, *On Organic Mechanochemistry and Its Practical Applications*, CRS, Taylor and Francis Group, LLC Boca Raton-London-New York (2006).
2. G. Kaupp, Mechanochemistry: The varied applications of mechanical bond-breaking, *Cryst. Eng. Commun.*, 11 (2009) 388–403.
3. V. V. Boldyrev, K. J. Tkacova, Mechanochemistry of solids: Past, present, and prospects, *J. Mater. Synth. Process.*, 8(3–4) (2000) 121–132.
4. M. K. Beyer, H. Clausen-Schaumann. Mechanochemistry: The mechanochemical activation of covalent bonds, *Chem. Rev.*, 105(8) (2005) 2921–2948.
5. W. Ostwald, *Lehrbuch der Allgemeinen Chemie*, Board 1 Stoechiometrie. Engelmann: Leipzig, Germany (1885).
6. W. Ostwald, Lehrbuch der Allgemeinen Chemie, Board 2.1 Verwandtschaftslehre, 2nd ed.; Engelmann: Leipzig, Germany (1896) (1902).
7. W. Ostwald, W. Lehrbuch der Allgemeinen Chemie, Board 2.1 Chemische Energie, 2nd ed.; 3rd print; Engelmann: Leipzig, Germany (1911).
8. G. Heinicke, *Tribochemistry*, Academy-Verlag, (1984).
9. J. J. Gilman, Mechanochemistry, *Science*, 274(5284) (1996) 65.
10. Y. Jin, S. Li, in: *Superlubricity* (ed. A. Erdemir, J.-M. Martin), Elsevier, Amsterdam, Chapter 24 (2007).
11. S. Li, A. Erdemir, Y. Jin, Mechanistic modeling of tribo-induced chemical reactions at engineered tribological surfaces. In: *Proceedings of WTC2005, World Tribology Congress III*, September 12–16, Washington DC (2005).
12. V. V. Boldyrev, Mechanochemistry and mechanical activation of solids, *Solid State Ionics*, 63–65 (1993) 537–543.
13. C. Kajdas, A. Kulczycki, A new idea of the influence of solid materials on kinetics of chemical reactions, *Mater. Sci. Poland*, 26(3) (2008) 787–796.
14. R. Dante, C. Kajdas, A. Kulczycki, Theoretical advances in the kinetics of tribochemical reactions, *Reac. Kinet. Mech. Cat.*, 99(2010) 37–46.

15. C. Kajdas, A. Kulczycki, K. J. Kurzydlowski, G. J. Molina, Activation energy of tribochemical and heterogeneous catalytic reactions, *Mater. Sci. Poland*, 28(2) (2010) 523–533.

16. M. Baron, A. Chamayou, L. Marchioro, J. Raffi, Radicalar probes to measure the action of energy on granular materials, *Adv. Powder Technol.*, 16 (2005) 199–211.

17. B. Vick, M. J. Furey, C. Kajdas, An examination of thermionic emission due to frictionally generated temperatures, *Tribol. Lett.*, 13 (2002) 147–153.

18. C. Kajdas, On a negative-ion concept of EP action of organo-sulfur compounds, *ASLE Trans.*, 28 (1985) 21–30.

19. J. F. Fernandes-Bertran, Mechanochemistry: An overview, *Pure Appl. Chem.*, 71(4) (1999) 581–586.

20. L. Takacs, M. Carey Lea, The first mechanochemist, *J. Mater. Sci.*, 39 (2004) 4987–4993.

21. C. Suryanarayana, *Mechanical Alloying and Milling*, Marcel Dekker, New York (2004).

22. J. S. Benjamin, Dispersion strengthened superalloys by mechanical alloying, *Meter. Trans.*, 1 (1970) 2943–2951.

23. A. K. Bhattacharya, E. Arzt, Temperature rise during mechanical alloying, *Scrip. Metallurgica Mater.*, 27(6) (1992) 749–754.

24. P. Balaz, E. Dutkova, Fine milling in applied mechanochemistry, *Minerals Eng.* 22 (2009) 681–694.

25. T. Hungria, M. Alquero, A. Castro, Grain growth control in $NaNbO_3$–$SrTiO_3$ ceramics by mechanosynthesis and spark plasma sintering, *J. Am. Ceram. Soc.*, 99(7) (2007) 2122–2127.

26. K. Komatsu, The mechanochemical solid-solid state reaction of fullerenes, In: *Organic Solid State Reactions, Topics in Current Chemistry* (ed. F. Toda), 254, Springer (2005), pp. 185–206.

27. K. Komatsu, K. Fujiwara, Y. Murata, T. Braun: Aqueous solubilization of crystalline fullerenes by supramolecular complexation with γ-cyclodextrin and sulfocalixarene under mechanochemical high-speed vibration milling, *J. Chem. Soc. Perkin*, 1(20) (1999) 2963–2966.

28. A. Shahverdi, G. Soucy, Thermogravimetric analysis of single-walled carbon nanotubes synthesized by induction thermal plasma, *J. Thermal Anal. Calorimetry*, 110 (2012) 1079–1085.

29. S. M. Hick, C. Griebel, D. T. Restrepo, J. H. Truitt, C. Bylda, R. G. Blair, Mechanocatalysis for biomass-derived chemicals and fuels, *Green Chem.*, 12 (2010) 468–474.

30. X. Guo, D. Xiang, G. Duan, P. Mou, A review of mechanochemistry applications in waste management, *Waste Manag.*, 30 (2010) 4–10.

31. K. Hiratsuka, M. Kuzuya, T. Sasada, Friction catalysis in the synthesis reaction of H_2O during adhesive wear, *Proc. 33rd Japan Cong. Mater. Res.*, 33 (1990) 191–195.

32. K. Hiratsuka, M. Kuzuya, T. Sasada, Friction catalysis in the synthesis reaction of CO_2 during adhesive wear, *Proc. 34th Jpn. Cong. Mater. Res.*, 34 (1991) 119–123.

33. K. Hiratsuka, C. Kajdas, M. Yoshida, Tribo-catalysis in the synthesis reaction of carbon dioxide, *Tribol. Trans.*, 47 (2004) 86–93.

34. K. Hiratsuka, T. Abe, C. Kajdas, Tribocatalytic oxidation of ethylene in the rubbing of palladium against aluminum oxide, *Tribol. Intern.*, 43 (2010) 1659–1664.

35. S. Ikeda, T. Takata, M. Komada, M. Hara, J. N. Kondo, K. Domen, A. Tanaka, H. Hosono, H. Kawazoe, Mechanocatalysis: A novel method for overall water splitting, *Phys. Chem. Chem. Phys.*, 1 (1999) 4485–4491.

36. S. Ikeda, T. Itani, K. Nango, M. Matsumura, Overall water splitting on tungsten-based photocatalysts with defect pyrochlore structure, *Catal. Lett.*, 98(4) (2004) 229–233.

37. E. McCafferty, *Introduction to Corrosion Science*, Springer (2010).

38. D. Landolt, S. Mischler, *Tribocorrosion of Passive Metals and Coatings*, Woodhead Publishing Series in Metals and Surface Engineering, No. 47 (2011).

39. V. V. Boldyrev, K. Tkacova, Mechanochemistry of solids: Past, present, and prospects, *J. Mater. Synth. Proc.*, 8(3/4) (2000) 121–132.

40. V. V. Boldyrev, Mechanochemistry and mechanical activation of solids, *Russ. Chem. Rev.*, 75(3) (2006) 177–189.

41. S. M. Hsu, J. Zhang, Z. Yin, The nature and origin of tribochemistry, *Tribol. Lett.*, 13 (2002) 131–139.

42. C. Kajdas, General approach to mechanochemistry and its relation to tribochemistry, in *Tribology in Engineering*, InTech, Chapter 11 (2013).

Chapter 2

Tribochemistry

Czesław Kajdas,[a] Ken'ichi Hiratsuka,[b] Gustavo Molina,[c] Akira Sasaki,[d] and Roberto C. Dante[e,f]

[a]*Institute of Chemistry in Płock, Warsaw University of Technology, Poland*
[b]*Department of Innovative Mechanical and Electronic Engineering, Chiba Institute of Technology, Japan*
[c]*Department of Mechanical Engineering, Georgia Southern University, USA*
[d]*Maintek Consultant, Yokohama, Japan*
[e]*Department of Rheology and Complex Materials, Instituto de Investigaciones en Materiales, Universidad Nacional Autónoma de México, Mexico*
[f]*Laboratorio de Materiales Avanzados ETSIIAA, University of Valladolid, Spain*

2.1 Tribochemistry

Czesław Kajdas[a] and Ken'ichi Hiratsuka[b]

[a]*Institute of Chemistry in Płock, Warsaw University of Technology, Poland*
[b]*Department of Innovative Mechanical and Electronic Engineering, Chiba Institute of Technology, Japan*

khiratsuka@it-chiba.ac.jp

2.1.1 Background

Tribochemistry is a subset of mechanochemistry. Although both of these terms are well known and their specific reaction

Tribocatalysis, Tribochemistry, and Tribocorrosion
Edited by Czesław Kajdas and Ken'ichi Hiratsuka

ISBN 978-981-4316-95-8 (Hardcover), 978-1-315-36498-8 (eBook)
www.panstanford.com

effects, especially mechanochemical ones [1–5], have been widely discussed for over a century, even in the past decade, it was stated that the precise nature of these reactions is still not well understood [6].

Works by Lea [1–5] were greatly interested in the relations of energy to the chemical changes in matter. A recent overview paper [7] documents unanimously that Lea was the first mechanochemist because his experiments relate to the first systematic investigations on the chemical effects of mechanical action. The most significant achievements by Carey Lea is that he differentiated the effect of mechanical stress from that of heat and started a new branch of chemistry "mechanochemistry" where materials react differently from thermal reactions.

Matthew Carey Lea suffered from weak health throughout his life and was educated at home by a tutor. Then he worked in the private laboratory of his home in Philadelphia and was primarily a photochemist. His next research was focused on investigating the effect of various kinds of energy: heat, light, mechanical action on allotropic (colloidal) silver and silver halides [7]; between 1892 and 1894, Lea studied the mechanochemical decomposition of dozens of stable compounds. He observed the decomposition of silver and mercuric chlorides by trituration in a mortar, although the same compounds are known to melt or sublime undecomposed when heated. For these achievements, M. Carey Lea was elected a member of the National Academy of Sciences in 1892. Lea's sound statement saying "What causes the reactions to take place is also a subject of speculation" was discussed two years later also in paper [5]; the assumed reason was assigned to insufficient reaction products yields and thereby insufficient information for a full understanding of the nature and origin of the complex chemical reactions, quantity of which was very small.

Historical retrospective of mechanochemistry beginning from the early observations by alchemists to the establishment of mechanochemistry as a branch of science is reviewed by Boldyrev and Tkacova [8]. The changes in structure and chemical properties of solids under three-axes loading and by combined action of pressure and shift had been presented. Among decisive factors influencing the mechanochemical synthesis, the formation and renewal of contact area between reacting compounds, the explosive evolution of heat, and the feedback phenomenon are specified. The perspective directions of practical application in the area known as mechanical alloying, as well as in preparation of functional ceramics and catalysts, and in pharmacy are also discussed in [8]. Boldyrev also wrote a paper that contains vast amounts of information concerning mechanochemistry and mechanical activation of solids [9]. It is better to understand in such a way that there are two mechanisms of mechanical activation: One is short-lived with flash temperature, electron emission, at the interface or vicinity of the contact area, while the other is long-lived associated with fresh surface with kinks and steps, which lasts until it is covered by the chemisorbed species.

On the other hand, paper [6] is of particular importance due to emphasis on the triboemission speculation saying "Recent observation of surface emissions of electrons, charged particles, from rubbing surfaces invites that this emission provides the source of energy causing tribochemical reactions to take place" and one of the conclusions related to the next speculation: "It may be further speculated that the emission of electrons, charged particles, would indeed induce tribochemical reactions that otherwise would not take place." Accordingly, tribochemistry can be defined as the chemical reactions that occur between the environment and the surfaces under boundary lubrication conditions.

2.1.2 Recent Results on Tribochemistry

Work [6] reviews some knowledge of tribochemistry until 2002 and examines the issue of the source of tribochemistry in boundary lubrication. Experiments were designed to examine the issue of whether mechanical disruption of surface bonds can lead to the formation of reaction products, and to find whether these reaction products correspond to those observed under normal rubbing conditions. Thermally induced reaction products were also examined for comparison. Specific tests in that work were as follows. Surfaces covered with a monolayer of stearic acid molecules were scratched by a glass lens covered with a layer of diamond particles. Unidirectional scratches were made at 20–30 nm depth.

The reaction products were monitored by a grazing angle FTIR capable of detecting the spectrum of a monolayer. The spectrum was then compared with the spectrum from a thermally heated case and a rubbed case. In WE-Heraeus Seminar in ILMENAU May 27–29, 2002, the topic was discussed much wider [10]. Over 20 invited speakers presented their research results, ideas, and/or hypothesis on this entangled subject. Examples of later published works [11–13] also relate to this subject. Paper [11] focuses on the wear transition from the thermal energy point of view. However, it does not refer to any chemical activations that would be the main subject of this book. Paper [12] reviews eight models to calculate the average and maximum temperature rise at sliding surfaces. The calculated flash temperature is also varied by the ratio of the real contact area to Hertzian contact area. Tribochemical reactions are discussed from the viewpoint of mechanical and thermal effects. Paper [13] deals with the relationship between wear and the entropy generation within the mechanically affected zone. The theoretical approach presented here does not include the atmospheric effect, which would play a crucial role in wear particle generation.

Most recent results on mechanochemistry were presented in 2011 at VII International Conference on Mechanochemistry and Mechanical Alloying INCOME 2011. In the general information of the INCOME 2011 Conference, it was emphasized that since the classical papers by M. Carey Lea on mechanical activation of solids over hundred years back (1893–1894), mechanochemistry and mechanical activation have moved a long way and evolved as a frontier area of research and applications, during the last two to three decades.

Notable developments had taken place in terms of scientific understanding and applications in diverse fields, encompassing all facets of metallurgy including metal extraction, alloying, composites; traditional and advanced ceramics processing; paints, pigments, and dyes; catalysis, pharmaceuticals, agriculture, utilization of wastes, energy and environment, and many more [14]. Significant progress of developments had also taken place in terms of characterization techniques of activated solids; quantification and control of activation processes; and mechanical activation devices for specific needs.

2.1.3 Are Triboelectrons of Low-Energy Important in Initiation of Mechanochemical/Tribochemical Reactions?

Section 2.2, "Triboemission of Exoelectrons," describes and reviews in detail triboelectrons of low-energy. At this point, we are mentioning only selected information along with some references. Triboemission of electrons is a particular case of the general phenomenon of exoelectron emission (EE), which is observed when a material surface is disturbed and whose origin is still unclear. EE has been observed from both metals and non-metals and there is strong evidence that oxides or other non-metallic surface layers are needed for EE to occur. Triboemitted electrons are thought to play a significant role in tribochemical reactions under boundary lubrication conditions [15]. Kim et al. [16] investigated photon emission, electron emission intensities, and electron kinetic energy from reciprocating scratching of MgO with diamond. The work of Molina et al. [17, 18] characterized triboemission of electrons from diamond-on-alumina, diamond-on-sapphire, alumina-on-alumina, and diamond-on-aluminum. The three ceramic-materials consistently showed burst-type negatively charged triboemission during contact at constant load and speed, while the aluminum system produced no significant emission. Decaying emission after contact was detected by Molina from the three ceramic systems [19]. For the cases of diamond-on-alumina and diamond-on-sapphire, energy spectrometry showed that a large fraction of the triboemitted negative charges were of low-energy (e.g. 1–5 eV). This finding was of significant importance because in the NIRAM approach, it was hypothesized that the energy level of triboelectrons to initiate tribochemical reaction should be 1–4 eV [20].

Presently it is well known and accepted that chemical reactions under boundary lubrication conditions are distinct from those of regular thermochemical reactions. The mechanically or tribochemically initiated reactions at the contact of asperities might be caused by the flash temperature effect. However, flash temperature is very short and it can also be expressed in terms of thermionic emission as evidenced by Vick et al. [21]. At this point, the following question arises: Can this reaction initiation process be considered in terms of overcoming the activation energy by heat or by another form of energy?

The common denominator of tribochemical reactions is that they might be initiated by low-energy electrons. This is consistent with the negative-ion-radical action mechanism (NIRAM) approach assuming that tribochemical reactions are initiated by electrons in the energy-range of 1–4 eV. The hypothesis is also relevant to the next hypothesis saying that the intermediate reactive species of both tribochemical reactions and heterogeneous catalytic reactions are produced by the same mechanism, governed by the NIRAM approach. Some 10 years ago, tribochemistry was reviewed in [22]. Another review [23] was on tribopolymerization.

2.1.4 Approach to Mechanochemistry Focused on Mechanochemical Synthesis/Mechanolysis

The term mechanolysis is rather rarely used. The only book [24], published in 2006, which presents and discusses the mechanolysis process, concerns organic mechanochemistry and its applications. According to author of the book, Z. V. Todres, the term *mechanolysis* denotes bond scission under mechanical activation. Such a reaction is reversible in principle, and as this takes place, the free radicals formed enter the usual free-radical reactions: recombination, decomposition, addition, and substitution [24]. It is of note here that the present authors consider that tribochemistry is a subset of mechanochemistry. The most recent overview paper on mechanochemistry as a key to understand the mechanisms of boundary lubrication, mechanolysis, and gas evolution during friction [25] discusses details and provides a novel understanding of mechanochemistry.

2.1.5 Brief Information on Mechanochemical Synthesis

Work [26] presents an interesting classification of mechanochemical syntheses occurring in various states of aggregation. The focus is on peculiarities of phenomena that take place under the action of mechanical impulses taking place in high-energy ball mills. In that context, both mechanochemical activation and mechanochemical reactions are discussed along with mechanical alloying and reactive milling. Definition of mechanochemistry is taken from Heinicke's book [27], defined as "*a branch of chemistry concerned with the chemical and physicochemical transformations of substances in all states of aggregation induced by mechanical energy.*" On the other hand, paper [28] clearly states that presently mechanochemistry is the subject of significantly increased interest in the theory and practical application. It also stresses that the mechanical treatment of solids using high-energy impulses causes both *mechanical activation and chemical reactivity.*

Usually, shredding and crushing concern the first step of those processes. Work [26] underlines that grinding in special mills is several times more effective than in regular devices, e.g., diminishing tungsten carbide particles from 2–3 mm pieces to particles of 3 μm conventionally takes 70 h, while the process realized in a planetary mill takes only 3 min. Paper [28] evidences that using mechanical treatment, it is possible to make a significant impact on the kinetics and extent of chemical processes. However, the relationship between the conditions of the action and the extent to which the physical and chemical properties of a solid substance are changed, still is not adequately known. In summary, the type of action and its parameters have different effects on the solid phase particles.

Nanoparticles can also increase the mechanical strength of a metal matrix. Work [29] demonstrates that nanoparticles increase the mechanical strength of a metal matrix more effectively than do microparticles, for instance, a uniform dispersion of nanoparticles ensures a good balance between the strengthening and the interparticle spacing effects. Accordingly, it is an interesting way to significantly improve mechanical properties of metal matrix materials.

2.1.6 Tribochemistry in Terms of Mechanochemical Reactions

Reactions in the solid state are very complex and their characteristic feature is that they involve product formation at the interfaces of reactants [26]. Additionally, such reactions are slow, therefore, a solid phase reactions require charge transfer to initiate them. It is known that mechanochemical reactions can proceed fast in high-energy ball milling processes. Namely, powder or powder mixtures may significantly make faster/

accelerate chemical reactions between two solids, a solid and a liquid, or a solid and gas, making it possible for the reaction to occur at temperatures lower than those of conventional processes [30, 31]. It is strictly combined with the 3-2-2 book Chapter [32].

In solid-solid systems, paper [26] indicates synthesis applications in $Pb(Zr,Ti)O_3$ and lead zirconate titanate (PZT). The latter of a perovskite structure is widely used in various sensing and actuating devices. It is of note that traditional method of obtaining this material is a solid-state reaction between stoichiometric mixtures of the constituent oxides: PbO, ZrO_2, and TiO_2 followed by calcination at high temperature, and such procedure often leads to an incomplete reaction [26]. Conversely, it is possible to obtain the PZT phase from $Pb(NO_3)_2$, $TiCl_4$, and $ZrO(NO_3)_2$ by mechanical activation, using a high-energy shaker mill operated at 900 rpm for 20 h [33]. The particles in the nanometer range are of particular importance, and they can be easily produced by using chemical synthesis routes under specific conditions [34].

Work [35] studied mechanochemical reactions of Ag_2O with V_2O_5 with the aim to generate/produce crystalline silver vanadates; it was proposed that compounds of the Ag–V–O system doped with lithium could be applied for positive electrode materials. On the other hand, while testing the Cr–V–O system, Wieczorek-Ciurowa et al. [36] suggested a number of applications in the field of heterogeneous catalysis, sensors, magnetic, and ceramic technologies. It should be emphasized, however, in these two cases, it was found that pure oxides—Ag_2O, V_2O_5, and Cr_2O_3—reveal strong stability during mechanical action. Their milling in mixtures composed of Ag_2O with V_2O_5 and Cr_2O_3 with V_2O_5 involved the formation of different spinel phases [36].

Extensive MDMO (molecular dynamic–molecular orbital) study on the solid state interfacial reaction under mechanical stressing, demonstrates that materials, prepared by applying the mechanochemical route, possess a lower density of acid sites as compared to samples obtained using the sol-gel method. Interestingly, mechanically synthesized zirconium phosphates (NZP) reveal the presence of the strongest Lewis centers [37].

Since mechanochemical synthesis concerns catalytic processes, mechanochemistry of catalysts combined with mechanochemical activation (MCA), should be here mentioned, especially in terms of factors that influence the specific catalytic activity [38]. It is detailed overview work emphasizing that degree of the catalyst increase in activity depends on the amount of excess energy that is stored in defects, and the energy amount is determined by the physicochemical nature of the catalytically active sites. Work [38] also underlined that the progress in this field might be achieved by the application of novel MCA methods/techniques.

Coming back to very small Cr_2O_3 particles, it is of note that study [39] on mechanochemical nanomaterial treatment shows the mechanochemically realized process occurs according to the following reaction:

$$Na_2Cr_2O_7 + S \rightarrow Cr_2O_3 + Na_2SO_4$$

It was found that 10 min of milling of the reactant mixture involved an abrupt increase in the vial temperature, confirming that Cr_2O_3 particles are formed in a combustion process during milling (with a large negative enthalpy change of $\Delta H = -562$ kJ/mol) [39].

Special ball milling is adequate method for preparing nano-sized metal matrix composites (MMCs) due to possibility of the composite generation with uniform distribution of grain sizes. Other works by Wieczorek-Ciurowa et al. [40, 41] show that materials consisting of ceramic and intermetallic phases, namely Cu–Al/Al_2O_3 and Ni–Al/Al_2O_3, can be obtained by milling mixtures of Cu(Ni) hydroxycarbonates with aluminum. Generation of such materials is caused by complex chemical reactions proceeding under milling conditions in air. These specific mechanochemical processes are described as follows [26]:

- mechanical decomposition: $M_2(OH)_2CO_3 \rightarrow 2MO + H_2O\uparrow + CO_2\uparrow$
- aluminothermic reduction: $3MO + 2Al \rightarrow 3M + Al_2O_3$
- mechanical alloying: $xM + yAl \rightarrow M_xAl_y$,

where M is Cu or Ni.

The above paper concludes/suggests that the heat generated ($\Delta H_{298} = -1675$ kJ/mol) speeds up the decomposition of hydroxycarbonates, induces the aluminothermic reaction, enhances the alloying of the two metallic phases, and the initiation temperature of the reaction can be monitored by differential thermal analysis (exo-effects) [26].

2.1.7 Computer Head–Hard Disk Lubrication

2.1.7.1 General Information

The computer lubrication system includes writing and reading magnetic flux patterns from rotating disks coated with magnetic materials, invented by the International Business Machines (IBM) Corporation in the late 1950s. Information is stored as bits in the magnetic layer, deposited onto a metallic support structure and coated with amorphous hydrogenated carbon. Very thin layer (1 nm) of selected adequate lubricant is then deposited over the carbon overcoat (COC) to further dampen the tribological impact of the slider. Then, a lubricant is applied to the COC surface.

Very significant progress in the technology of magnetic media has brought about a remarkable increase in recording density information storage. The magnetic layer of hard storage disks is coated with thin (few to 20 nm) film of sputter-deposited carbon to protect the magnetic layer from abrasive impacts and corrosion of the head slider.

Well-elaborated tribochemistry concerns also head-disk interface interactions for selected tribological systems. This is because chemistry controls head and disk surfaces along with the specific lubricant chemistry. PFPEs (perfluoropolyethers) are widely used to lubricate hard disks. In a magnetic rigid-disk drive, recording or playback is accomplished by relative motion between a magnetic medium against a stationary read-write magnetic-head slider [42, 43]. During starting and stopping between the slider and medium, physical contact-start-stop (CSS) occurs.

The need for increasing high recording densities requires that surfaces should be as smooth as possible and flying heights as low as possible; smoother surfaces lead to increase in adhesive interaction between the mating surfaces controlling both friction and interface temperatures [42]. Closer flying heights lead to occasional contact of asperities and increased wear, combined with different physical processes.

From the viewpoint of chemical reactions at surfaces, very comprehensive information of mechanically initiated reactions is described in [27], and it is emphasized that many tribological problems are concerned with various tribochemical reactions that cannot be accounted for by frictional heat alone. An earlier review on mechanochemical reactions [45], pointed out the technological importance of mechanically initiated chemical reactions by grinding, drilling, cutting or crushing. These mechanical processes facilitate the chemical compound formation at the worked interface. Tribochemical reactions are distinct from those of thermochemical ones because activation energy (E_a) for the latter is much higher, e.g., E_a for iron oxide formation from iron and oxygen differs very significantly if the thermochemical reaction (E_a = 54 kJ/mol) is compared with tribochemical reaction (E_a = 0.7 kJ/mol) [27]. The occurrence of reaction velocity independent of temperature evidences that the energy for releasing the reactions is applied to the solid through the mechanically activated regions only. The same is due to catalysis and tribocatalysis. It is of note, that chemically stimulated exoelectron emission (EEE) process occurs from silver catalyst during partial oxidation of ethylene and the emission rate is proportional to the rate of ethylene oxide formation [46].

New sophisticated techniques for measuring wear and friction, surface topography, lubricant thickness on a micro- to nanoscale have led to the development of a new field of research called microtribology, nanotribology, molecular tribology, or atomic-scale tribology [47]. More recent book [48], focused on dissipative mechanisms and hierarchical surfaces, covers fast developing topics of roughness-induced surfaces, the advances in the nanosciences and applications in nanotechnology. The converse of macrotribology, micro/nanotribology or components wear behavior studies, at least one is of extremely small mass, as only negligible wear occurs in such circumstances. At this small scale, the tribological performance is mostly controlled by the lubrication of component surfaces [49], and at these nanoscale boundary lubrication conditions, lubricant films are very thin.

The central thrust of the present Section is to review and discuss tribochemical reactions related to computer lubrication systems, starting from chemistry of selected fluorine-containing compounds.

2.1.7.2 Types of Major Lubricant

To reduce wear on the disk, from both sliding and flying contacts, it must be adequately lubricated. Fomblin Z-derivatives [50] are widely used as lubricants in the magnetic disk drive industry to decrease the friction between the head and disk. Lubricants for this application are based on a linear perfluoropolyether backbone end combined with two functional groups specifically designed to have a strong interaction with the disk surface. Such chemical structure enables the molecule to sit flat with a good adhesion on the surface. Usually, these lubricant molecules are known by the brand names: Fomblin Z-DOL, Fomblin Z-DOL TX, Fomblin Z-TETRAOL. PFPE lubricants of Fomblin-Z type molecular chains terminated with alkyl moieties having hydroxyl groups are presently widely applied in lubrication of computers.

The perfluoropolyether (PFPE) class of lubricants is widely used on magnetic recording disks for computer data storage. In computers, commonly used lubricants include selected PFPEs as well as their polar derivatives containing various reactive end groups. In a magnetic rigid-disk drive is accomplished by relative motion between a magnetic medium against a stationary read/write magnetic-head slider. Presently, diamond-like carbon (DLC) coatings lubricated with PFPE lubricants are widely used as the protective coating for thin-film magnetic storage disks. Recent developments in DLC films for magnetic media, and their chemical, mechanical, and tribological characteristics, are described elsewhere [51].

Another paper has reviewed the literature on interaction and degradation mechanisms of PFPE lubricants with protective carbon coatings used on magnetic media [42]. The computer head/disk interface is one of the most important factors controlling the performance and life of hard disk drives. To reduce friction on a thin-film magnetic recording disk used with a read/write head, the disk surface is usually lubricated with fluorinated or perfluorinated organic compounds. Presently, the most widely used lubricants for disk drives are PFPEs with hydroxyl functional groups, for instance Z-DOL.

The need to increase recording density has recently led to smoother disk surfaces and lower flying heights, but these changes have made the tribological environment at the head/disk interface much more severe. Therefore, new lubricants and/or additives have been developed, which encompass cyclic phosphazene compounds that are environmentally stable and have better tribological performance than polar PFPEs. The literature comprises many reports about PFPE lubricant interaction with carbon coatings under both static and sliding conditions. Other publications relate to PFPE degradation mechanisms. The mechanisms include thermal decomposition, catalytic degradation, electron mediated degradation, and mechanical degradation processes. All the processes have recently been reviewed [42].

2.1.7.3 Lubrication Process

The primary goal of this Section is to provide a better understanding of the PFPE lubricant interaction with carbon overcoats, under sliding conditions. The next aim is focused on PFPE degradation mechanisms. Emphasis is on importance of low-energy electrons in the degradation processes of head-hard disk lubricants PFPE, X-1P, and the presently most sophisticated lubricants, e.g., A20H chemically linking the cyclotriphosphazene moiety to the PFPE molecular chain [52]. In the triphosphazene sector of A20H, two of the phosphorus atoms have two phenoxy groups attached to them, while the third phosphorus atom has one phenoxy group and the Fomblin chain attached to it. Figure 2.1 shows the formulae of X1P and A20H [53].

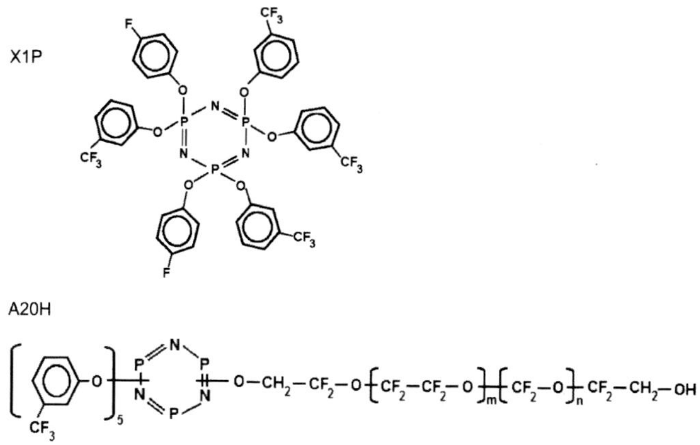

Figure 2.1 Chemical formulae of X1P and A20H. *Tribol. Lett.*, Disk lubricant additives, A20H and C2: Characteristics and chemistry in the disk environment, 31, 2008, 25–35, P. H. Kasai, A. Wakabayashi, with permission of Springer.

To reduce the bulk size of A20H, alternative lubricant C2 was developed by replacing the phenoxy groups in A20H by trifluoro-ethoxy groups [54]. The formula of C2 lubricant is shown in Fig. 2.2 [53].

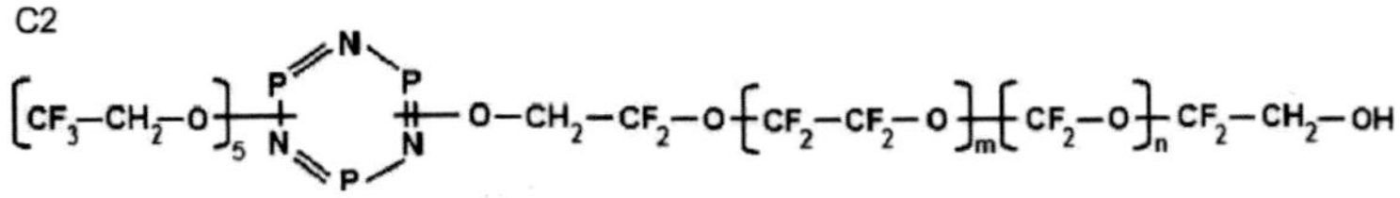

Figure 2.2 Chemical formula of C2. *Tribol. Lett.*, Disk lubricant additives, A20H and C2: Characteristics and chemistry in the disk environment, 31, 2008, 25–35, P. H. Kasai, A. Wakabayashi, with permission of Springer.

The adhesive interaction occurs between the benzene rings in contact with the surface and the carbon overcoat (Fig. 2.3).

Referring to Kasai's earlier paper [55], it is of note at this point that the electron donating property (the nucleophilic property) of X1P stems from the triphosphazene ring and not from the phenoxy sectors. A20H is topically used as an additive to Z-DOL or Z-TETRAOL, and regularly used with UV 185 nm photo-curing. It has been reported [56] that the UV-curing results in enhancement of the bonded fraction, which relates to the lubricant that cannot be removed by solvent. The reaction that occurs under the UV-curing process has not been accounted for in detail yet. Wakabayashi et al. [56] found that A20H was completely miscible with benzene while C2 was immiscible. Then they evidenced that the UV-curing did enhance bonding of the A20H lubricant to the disk COC but did not enhance the bonding of the C2 lubricant.

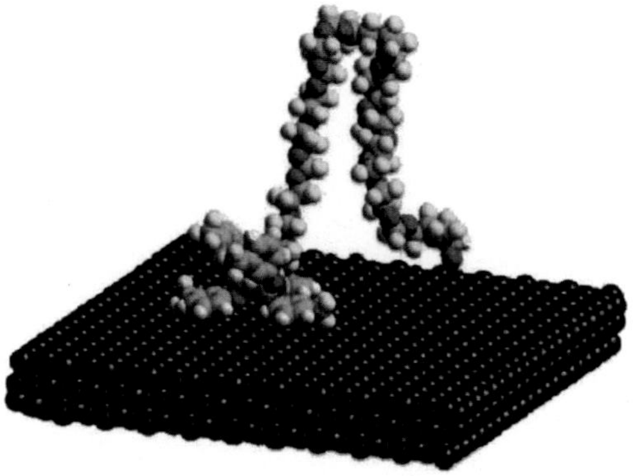

Figure 2.3 An A20H molecular chain anchored to the carbon overcoat by the phosphazene end-group. *Tribol. Lett.*, Disk lubricant additives, A20H and C2: Characteristics and chemistry in the disk environment, 31, 2008, 25–35, P. H. Kasai, A. Wakabayashi, with permission of Springer.

2.1.7.4 Degradation of PFPE Lubricants

PFPE lubricants are widely applied in very special tribological applications. It is due to their superb oxidative and thermal stability properties and low vapor pressure. A good example of applications relates to magnetic recording disks. Commonly used lubricants for topical lubrication include Fomblin Z and Demnum as well as their polar derivatives containing various reactive end groups, e.g., hydroxyl (Fomblin Z-DOL, Demnum SA), and piperonyl (Fomblin AM 2001). The areal storage density, the number of bits stored per unit area, has been steadily increasing. This trend relates to a decrease in the flying height or spacing between the read-write head and the surface of magnetic storage disk, which presents tribological challenges [57]. The magnetic media is protected with a hard overcoat that is mostly a hydrogenated amorphous carbon film. Most recent developments and/or applications are described in [58]. An earlier paper [42] reviewed and discussed/analyzed the literature on interaction and degradation mechanisms of perfluoropolyether (PFPE) lubricants with carbon protective overcoats used for magnetic media.

Lubricants deposited on carbon coatings under static conditions primarily interact with the surfaces by physical adsorption forces. The adsorbed lubricant film can also be chemically bonded by a specific treatment of the deposited film. To produce chemically bonded films, exposure of the lubricated disk to specific radiation is used. There are several modes: (a) low-energy X-ray [59], (b) nitrogen plasma [60], (c) high-energy ion beam [61], and (d) low-energy electrons and far ultraviolet radiation [62].

During sliding both physically adsorbed and chemisorbed PFPE lubricant molecules undergo degradation. The degradation process is very complex. Although a number of papers aim at accounting for the PFPE degradation mechanism, e.g., Zhao et al. [63], it is still not yet fully understood. Broad, detailed, and sophisticated research resulted in the development of several approaches. They include [42] (i) thermal decomposition processes, (ii) catalytic degradation processes, (iii) electron mediated degradation, and (iv) mechanical degradation by a shearing process. The following subsections will discuss all the approaches of PFPE degradation mechanisms.

Thermal decomposition processes

Although PFPE lubricants exhibit excellent thermal stability and they are thermally stable up to 350°C [64], there is clear evidence that their actual thermal stability is adversely affected by the presence of metals. The presence of metal alloys drastically increases the rate of degradation [65]; two inhibitors, a perfluoroalkyl ether substituted monophospha-s-triazine and a perfluorophenylphosphine, were found to be highly effective in degradation reduction at 288°C, but they had only limited effectiveness at 316°C. Work [66] showed that complete degradation of Fomblin Z takes place at 185°C in the presence of iron oxide (Fe_2O_3).

Z-DOL-containing hydroxyl groups strongly interact with dangling bonds and functional groups on the carbon surface, apart from the overcoat interaction with the lubricant ether groups. Additionally, hydroxyl groups of the Z-DOL molecule can form hydrogen bonding with oxygen and hydrogen atoms of the carbon surface film functional groups [63]. Heating enhances the bonded Z-DOL film significantly. Prebaking at 150°C doubles the bonded film of Z-DOL lubricant; this effect was accounted for in terms of the water molecules' removal from the carbon overcoat [67], and thereby exposing more dangling bonds and the surface functional groups to an interaction with the lubricant molecules (see Fig. 2.4).

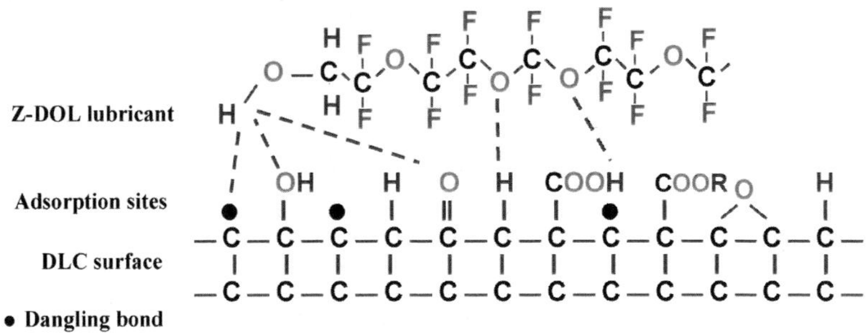

Figure 2.4 Original Yanagisawa model of amorphous carbon surface. Reproduced with permission from Society of Tribologists and Lubrication Engineers. M. Yanagisawa, Adsorption of perfluoro-polyethers on carbon surfaces, SP-36, *Tribol. Mech. Magn. Storage Syst.*, 9 (1994) 25–32.

Work [68] investigated the adsorption of water and small perfluorinated compounds on hydrogenated amorphous carbon films (a-C:H). They found changes in hydrogen content of the films and concluded that hydrogenated carbon films exhibiting the highest surface free energy provided a greater attractive interaction for the model lubricants and may provide greater stability of thin lubricant films on these surfaces.

Carbon surface chemistry of magnetic disks is of particular importance from the viewpoint of lubricant interaction with the disk or head surface. A variety of functional groups has been found to exist on carbon surfaces [69]. Applying X-ray photoelectron spectroscopy, Yanagisawa determined many oxygen-containing groups along with the conjugated group on the sputtered carbon surface [70], and proposed an amorphous carbon surface model. Figure 2.4 depicts the original Yanagisawa's model [70].

Figure 2.5 illustrates a modified model [42]. In the modified model, instead of the C five valence bond combined with C=O functional group, on its right side, the C–C bonding should be broken to form another dangling bond.

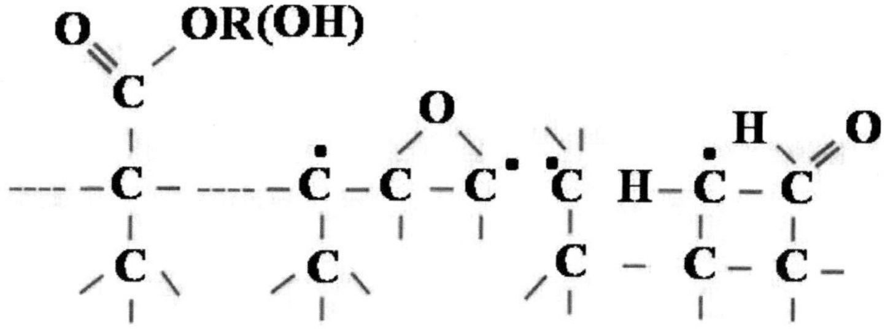

Figure 2.5 Modified Yanagisawa's carbon overcoat model. *J. Info. Storage Proc. Syst.*, Mechanism of interaction and degradation of perfluoropolyethers with a DLC coating in thin-film magnetic rigid disks: A critical review, 1 (1999) 303–320, C. Kajdas, B. Bhushan, with permission of Springer.

In this model there are four dangling bonds (4 dots in the modified model) and several oxygen-containing functional groups out of five possible ones: hydroxyl, carbonyl, carboxyl, oxide, and ester. Z-DOL lubricant molecule is adsorbed on the carbon surface at 20°C. As the carbonyl group cannot be combined with a five-valence carbon, work [42] considered the presence of the simplest aldehyde group (methanal) and it was substantiated by the fact that mostly carbon overcoats are hydrogenated. According to paper [71], even if the carbon is hydrogenated, the density of unpaired electrons (dangling bonds) in the hydrogenated carbon decreases only from 10^{18} to 10^{16} cm^{-3}. The graphite surface layer plane with various functional groups located at the periphery of the plane is illustrated in Fig. 6 of ref. [72].

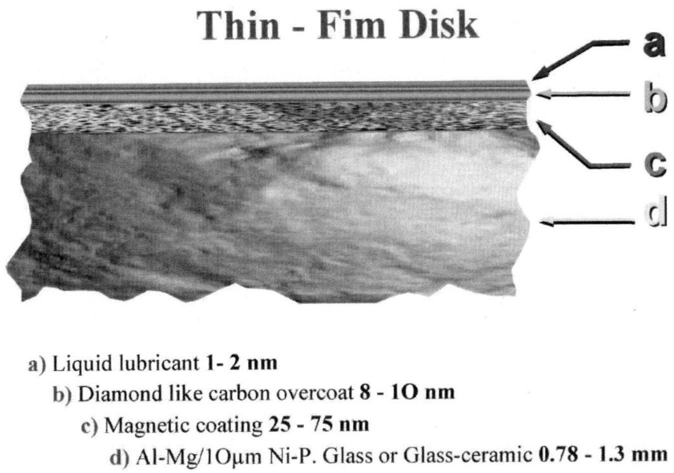

Figure 2.6 Simplified diagram of magnetic storage disk redrawn from *J. Info. Storage Proc. Syst.*, Mechanism of interaction and degradation of perfluoropolyethers with a DLC coating in thin-film magnetic rigid disks: A critical review, 1 (1999) 303–320, C. Kajdas, B. Bhushan, with permission of Springer.

Carbon overcoats lubricated with PFPE are applied as the protective coating for magnetic storage disks and metal evaporated (ME) tapes. The thickness of the protective overcoat is 8–10 nm, as presented in Figs. 2.6 and 2.7 (redrawn from [42]). The addition of a limited amount of hydrogen in the amorphous carbon overcoat improves the CSS durability of thin-film disks lubricated with PFPE's.

Interestingly, such degradation was not observed with Krytox lubricant $F-(CF(CF_3)-CF_2-O)_n-CF_2CF_3$ (polyhexafluoropropylene oxide). The complete degradation of Fomblin Z was ascribed to the presence of acetal groups (—O—CF_2—O—) in the polymeric chain, as suggested in 1983 by Jones et al. [65]. Under sliding conditions, particularly at high sliding speeds, the temperature on hard carbon coatings can be sufficiently high to initiate PFPE thermal degradation [67]. Investigating friction and wear of hard carbon coatings at sliding speeds in the range 30–35 m/s in air, Huu et al. [73] presented evidence that the temperature increase produced by friction in the contact zone was high enough (about 285°C), and emphasized that the transition from the sp^3 to sp^2 phase occurs at this temperature. Since the transition consumes a part of the friction energy in the contact, apart from creating a thin graphite layer on the track, it is also reasonable to suggest that in such specific situation the thermal degradation of PFPE lubricant can take place.

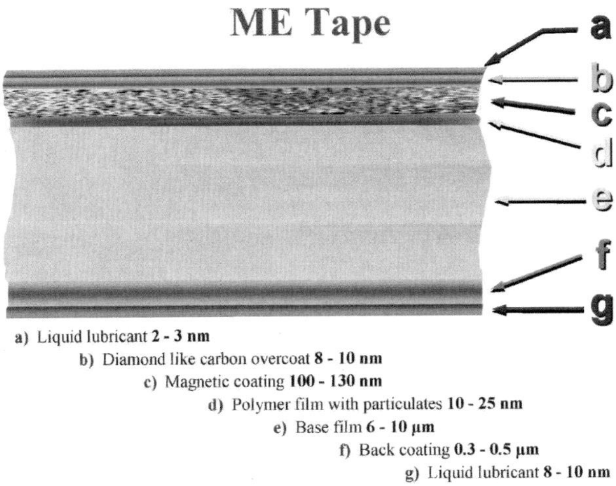

Figure 2.7 Simplified diagram of metal evaporated tape redrawn from *J. Info. Storage Proc. Syst.*, Mechanism of interaction and degradation of perfluoropolyethers with a DLC coating in thin-film magnetic rigid disks: A critical review, 1 (1999) 303–320, C. Kajdas, B. Bhushan, with permission of Springer.

Figure 1 of ref. [55] illustrates weight loss observed on PFPE lubricants when heated at different material/temperature combinations. It is of note here, that the presence of oxygen in the environment may also lead to the thermal oxidative degradation process generating very reactive peroxide radical, which is the compound containing the peroxide anion $(O–O)^{2-}$. Usually, O–O group is called the *peroxide group* or *peroxo group*.

Catalytic degradation processes

Considering the fact that alpha-alumina exists as the structural element in the construction of head slider for magnetic disk drive, work [74] examined the catalytic degradation mechanism of PFPE's by that alumina material. Then, the experiment of the degradation process was followed up in the presence of $AlCl_3$, and at higher temperatures [74]. Kasai's earlier study [55] conclusively suggested that the degradation of PFPE's in the presence of Lewis acids (metal oxides or halides) is singularly dominated by the intra-molecular disproportionation reaction,

$$R_1—CF_2—O—CF_2—O—CF_2—R_2 \rightarrow R_1—CF_3 + FC(O)—R_2,$$

where R_1 is the left part of Z-DOL molecule from the acetal sector, and R_2 is the right portion of Z-DOL molecule.

This reaction type occurs at acetal sector ($O—CF_2—O$), and in the presence of a stronger Lewis acid, it occurs also at other linkages. The fluorine transfer was always from the terminal group into the internal sector. The degradation mechanism of a Z-type lubricant in the presence of alumina it is catalyzed by $A1F_3$, *formed during the induction period,* and involves the acetal moiety ($—O—CF_2—O$) of the polymer chains producing methoxy end groups ($—O—CF_3$). Concluding, this degradation mechanism results in fragmentation of polymer chains but does not involve an unzipping process [63]. The reaction sequence is initiated by a bidentate linkage between an acidic aluminum on $A1F_3$ and the two oxygen atoms of an acetal unit, as shown in the upper part of Fig. 2.8. The partial positive charge developed at the acetal carbon induces a fluorine atom transfer from the adjacent CF_2 unit leading to chain scission with transformation of the acetal unit into a methoxy ($—O—CF_3$) end group, and the adjacent unit into either a fluoro-formate $F—CO—O—CF_2$ end group or an acylfluoride $F—CO—CF_2$ end group from the adjacent unit, depending upon whether the adjacent unit was originally a methylene oxide unit or an ethylene oxide unit.

Weight loss of major PFPE lubricant types during heating in the presence of various catalysts [55] provides good information on the thermocatalytic stability of these lubricants. The degradation was induced by placing 5 g of lubricant and

1 wt% of Al_2O_3 (or $AlCl_3$) in a test tube, and immersing the tube in an oil bath maintained at a desired temperature. For each lubricant, the weight loss was measured after the heat treatment for a given period of time [55]; under the applied test conditions (Al_2O_3/200°C) all the tested lubricants apart from the Z-lubricant type were very stable. On the other hand, even the least stable lubricant started to degrade after 50 min of heating.

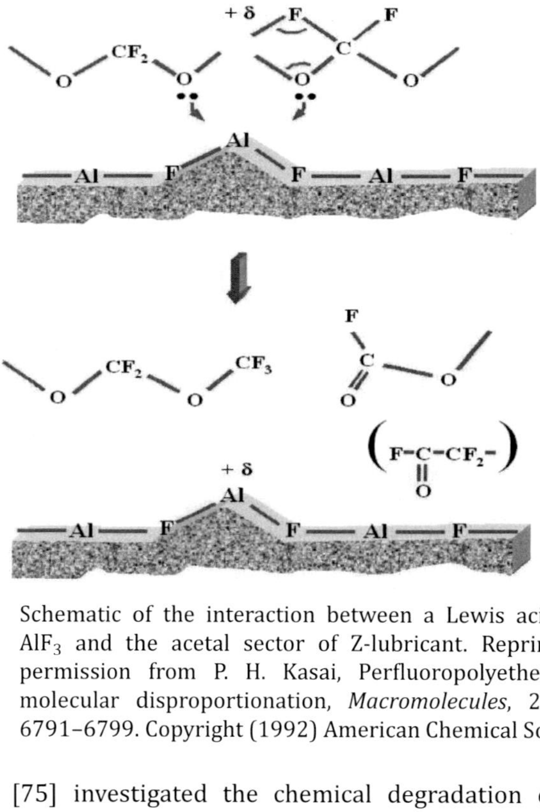

Figure 2.8 Schematic of the interaction between a Lewis acid site on AlF_3 and the acetal sector of Z-lubricant. Reprinted with permission from P. H. Kasai, Perfluoropolyethers: Intramolecular disproportionation, *Macromolecules*, 25 (1992) 6791–6799. Copyright (1992) American Chemical Society.

Work [75] investigated the chemical degradation of major PFPE lubricant fluids, such as Z-DOL, Fomblin AM-2001, Z–25, Demnum SP-3, and Demnum SA, with bivalent metal ions. They were performed to react with a variety of metal sulfates under identical heating conditions; the resulting fluoride ion concentrations were measured. The respective reaction mechanisms involve an ether cleavage that is caused by the coordination of metal ions with oxygen atoms in the main chain of the PFPE; each lubricant with a given sulfate was heated at 250°C in an oven. Heating times were 30, 60, 90, or 150 min. The evaluated ability of metal ions to degrade a lubricant is the following sequence [76]: Cu^{+2} > Ni^{+2} > Co^{+2} > Fe^{+2} > Mn^{+2} > Mg^{+2}; this trend has been explained by the ionization potentials of the metal atoms.

Mechanical scission/degradation processes

Removal of PFPE lubricant molecules from sliding and flying tracks is controlled by its film thickness, chemical structure, and molecular weight. The friction and properties of molecularly thin liquid films were discussed in detail some 15 years ago [77]. It is known that under flying and sliding operation, the lubricant removal rate from monolayer films is smaller than from multilayer films [78]. It is of note that similarly to the thermal degradation process, mechanical scission of PFPE bonds also leads to the formation of free radicals. Figure 2.9 shows the PFPE degradation process initiated by shearing.

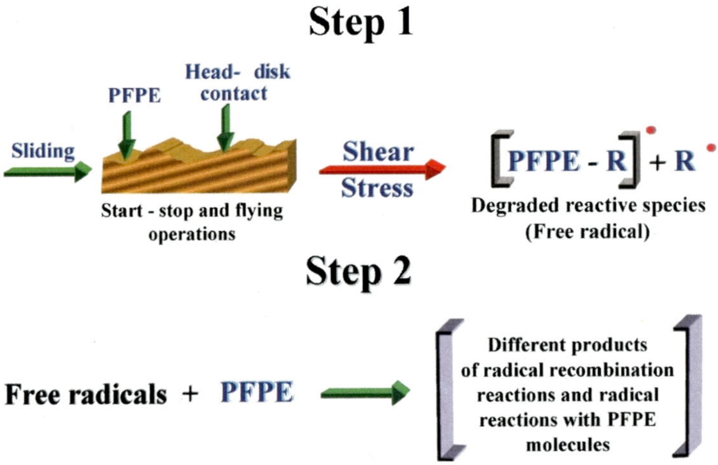

Figure 2.9 PFPE degradation process caused by shearing. *J. Info. Storage Proc. Syst.*, Mechanism of interaction and degradation of perfluoropolyethers with a DLC coating in thin-film magnetic rigid disks: A critical review, 1 (1999) 303–320, C. Kajdas, B. Bhushan, with permission of Springer.

Electron-induced degradation mechanism

High-vacuum test results performed by Zhao and Bhushan evidence that the degradation process of the 1.5-nm-thick Z-DOL lubricant film generates CFO, HCF_2, and CF species during sliding [79], but CF_2O fragment has not been detected. On the other hand, experiments performed by Zhao et al. [63], also including the Z-DOL lubricant degradation process, has clearly demonstrated the importance of CF_2O and CFO species along with HCF_2, CF_3, and other fragments. The Z-DOL degradation fragments have been accounted for in terms of the electron-induced degradation mechanism. For instance, to account for the CF_2H and $CFCF_3$ Z-DOL degradation fragments detected in mass spectra under UHV sliding conditions, the following reactions were proposed:

$$—CF_2—CF_2—O—CF_2—O—CF_2— + e \rightarrow$$

$$—CF_2—CF_2—O^- + {}^{\bullet}CF_2—O—CF_2$$

$$\mathbf{A^- \text{ (anion)} + R^{\bullet} \text{ (free radical)}} \qquad (2.1)$$

The negative ion A^- reacts with the DLC surface. The formed free radical $R^{\bullet}$ can further react in various ways. The following reactions lead to the formation HCF_2 and CF_3 of species found in mass spectra:

$$R^{\bullet} + H^{\bullet} \rightarrow HCF_2—O—CF_2—$$

$$\textbf{Compound I} \qquad (2.2)$$

$$R^{\bullet} + \text{Z-DOL} \rightarrow CF_3—O—CF_2— + [Z—DOL—F]^{\bullet}$$

$$\textbf{Compound II} \qquad (2.3)$$

Compounds I and II, interacting with emitted low-energy electrons, generate the fragments detected in mass spectra in the following reactions:

$$HCF_2—O—CF_2— + e \rightarrow {}^{\bullet}CF_2H + {}^{-}O—CF_2— \qquad (2.4)$$

$$CF_3—O—CF_2— + e \rightarrow {}^{\bullet}CF_3 + {}^{-}O—CF_2— \qquad (2.5)$$

Negative ions formed in reactions (2.4) and (2.5) react with the DLC positively charged spots generated during sliding. There are several possibilities to produce CF and CF_2 fragments during the Z-DOL degradation process, e.g., CF_2 species may relate to the decomposition of CF_2CF_2O species into CF_2 and CF_2O by friction shear as well as electron impact [80]. CF_2 may be further cleaved into the F atom and CF species.

The detailed Z-DOL lubricant degradation mechanism, as described by Zhao et al. [63], emphasizes the importance of the negative-ion-radical action mechanism for the PFPE lubricant degradation process. The experimental results of that study also show that the decomposition of the lubricant molecule is greatly affected by the mechanical shearing, as illustrated in Fig. 2.9. Any Z-DOL lubricant molecule presents a linear long chain structure, including weak C—O and C—C bonds. In accord with that, the lubricant molecules can be degraded during sliding by microasperities of the rubbing surfaces, and this results in the decomposition of the lubricant. No specific effect of temperature on the decomposition of the lubricant was observed [65]. Thus, the shear stress influence on the Z-DOL lubricant degradation process should not be neglected.

Interestingly, triboemitted low-energy electrons (1–4 eV) can also be applied in accounting for experimental results concerning chemical bonding of PFPE lubricant films with DLC under sliding conditions. Figure 2.10 shows the steps through which lubricant Z-DOL molecules are chemically bonded to the DLC surface under sliding/contact operating conditions.

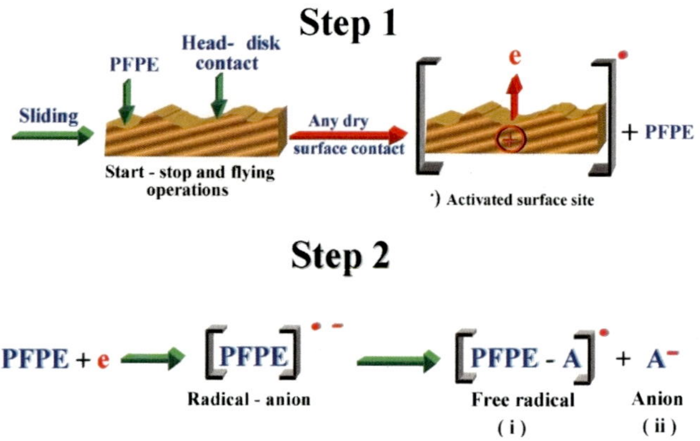

Figure 2.10 Electron-induced PFPE degradation process. *J. Info. Storage Proc. Syst.*, Mechanism of interaction and degradation of perfluoropolyethers with a DLC coating in thin-film magnetic rigid disks: A critical review, 1 (1999) 303–320, C. Kajdas, B. Bhushan, with permission of Springer.

Under sliding conditions, low-energy electrons are emitted at the interface (Step 1); these electrons then attach to lubricant molecules forming reactive species (Step 2). The generated negative ions are presented in Eq. (2.4) and (2.5). Figure 2.11 illustrates three steps through which lubricant Z-DOL-15 or Z-DOL-25 molecules are chemically bonded to the DLC surface after sliding.

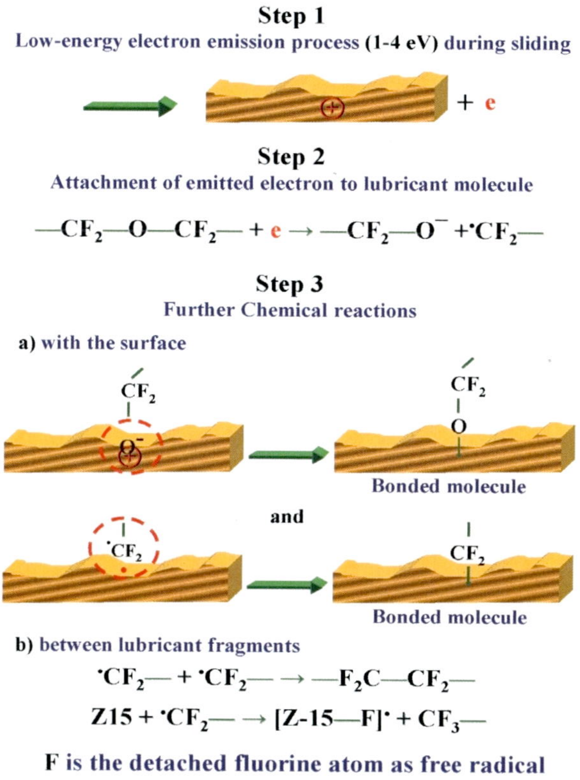

Figure 2.11 Interaction of emitted electrons with Z-15 (Z-DOL-15) or Z-25 (Z-DOL-15) lubricant molecules and subsequent interactions with the surface. *J. Info. Storage Proc. Syst.*, Mechanism of interaction and degradation of perfluoropolyethers with a DLC coating in thin-film magnetic rigid disks: A critical review, 1 (1999) 303–320, C. Kajdas, B. Bhushan, with permission of Springer.

Step 1 is similar to that in Fig. 2.10. The low-energy electrons then attach to lubricant molecules and form the reactive species (Step 2). The produced negative-ion reactive species allow chemical reaction with the surface and form bonded molecules. Free-radical species can react with the surface dangling bonds (Step 3a). They can also recombine, forming a PFPE molecule, or they can interact with Z-15 lubricant molecules, producing a smaller PFPE compound and generating another free radical [Z-15-F]$^{\bullet}$. The latter radical can recombine with $^{\bullet}CF_2$-free radical, producing higher molecular-weight lubricant compounds.

2.1.8 Present State of the Computer Head–Hard Disk Tribochemistry

2.1.8.1 Brief introductory background

We consider tribochemistry as a subfield of mechanochemistry. The same is due to mechanocatalysis and tribocatalysis. All the mechano-activated processes are very complex and are governed by lower activation energies than those thermochemical or catalytic processes. Processes initiated by mechanical action are also widely known as tribochemical reactions. Although start-stop contacts in computers somehow recall very low-loaded specific mechanochemistry, some catalytic and shearing processes might take place as well (see Fig. 2.5 and ref. [42]). Very extensive historical review on tribochemistry by T. E. Fischer [81] points out the term mechanical catalysis. All the information and discussion of Section 2.1.7, "Computer Head–Hard Disk Lubrication," allows us to say that tribochemistry of the presently available computers should be especially considered in terms of the tribochemistry mechanisms. Emphasis will be on the reaction mechanisms initiated by tribo- and photoelectrons. Interaction of low-energy tribo electrons with different PFPE lubricant molecules and subsequent interactions with the surface was proposed and discussed as early as 1985 [42, 79, 82–84]. Initial works [20, 22, 83, 84] evidence recent efforts to improve understanding of the mechanism factors controlling computer lubricant degradation [15, 43, 44, 49].

This part of Section 2.1.8 describes new concepts, prospects, and also shows how the molecular dynamics calculation revealed that alkyl-phenyl ether, on capture of an electron, would dissociate to yield the phenoxide anion and the alkyl radical [85]. Thus, it was predicted that PFPE lubricants with an end-group possessing a phenoxy unit would spontaneously adsorb on the carbon overcoat, and that irradiation of disks coated with such lubricant leads to grafting PFPE molecular chains to the carbon surface at the chain terminus. Four new PFPE lubricants, Z-SA1 and Z-SA2 based on the Fomblin Z type backbone, and D-SA1 and D-SA2 based on the Demnum backbone were synthesized, where SA1 and SA2 indicate end-groups possessing a phenoxy unit and a p-methoxy-phenoxy unit, respectively [85].

2.1.8.2 Novel mechanisms of computer tribochemistry

Photo-grafting by UV radiation has recently been demonstrated by Kasai and Wakabayashi [85]. They showed that a photoelectron produced by UV (185 nm) readily migrates into the phenoxy unit of the A20H molecule, resulting in its detachment. Detached phenoxide anions react with the carbon overcoat forming chemical bonds. This approach confirms well previously discussed tribo electron initiated degradation of Z-DOL type lubricants [42, 43, 79]. How A20H molecule adsorbs on the carbon surface by the interaction of the phenoxy group and the carbon substrate depicts Fig. 2.12.

Figure 2.12 clearly shows that photoelectron generated by UV (185 nm) migrates into the phenoxy unit resulting in its detachment and generation of PFPE molecular chain with a radical end. Both species can react with the carbon overcoat forming regular chemical bonds. The grafting reaction sequence of the Demnum-COOH lubricant type –CF2–CF2–O–[-CF2-CF2-CF2–O-]$_m$ –CF2–CF2–C(O)–OH irradiated with long UV (254 nm) is depicted in Fig. 2.13. The middle part of the figure shows the decarboxylation process combined with generation the two free radicals, which in turn react with the carbon overcoat double bond.

During testing PFPE-based disk lubricants, Kasai suspected an additional liability for a hydrogen bonded pair of hydroxyl end-groups in the presence of an electron donor. The plausible reaction sequences upon electron capture of hydrogen-bonded methanol dimer, hydrogen-bonded ethanol dimer, and 1,2-propanediol were first examined by a molecular dynamics (MDs) method, and then by the matrix isolation electron spin resonance (ESR) spectroscopy [86].

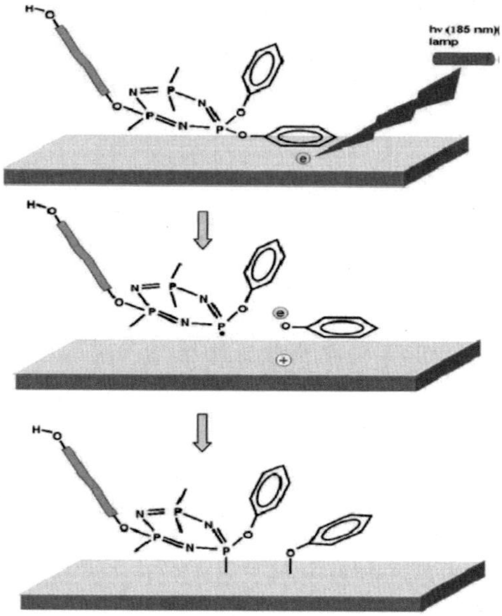

Figure 2.12 Adsorption of A20H molecule on the carbon surface. *Tribol. Lett.*, Disk lubricants for spontaneous adsorption and grafting to carbon overcoat by UV irradiation, 38, 2010, 241–251., P. H. Kasai, A. Wakabayashi, with permission of Springer.

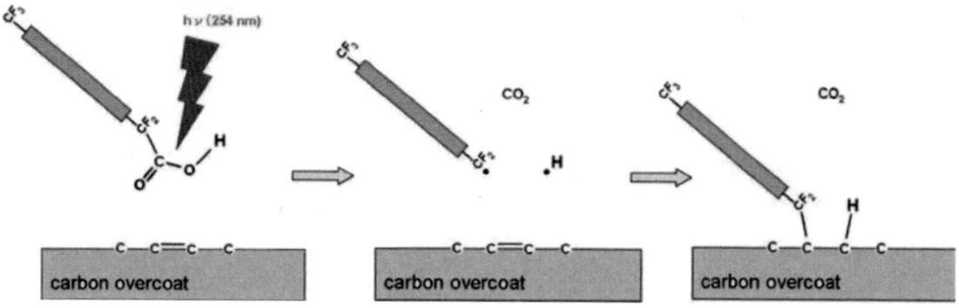

Figure 2.13 Irradiation of a disk coated with a PFPE molecular chain terminated with the carboxylic group with long UV (254 nm). *Tribol. Lett.*, Disk lubricants for spontaneous adsorption and grafting to carbon overcoat by UV irradiation, 38, 2010, 241–251, P. H. Kasai, A. Wakabayashi, with permission of Springer.

Alcohol dimer anions were generated by co-trapping Na atoms and alcohol molecules in argon matrices and inducing electron transfer by photo-irradiation. In methanol and ethanol dimers, the MDs study predicted that on capture of an electron, the hydroxyl hydrogen of the acceptor side would cleave as an atom, and then abstract a hydrogen atom from the nearest carbon of the proton donating alcohol [86]. The calculation revealed the reaction sequence in which the hydroxyl hydrogen of the acceptor side cleaved as an atom, and then abstracted a hydrogen atom from the alcohol donor CH_3 group, as presented in Fig. 2.14.

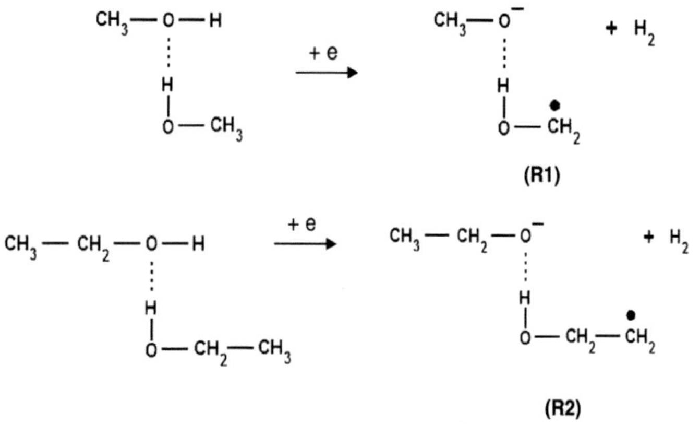

Figure 2.14 Dissociation sequence of methanol dimer upon electron capture predicted by the MDs study calculation. Reproduced with permission from P. H. Kasai, Dissociative electron capture of hydrogen-bonded hydroxy groups: Molecular dynamics and matrix isolation ESR study, *Bull. Chem. Soc. Jpn.*, 83(3) (2010) 254–260.

Summarizing the above information, it is necessary to say that very specific and sophisticated research combined with MDs calculation, described in papers [85, 86], provide further good evidence for the earlier NIRAM-related works [42, 43, 79, 82–84]. The novelty is that presently photo-emitted electrons are documented to work similarly as triboelectrons of low-energy (1–4 eV). Thus, research by Kasai and Kasai et al. might be compared with very extensive recent investigation by Zheng et al. [87], which not only confirmed experimentally the NIRAM approach but also extended and modified it.

2.1.8.3 Present extension of the NIRAM approach

The NIRAM approach proposed/hypothesized that the negative ion (A^-) attaches to rubbing material surface (hole) under the boundary friction process conditions. In work [87], the negative ion is the monomer fragment bearing a double bond as shown in Fig. 2.15.

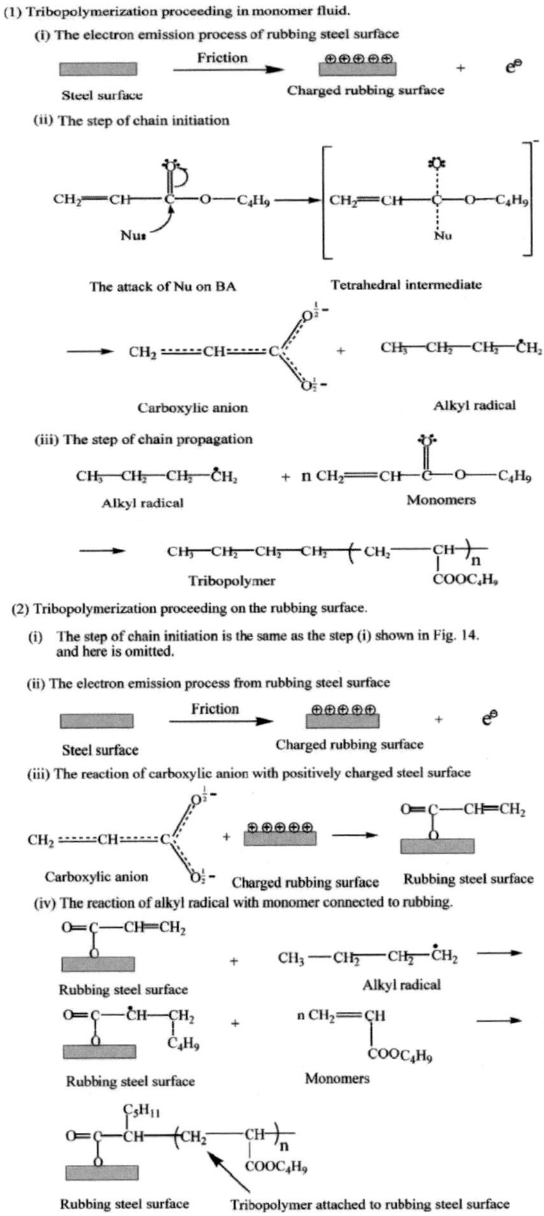

Figure 2.15 The tribopolymerization processes proceeding on the rubbing surface. Reprinted from *Tribol. Int.*, 41, J. Zheng, L. Zhang, Z. Du, C. Zhang, H. Li, Tribopolymerization of n-butyl acrylate on the steel–steel rubbing surface, pp. 769–777, Copyright (2008), with permission from Elsevier.

The tribopolymerization process initiated by low-energy electrons occurs in the bulk fluid, however, temperature and probable concentration of alkyl radical near rubbing surface should be higher than in the bulk fluid. Since the tribopolymers and the wear debris are connected by chemical bonds, such tribopolymer particles cannot be easily removed from the surface, for instance, by regular ultrasonically washing. The TG curve of wear debris proves it by exhibiting a significant weight loss in the temperature range from 308°C up to 480°C [87]. It is of note that the IR spectrum of the washing solution shows a broad absorption peak at around 1600 cm^{-1}, most probably due to carboxylic anions (see FTIR spectrum, Fig. 7 of ref. [87]).

2.1.8.4 Brief summary of the modified NIRAM tribopolymerization mechanism

In the research work discussed above, the AW property of n-BA in hexadecane for the steel–steel friction system was studied in detail. SEM images of worn scars of the test balls, and width measurements of the worn disk tracks indicated that n-BA has good AW property. Tribopolymerization tests of n-BA used as the lubricant were conducted to determine AW mechanisms.

Infrared spectroscopy (IR) of washing solution and thermogravimetry (TG) traces of wear debris confirmed that tribopolymers were generated on steel–steel interface in situ. Also considerable wear products were precipitated from vacuum-condensed worn fluids with methanol as the non-solvent. These wear products were analyzed by IR, gel permeation chromatography and nuclear magnetic resonance spectroscopy. The obtained analyzes results show that the precipitates were poly(n-butyl acrylate) (PBA) with very high molecular weight.

Detailed research presented in work [87], studied also the tribopolymerization mechanisms of n-BA under various sliding velocities, and the obtained results demonstrated that *tribopolymer was generated just due to the friction process.* Interestingly, *no substantial worn products were precipitated from vacuum-condensed worn fluids.* New experimental results allowed Zheng et al. to propose an extended NIRAM model and conclude that the exo-electron-radical-tribopolymerization model is consistent with Kajdas' tribochemistry theories.

Some extended work of the Zheng et al. research was on the n-BA polymerization mechanism. To test/determine the polymerization mechanism, parallel experiments were carried out using as the lubricant n-BA containing 1 wt% *n-butyl alcohol,* added as the inhibitor for ionic polymerization. Interestingly, it was found that the incorporated *n-butyl alcohol did not affect the amount of tribopolymer.* This experimental result indicated that the tribopolymerization mechanism of n-BA in the fluid phase (used as lubricant) was the free radical mechanism.

2.1.9 Summary Information on the Present Knowledge State of the Tribochemistry Discipline

In this book, we consider and discuss tribochemistry processes as a subfield of mechanochemistry because they are activated by the mechanical energy. This is because such processes are governed by lower activation energies than those of regular thermochemical or catalytic processes. Thus, mechanical energy alters significantly chemical reactions. Actually, processes initiated by mechanical action are also widely known as tribochemical reactions. For instance, book [24] provides a good evidence for that. Another work [25] that focused on better understanding of some tribochemical reactions from the mechanochemical point of view shows that tribochemical reactions taking place between frictional materials and surrounding chemicals enhance the decomposition of the solid and molecules.

From the viewpoint of terms [88], tribochemistry and mechanochemistry may be compared with the terms chemical physics and physical chemistry. In the latter term, chemistry is first and physics second. In tribochemistry, friction (*tribos*) is the first. Because mechanics includes friction, tribochemistry should be included in mechanochemistry. To initiate thermochemical reactions an adequate heat amount has to be supplied to overcome the activation energy. Good information relating to tribochemistry can be found in the most recent overview paper [89] and in the very specific experimental research work [90].

It is of note at this point that major basic knowledge on presently accepted tribochemistry is well described in the book on organic mechanochemistry and its practical applications [24], papers [89, 90], book chapters [92–96], reviews [15, 22,

97] and the book describing tribochemistry of lubricating oils [98]; therefore, the present chapter is mostly focused on novel tribochemistry approaches. These novel approaches include the modified/enhanced NIRAM theory, action mechanisms and the tribochemistry of extreme pressure additives, computer head–hard disk tribochemistry, degradation of perfluoro-polyether lubricants, and tribochemistry of soap formation.

2.1.10 Tribochemistry of Soap Formation

The chemical reaction between chemisorbed fatty acid and surface metal atom to form soap is a fundamental process for boundary lubrication. However, the precise mechanisms of soap formation have not been fully elucidated. A freshly generated metal surface has dangling bonds, which are active in chemical reactions. However, it is not active enough to cleave O–H bond in carboxyl group to form oxygen-metal bond, because fatty acid is not chemisorbed on metal surface as pointed out by Bowden and Tabor in *The Friction and Lubrication of Solids.*

Electrons emitted from the surface can be the energy source for soap formation. According to the model by [91], the fatty acid is dissociated into negative ion and hydrogen radical by the attachment of the emitted electron from the metal onto the fatty acid. Then, the negatively charged fatty acid is chemisorbed on the positively charged metal surface.

The complete reaction is written as follows:

$$\frac{1}{2}\mathrm{Cu} \rightarrow \frac{1}{2}\mathrm{Cu}_2^+ + e^-$$

$$e^- + \mathrm{H(Nd)} \rightarrow (\mathrm{Nd})^- + \frac{1}{2}\mathrm{H}_2$$

$$(\mathrm{Nd})^- + \frac{1}{2}\mathrm{Cu}^{2+} \rightarrow \frac{1}{2}\mathrm{Cu(Nd)}_2,$$

where Nd denotes nonadecanoic acid.

This model well explains the reaction scheme between metal and acid with the help of electron emission. However, it is also evident that soap is not formed on the genuine metal surface. Bowden and Tabor wrote, "For effective lubrication it is apparently necessary for water to have been present as well as oxygen." Then, the point is, how is fatty acid facilitated by oxygen and water to react with metal?

Our model is presented in Fig. 2.16, where a fatty acid is decomposed by water molecule plus the emitted electrons, which results in soap formation on the metal surface [89]. The mechanism is described as follows:

In the friction process, both metal–metal and metal–oxygen bonds are cleaved. The broken metal bond remains as a dangling bond. On the other hand, the scission of oxygen-metal bond emits electrons leaving a positively charged site and an oxygen dangling bond on the surface. When an electron attacks a water molecule, it dissociates the molecule into a hydroxyl anion (HO^-) and a hydrogen radical ($H^\bullet$). The hydrogen radical is then combined with another hydrogen radical to form a hydrogen molecule, whereas the "basic" HO^- reacts with the fatty acid resulting in the formation of water and a carboxyl anion. Finally, the carboxyl anion reacts with the positively charged metal cation to complete the soap formation.

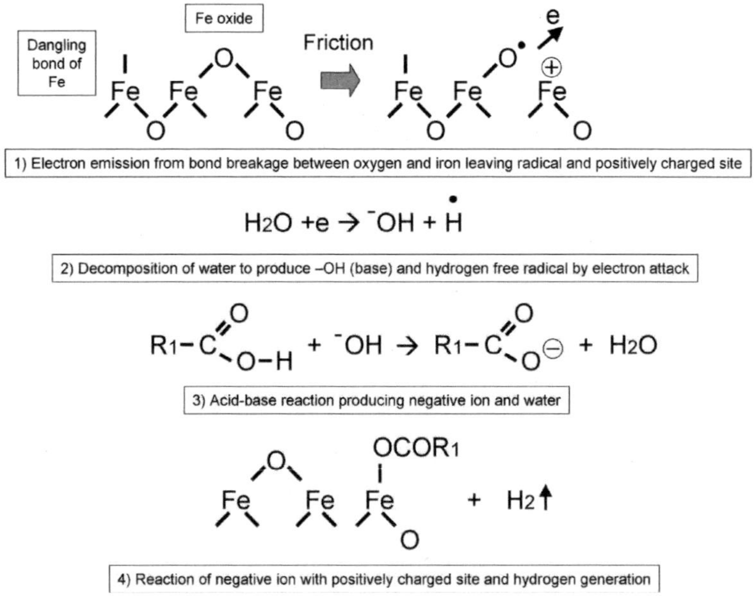

Figure 2.16 Mechanism of soap formation on metal surface. Reproduced with permission from K. Hiratsuka, C. Kajdas, Mechanochemistry as a key to understand the mechanisms of boundary lubrication, mechanolysis and gas evolution during friction, *Proc. Inst. Mech. Eng. Part 7 J. Eng. Tribol.*, 227 (2013) 1191–1203.

2.1.11 Action Mechanisms and Tribochemistry of EP Additives

Extreme pressure (EP) additives operate when the tribological parts are at elevated temperatures due to high pressure, where boundary lubrication additives become ineffective under these conditions. It has been believed that chemical reactions between metal surface and EP additives are driven by the heat produced from friction. However, when temperature is raised, thermionic electrons are emitted, which can be another promoter of the chemical reactions between EP additives and surface. To elucidate the detailed reaction mechanisms, chemisorption kinetics of one of the model compounds of EP additives was demonstrated on a steel surface. When steel was cut in diethyl disulfide gas atmosphere, its pressure was decreased, producing hydrogen and ethylene [90]. It means that the remaining sulfur compounds were left behind on the surface.

We propose the reaction processes in Fig. 2.17 [89]. When an electron is emitted and it attacks diethyl disulfide, the dissociation of a radical and an anion takes place. Both species react with the dangling bond and the positive site on the freshly generated surface, respectively. Because S–C bond is the weakest in the chemisorbed molecule, bond scission takes place by shear stress, leaving the dangling bond on sulfur and the carbon radical of $^{\bullet}CH_2CH_3$. $^{\bullet}CH_2CH_3$ radical is so unstable that it becomes ethylene by releasing a hydrogen radical. A hydrogen radical stabilizes to a diatomic hydrogen molecule or bonds with sulfur. Further shear stress applied onto the –SH group separates sulfur and hydrogen. Finally, sulfur sits on the surface with a double bond and free hydrogen is released.

In these processes, the time constants of chemisorption and desorption are totally different; chemisorption of diethyl disulfide is a slow process while desorptions of hydrogen and ethylene are rapid. The difference should be due to the difference in reaction area. There are two kinds of frictional interface between the tool and the material; flank face and rake face. Diethyl disulfide is chemisorbed by the newly generated surface after cutting. The large molecular weight of diethyl disulfide does not allow its fast chemisorption. On the other hand, in front of the tool, as the surface material is under shear stress, ethylene and hydrogen are quickly desorbed because friction force acts directly on the molecules on the surface after diethyl disulfide has been dissociatively chemisorbed.

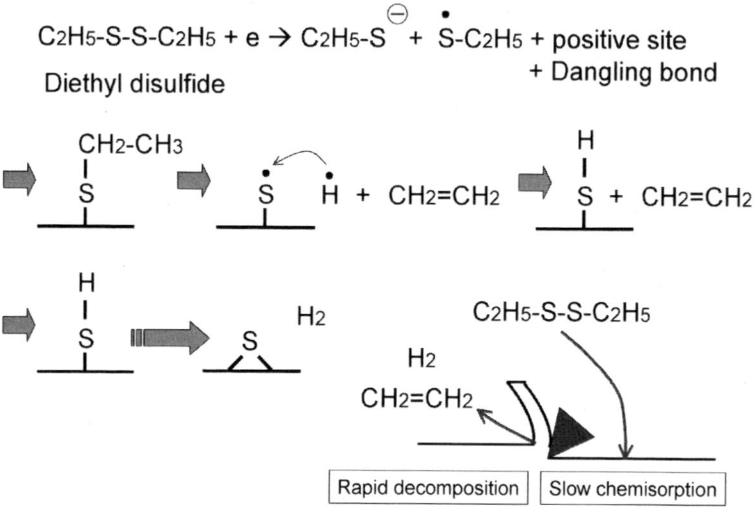

Figure 2.17 Decomposition of diethyl disulfide on cut steel and evolution of hydrogen and ethylene. Reproduced with permission from K. Hiratsuka, C. Kajdas, Mechanochemistry as a key to understand the mechanisms of boundary lubrication, mechanolysis and gas evolution during friction, *Proc. Inst. Mech. Eng. Part 7 J. Eng. Tribol.*, 227 (2013) 1191–1203.

2.1.12 Concluding Remarks

As introduced widely in this chapter, electron emission or transfer is a key to understanding tribochemistry. However, the number of electrons involved in oxidation/reduction processes cannot be measured from the outside of the system. This keeps tribochemical reactions apart from quantitative description.

Triboelectricity in the meaning of charges generation with rubbing (friction) is widely known due to the fact that two objects rubbing together usually create electrostatic charges. Triboelectricity is combined with electronegativity because various materials have various desire for electrons. Simply, some materials have tendency to take electrons from other ones that are ready to render (emit) electrons. Since we consider tribochemistry processes as the subfield of mechanochemistry, we expect that electrochemical approach to tribochemistry, similarly as in the mechanochemistry-derived tribocorrosion, in the future electrochemistry will open a new horizon to better understand and control also tribochemical reactions.

References

1. M. Carey Lea, On picric acid and some of its salts, with an advantageous modification of one of the processes for obtaining it, *Am. J. Sci., II*, 26 (1858) 379.
2. M. Carey Lea, On numerical relations existing between the equivalent numbers of elementary bodies, *Am. J. Sci., II, xxix*, 98 (1860) 349.
3. M. Carey Lea, Notes on allotropic silver, *Am. J. Sci., 3rd Series*, 42 (1891) 312.
4. M. Carey Lea, On endothermic decompositions, obtained by pressure. Part II. Transformation of energy by shearing stress, *Am. J. Sci., 3rd Series*, 46 (1893) 413.
5. M. Carey Lea, Transformations of mechanical into electrical energy, 3rd paper, Action of shearing stress continued, *Am. J. Sci., 3rd Series* 47 (1894) 377.
6. S. M. Hsu, J. Zhang, Z. Yin, The nature and origin of tribochemistry, *Tribol. Lett.*, 13 (2002) 131–139.
7. L. Takacs, M. Carey Lea, the first mechanochemist, *J. Mat. Sci.*, 39, (2004) 4987–4993.
8. V. V. Boldyrev, K. Tkacova, Mechanochemistry of solids: Past, present, and prospects, *J. Mater. Synth. Proc.*, 8(3/4) (2000) 121–132.
9. V. V. Boldyrev, Mechanochemistry and mechanical activation of solids, *Russ. Chem. Rev.*, 75(3) (2006) 177–189.
10. M. Scherge, J. A. Schaefer, Integrating friction and wear research, 280. WE-Heraeus-Seminar in Ilmenau, Germany, May 27–29, 2002, *Wear*, 254 (2003) 799–800.
11. H. A. Abdel-Aal, On the interdependence between kinetics of friction-released thermal energy and the transition in wear mechanisms during sliding of metallic pairs, *Wear*, 253 (2002) 11–12.

12. M. Kalin, Influence of flash temperatures on the tribological behaviour in low-speed sliding: A review, *Mater. Sci. Eng.*, A 374 (2004) 390–397.
13. H. A. Abdel-Aal, Wear and irreversible entropy generation in dry sliding, *The Annals of University "Dunărea De Jos" of Galați Fascicle tribology VIII* (2006) 34–44.
14. VII International Conference on Mechanochemistry and Mechanical Alloying (INCOME) 2011, Herceg Novi, Montenegro, August 31–September 3 (2011).
15. C. K. Kajdas, Importance of the triboemission process for tribochemical reaction, *Tribol. Int.*, 38 (2005) 337–353.
16. M. Kim, M. S. C. Langford, J. T. Dickinson, Electron and photon emission accompanying the abrasion of MgO with diamond, *Tribol. Lett.*, 1 (1995) 147–157.
17. G. J. Molina, M. J. Furey, A. L. Ritter, C. Kajdas, Triboemission from alumina, single crystal sapphire, and aluminum, *Wear*, 249 (2001) 214–219.
18. G. J. Molina, M. J. Furey, B. Vick, A. L. Ritter, C. Kajdas, Triboemission from the sliding contact of alumina systems. Proceedings of the Second World Tribology Congress, Vienna, Austria, September 3–7 (2001).
19. G. J. Molina, Triboemission from ceramics: Charge intensity and energy distribution characterizations. PhD dissertation, Department of Mechanical Engineering, Virginia Polytechnic Institute and State University, Blacksburg; (2000).
20. C. Kajdas, On a negative-ion concept of EP action of organo-sulfur compounds, *ASLE Trans.*, 28 (1983) 21–30.
21. B. Vick, M. J. Furey, C. Kajdas, An examination of thermionic emission due to frictionally generated temperatures, *Tribol. Lett.*, 13 (2002) 147–153.
22. C. Kajdas, Tribochemistry, in: *Surface Modification and Mechanisms* (G. E. Totten, H. Liang, eds.), Marcel Dekker Inc, New York (2004), Chapter 6, pp. 99–164.
23. M. J. Furey, C. Kajdas, Tribopolymerization as a mechanism of boundary lubrication, in: *Surface Modification and Mechanisms* (G. E. Totten, H. Liang, eds.), Marcel Dekker Inc, New York (2004), Chapter 7.
24. Z. V. Todres, *Organic Mechanochemistry and Its Practical Applications*, CRC Press, Taylor & Francis Group, Boca Raton, USA Florida (2006).

25. K. Hiratsuka, C. Kajdas, Mechanochemistry as a key to understand the mechanisms of boundary lubrication, mechanolysis and gas evolution during friction, Proceedings of the Institution of Mechanical Engineers, *Part 7 J. Eng. Tribol.*, Online, April 24 (2013) DOI: 10.1177/1350650113483222.

26. K. Wieczorek-Ciurowa, K. Gamrat, Some aspects of mechanochemical reactions, *Mater. Sci. Poland*, 25 (2007) 219–232.

27. G. Heinicke, *Tribochemistry*, Academie-Verlag, Berlin (1984).

28. S. P. Bobkov, A. N. Frolov, Effect of parameters of mechanical action on its effectiveness while handling solid particles, *Izvestiya Vysshikh Uchebnykh Zavedeniy: Khimiya i Khimicheskaya Tekhnologiya*, 34(6) (1991) 93–96.

29. A. Rougier, S. Soiron, I. Haihal, L. Aymard, B. Taouk, J.-M. Tarascon, Influence of grinding on the catalytic properties of oxides, *Powder Technol.*, 128 (2002) 139–147.

30. F. Kh. Urakaev, V. V. Boldyrev, Mechanism and kinetics of mechanochemical processes in comminuting devices 2. Applications of the theory. Experiment, *Powder Technol.*, 107 (2000) 197–206.

31. T. H. Courtney, D. Maurice, Process modeling of the mechanics of mechanical alloying, *Scrip. Mater.*, 34(1) (1996) 5–11.

32. A. Kulczycki, C. K. Kajdas, Activation energy (Ea), Chapter 3.2. In: *Tribocatalysis, Tribochemistry and Tribocorrosion*: C. Kajdas and K. Hiratsuka, eds, Pan Stanford Publishing, (2018).

33. X. Junmin, J. Wang, T. Weiseng, J. Synthesis of lead zirconate titanate from an amorphous precursor by mechanical activation, *J. Alloys Compounds*, 308(1–2) (2000) 139–146.

34. H. Schmidt, Nanoparticles by chemical synthesis, processing to materials and innovative applications, *Appl. Organometallic Chem.*, 15 (2001) 331–343.

35. S. Kittaka, S. Nishida, T. Ohtani, Mechanochemical reactions between Ag_2O and V_2O_5 to form crystalline silver vanadates, *J. Solid State Chem.*, 169(1) (2002) 139–142.

36. K. Wieczorek-Ciurowa, A. Gomułczak, J. Rakoczy, J. Stoch, K. Kamrat, Z. Sawłowicz, The influence of mechanical activation by a high-energy ball milling on the physical, chemical and catalytic performance of V_2O_5, Cr_2O_3 and their mixture, *Polish J. Appl. Chem.*, 49(4) (2005) 253–262.

37. S. N. Pavlova, V. A. Sadykov, G. V. Zabolotnaya, R. I. Maximovskaya, V. I. Zaikovskii, S. A. Tsybulya, E. B. Burgina, M. V. Chaikina, D. Agrawal, R. Roy, Molecular dynamic–molecular orbital combined study on the solid state interfacial reaction under mechanical stressing, *Solid State Ionics*, 141–142 (2001) 683–688.

38. V. V. Molchanov, R. A. Buyanov, Mechanochemistry of catalysts, *Russ. Chem. Rev.*, 69(5) (2000) 435–450.

39. T. Tsuzuki, P. G. McCormick, Synthesis of Cr_2O_3 nanoparticles by mechanochemical processing, *Acta Mater.*, 48 (2000) 2795–2801.

40. K. Wieczorek-Ciurowa, K. Kamrat, K. Fela, Chemical reactions during high-energy ball milling of the $Cu_2(OH)_2CO_3$–Al^0 system, *Solid State Ionics*, 164(3–4) (2003) 193–198.

41. K. Wieczorek-Ciurowa, P. Julian, A. Nosal, J. Domagała, Effects of reagents' nature on mechanochemical synthesis of calcium titanate, *J. Therm. Anal. Calorim.*, 101 (2010) 471–477.

42. C. Kajdas, B. Bhushan B., Mechanism of interaction and degradation of prfluoropolyethers with a DLC coating in thin-film magnetic rigid disks: A critical review, *J. Info. Storage Proc. Syst.*, 1 (1999) 303–320.

43. X. Zhao, B. Bhushan, Comparison studies on degradation mechanisms of PFPE lubricants and model lubricants, *Tribol. Lett.*, 9(3–4) (2000) 187–197.

44. C. Kajdas, Importance of anionic reactive intermediates for lubricant component reactions with friction surfaces, *Lubrication Sci.*, 6(3) (1994) 203–228.

45. P. G. Fox, Review, Mechanically initiated reactions in solids, *J. Mat. Sci.*, 10 (1975) 340–360.

46. N. Sato, M. Seo, Chemically stimulated exoelectron emission from silver catalyst curing partial oxidation of ethylene, *J. Catal.*, 24 (1972) 224–232.

47. B. Bhushan, Micro/nanotribology using atomic force/friction force microscopy, *State Art. Proc. IMechE, Part J: J. Eng. Tribol.*, 212 (1998) 1–18.

48. M. Nosonovsky, B. Bhushan, *Multiscale Dissipative Mechanisms and Hierarchical Surfaces, Friction, Superhydrophobicity, and Biomimetics, Series: NanoScience and Technology*, Springer Verlag (2008).

49. C. Kajdas, Nanoscale boundary lubrication of diamond-like carbon coatings with fluorinated compounds, *J. Synthetic Lubricants*, 18(1) (2001) 17–38.

50. P. H. Kasai, V. Raman. Z-dol versus Z-tetraol: Bonding and durability in magnetic hard disk application, *Tribol. Lett.*, 16(1–2) (2004) 29–36.

51. B. Bhushan, Chemical, mechanical and tribological characterization of ultra-thin and hard amorphous carbon coatings as thin as 3.5 nm: Recent developments, *Diamond Rel. Mater.*, 8 (1999) 1985–2015.

52. R. J. Waltman, D. J. Pocker, H. Deng, N. Kobayashi, Y. Fujii, T. Akada, K. Hirasawa, G. W. Tyndall, Investigation of a new cyclotriphosphazene-terminated pefluoropolyether lubricant. Properties and interactions with a carbon surface, *Chem. Mater.*, 15 (2003) 2362–2375.

53. P. H. Kasai, A. Wakabayashi, Disk lubricant additives, A20H and C2: Characteristics and chemistry in the disk environment, *Tribol. Lett.*, 31 (2008) 25–35.

54. N. Kobayashi, A. Wakabayashi, Japanese patent publication of application, JP 2006, 307123A (2006).

55. P. H. Kasai, Degradation of perfluoropoly(ethers) and role of X-1P additives in disk files, *J. Info. Storage Proc. Syst.*, 1 (1999) 23–31.

56. A. Wakabayashi, T. Shimizu, N. Kobayashi, P. H. Kasai, UV bonding of hard disk drive: characteristic and mechanism. In: Proceedings of ASIATRIB Conference, Kanazawa, Japan, October 16–19 (2006).

57. B. Bhushan, C. Kajdas, The present state of the art on degradation models of perfluoropolyethers with DLC coatings in thin film magnetic rigid disks, in: *Fundamentals of Tribology and Bridging the Gap Between the Macro- and Micro/Nanoscales*, B. Bhushan (ed). Kluwer Academic Publishers, Dordrecht, The Netherlands. Published in cooperation with NATO Scientific Affairs Division (2001), pp. 735–745.

58. B. Bhushan, *Principles and Applications of Tribology*, John Wiley and Sons, Inc. (1999).

59. R. Heideman, M. Wirth, Transforming the lubricant on a magnetic disk into a solid fluorine compound, *IBM Tech. Disclosure Bull.*, 27 (1984) 3199–3205.

60. A. M. Homola, L. Lin, D. D. Saperstein, Process for bonding lubricant to a thin magnetic recording disk, U.S. Patent, 4,960,609, October 2 (1990).

61. Y. Li, B. Bhushan, Degradation of perfluoropolyether lubricants on magnetic recording disks during sliding, *Proc. Inst. Mech. Eng. Part J.*, 211 (1997) 279–288.

62. G. H. Vurens, C. S. Gudeman, L. J. Lin, J. S. Foster, Mechanism of ultraviolet and electron bonding of PFPEs, *Langmuir*, 8 (1992) 1165–1169.

63. X. Zhao, B. Bhushan, C. Kajdas, Lubrication studies of head-disk interfaces in a controlled environment: Part II. Degradation mechanisms of PFPE lubricants, *Proc. Inst. Mech. Eng. Part J.*, 214 (2000) 547–559.

64. L. S. Helmick, W. R. Jones, Deformation of the thermal stability of perfluoroalkylethers, NASA memo TM-102493 (1990), NASA, Washington, D.C.

65. W. R. Jones, K. J. L. Paciorek, T. L. Ito, R. H. Kratzer, Thermal oxidative degradation reactions of linear perfluoroalkyl ethers, *Ind. Eng. Chem. Prod. Res. Dev.*, 22 (1983) 167–170.

66. M. J. Zehe, O. D. Faut, Acidic attack of perfluorinated alkyl ether lubricant molecules by metal oxide surface, NASA Memo TM-101962 (1989), NASA, Washington, D.C.

67. Z. Zhao, B. Bhushan, C. Kajdas, Effects of thermal treatment and sliding on chemical bonding of PFPE lubricant films with DLC surfaces, *J. Info. Storage Proc. Syst.*, 1 (1999) 259–264.

68. S. S. Perry, G. A. Somorjai, C. M. Mate, R. White, The interaction of short-chain model lubricants with the surfaces of hydrogenated amorphous carbon films, *Tribol. Lett.*, 1 (1995) 47–58.

69. S. R. P. Silva, A. J. Amaratunga, C. P. Constantinou, Optical properties of amorphous carbon/diamond thin films, *J. Appl. Phys.*, 72 (1992) 1149–1153.

70. M. Yanagisawa, Adsorption of perfluoro-polyethers on carbon surfaces, SP-36, *Tribol. Mech. Magn. Storage Syst.*, 9 (1994) 25–32.

71. F. Jansen, M. Machonkin, S. Kaplan, S. Mark, The effects of hydrogenation on the properties of ion beam sputter deposited amorphous carbon, *J. Vac. Sci. Technol.*, A3 (1985) 605–609.

72. E. M. Dannenberg, L. Paquin, Carbon Black, *Encyclopedia of Chemical Technology*, 4th ed., vol. 4, Bearing Materials to Carbon, John Wiley & Sons, Inc. (1992) 1037–1074.

73. T. L. Huu, H. Zaidi, D. Paulmier, Friction and wear properties of hard carbon coatings at high sliding speed, *Wear*, 203–204 (1997) 442–446.

74. P. H. Kasai, W. T. Tang, P. Wheeler, Degradation of perfluoropolyethers catalyzed by alumina oxide, *Appl. Surf. Sci.*, 51 (1991) 201–211.

75. P. H. Kasai, Perfluoropolyethers: Intramolecular disproportionation, *Macromolecules*, 25 (1992) 6791–6799.

76. T. Fukuchi, Degradation of perfluoropolyether fluids with metal ions, *ASME J. Tribol.*, 121 (1999) 348–351.

77. B. Bhushan, *Principles and Applications of Tribology*, Micro/Nanotribology, Wiley, New York (1999), Chapter 11.

78. V. J. Novotny, M. A. Baldwinson, Lubricant dynamics in sliding and flying, *J. Appl. Phys.*, 70 (1991) 5647–5652.

79. Z. Zhao, B. Bhushan, Tribological performance of PFPE and X-1P lubricants at head-disk interface. Part I. Experimental results, *Tribol. Lett.*, 6 (1999) 129–139.

80. J. Wei, W. Fong, D. B. Bogy, C. S. Bhatia, The decomposition mechanisms of a perfluoro-polyether at the head/disk interface of hard disk drives, *Tribol. Lett.*, 5 (1998) 203–209.

81. T. E. Fischer, Tribochemistry, *Ann. Rev. Mater. Sci.*, 18 (1988) 303–323.

82. C. Kajdas, About a negative-ion concept of the antiwear an antiseizure action of hydrocarbons during friction, *Wear*, 101 (1985) 1–12.

83. C. Kajdas, About an ionic-radical concept of the lubrication mechanism of alcohols, *Wear*, 116 (1987) 167–180.

84. C. Kajdas, M. Majzner, The influence of fatty acids and fatty acids mixtures on the lubricity of low-sulphur diesel fuels, SAE Technical paper series, 2001-01-1929.

85. P. H. Kasai, A. Wakabayashi, Disk lubricants for spontaneous adsorption and grafting to carbon overcoat by UV irradiation, *Tribol. Lett.*, 38 (2010) 241–251.

86. P. H. Kasai, Dissociative electron capture of hydrogen-bonded hydroxy groups: Molecular dynamics and matrix isolation ESR study, *Bull. Chem. Soc. Jpn.*, 83(3) (2010) 254–260.

87. J. Zheng, L. Zhang, Z. Du, C. Zhang, H. Li, Tribopolymerization of n-butyl acrylate on the steel–steel rubbing surface, *Tribol. Int.*, 41 (2008) 769–777.

88. C. Kajdas, General approach to mechanochemistry and its relation to tribochemistry, *Tribology in Engineering*, InTech (2013), Chapter 11.

89. K. Hiratsuka, C. Kajdas, Mechanochemistry as a key to understand the mechanisms of boundary lubrication, mechanolysis and gas evolution during friction, *Proc. Inst. Mech. Eng. Part 7 J. Eng. Tribol.*, 227 (2013) 1191–1203.

90. S. Mori, Y. Imaizumi, Adsorption of model compounds of lubricant on nascent surfaces of mild and stainless steels under dynamic conditions, *Tribol. Trans.*, 31(4) (1988) 449–453.

91. H. A. Smith, R. M. McGill, The adsorption of n-nonadecanoic acid on mechanically activated metal surfaces, *J. Phys. Chem.*, 61(8) (1957) 1025–1036.

92. T. T. Tysoe, P. T. Kotvis, Surface chemistry of extreme pressure lubricant addtitives, in: *Surface Modification and Mechanisms* (G. E. Totten, H. Liang, eds.), Chapter 10, 299–351, Marcel Dekker, Inc., New York (2004), Basel (2004).
93. J. Vižintin, Additive reaction mechanisms. In *Surface Modification and Mechanisms* (G. E. Totten, H. Liang, eds.), Marcel Dekker, Inc., New York, Basel (2004), Chapter 9, pp. 243–298.
94. I. A. Buyanovsky, V. I. Zimaida, R. N. Zaslavsky, Tribochemistry of boundary lubrication processes, in *Surface Modification and Mechanisms* (G. E. Totten, H. Liang, eds.), Marcel Dekker, Inc., New York, Basel (2004), Chapter 11, pp. 353–404.
95. C. Kajdas C. Tribochemistry. In *Surface Modification and Mechanisms* (G. E. Totten, H. Liang, eds.), Marcel Dekker, Inc., New York, Basel (2004), Chapter 6, pp. 99–164.
96. C. Kajdas, Physical and chemical phenomena related to tribochemistry, in *Advances in Contact Mechanics: Implications for Materials Science, Engineering and Biology* (R. Buzio, U. Valbusa, eds.), Transworld Research Network, Kerala, India (2006), Chapter 12, pp. 383–412.
97. D. A. Fischer, Z. S. Hu, S. M. Hsu, Tribochemical and thermochemical reactions of stearic acid on copper surfaces in air as measured by Ultrasoft X-Ray Absorption Spectroscopy, X-Ray Absorption Spectroscopy, *Tribol. Lett.*, 3(10) (1997) 35–40. DOI: 10.1023/A:1019109407863.
98. Z. Pawlak, *Tribochemistry of Lubricating Oils*, Amsterdam: Elsevier (2003).

2.2 Triboemission of Exoelectrons

Gustavo Molina,[a] Czesław Kajdas,[b] and Ken'ichi Hiratsuka[c]

[a]*Department of Mechanical Engineering, Georgia Southern University, USA*
[b]*Institute of Chemistry in Płock, Warsaw University of Technology, Poland*
[c]*Department of Innovative Mechanical and Electronic Engineering, Chiba Institute of Technology, Japan*

gmolina@georgiasouthern.edu

2.2.1 Introduction

2.2.1.1 Importance of low-energy exoelectron emission

Friction and sliding contact of solid surfaces is known to produce a variety of material and energy outputs, e.g., material debris, heat, acoustic emission, charged particles and various radiations. These energy outputs result from transformation of the mechanical energy, which is input by rubbing, into other forms of energy. This energy outputs, the so-called triboemission among them, also provides the driving force for a number of physical and chemical changes on the surface.

The term triboemission refers to the emission of electrons, ions, neutral particles, photons, and radiation under conditions of tribological damage. Triboemission of charged particles, particularly low-energy electrons, and friction-generated heat are considered the most important initiating and controlling factors in tribochemical mechanisms; triboemitted electrons are thought to be connected to the heterogeneous catalyst action.

Heterogeneous catalysts are environmentally benign and could be easily operated in continuous processes. However, the heterogeneous catalyst mechanisms by which lower activation energy might be achieved are still mostly unknown. The work published in 1992 by Illenberger discussed how low-energy electrons can be effectively captured by molecule clusters, which then undergo rapid decomposition [1]. Illenberger focused on reactions in weakly bound van der Waals clusters that are induced by capture of 0–10 eV electrons, noting that electron-transfer reactions play a key role in many fields of both pure and applied sciences. Evidence of low-energy electron emission from mechanical work is abundant; for example, the early work of von Voss and Brotzen [2] on electron emission from plastically strained aluminum, and the investigations of Ramsey [3] on emissions during and after wear of metals.

The earlier research by Sato and Seo [4] demonstrated the production of EEE from a silver catalyst, followed by very detailed research work toward partial ethylene oxidation [5], and that thermally stimulated EEE occurred in the same range of temperatures as that of typical of heterogeneous catalysis. Molina et al. [6] discussed in 2003 the importance of low-energy triboelectrons for the initiation of some typical tribochemical reactions.

In recent years, good progress has been made to better understand how reactant molecules, which adsorb in the catalyst, could reach the activation energy [7]. It is generally accepted that

(i) heterogeneous catalytic reactions take place at sites (or centers) in the catalyst–substrate system, where electron exchange is facilitated, and (ii) that EEE can be produced from sites where electrons are weakly bound. There are many works demonstrating the importance of EEE for heterogeneous catalysis [8–13]. More details are presented and discussed in the present book's Section 3.1, "Tribocatalysis."

It should be stressed that better understanding the significance of charged particle emission, also helps explain the recently discovered enhanced catalytic performance of some noble metals clustered in nanocatalytic clusters. Nanoclusters of some 10 atoms (about 1 nm in size) exhibit enhanced catalytic activity when supported on metal oxides or as charged gas-phase clusters [14, 15]. Many other recent works support this concept [16–27], they deal with models of gold nanoparticles deposited on thin layers (e.g., on MgO) for oxidation reactions. Simulation of these processes based on density functional theory [22] has found that adsorption energies of CO on cationic clusters are generally greater than those on neutral and anionic complexes, and the findings of Molina and Hammer [23] concluded that some gold atoms, independent of the substrate, became positively charged and explained the catalytic activity of Au. These results correspond well to observed exchange of electrons with the reactants. Recent experimental evidence [14, 15] shows that for heterogeneous catalysis, active sites on the oxide support-surface are required, and that these active sites control the charge state and exchange of the catalytic clusters by promoting the activation of adsorbed reactant molecules.

2.2.1.2 General description of triboemission

The triboemission process, which occurs under conditions of tribological surface damage, encompasses the emission of electrons, photons, and ions. The conceptual view of triboemission is illustrated in Fig. 2 of ref. [26].

Being triboemission just a fraction of the energy released during tribological contact, possible correlations to other better known energy outputs can help understanding the phenomenon dynamics, as the study of the concomitant acoustic emission.

The pioneer work of Dickinson et al. [27] showed that during the tensile strain of anodized aluminum the observed bursts of acoustic emission, closely associate with cracking of the surface oxide, that acoustic emission and electron emission count rates versus strain rates are closely related, and that most electron emission occurs immediately after and following the acoustic emission event evolution. Figure 2.18 summarizes these findings.

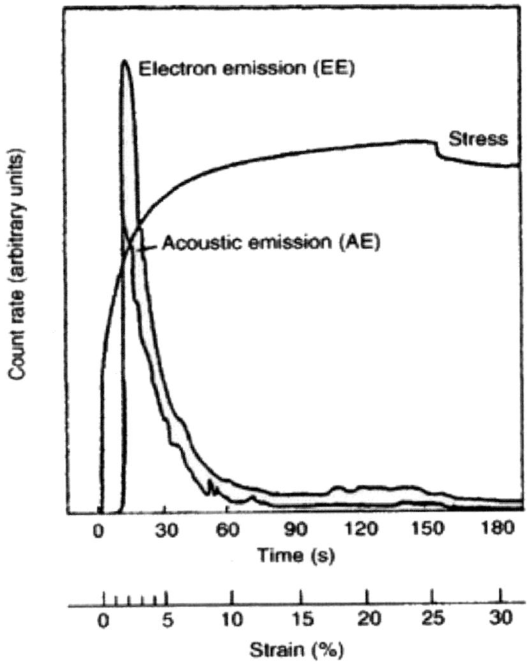

Figure 2.18 Acoustic emission, exoelectron emission, and stress vs. strain or time, for a sample of anodized Al 13450 under tension. Reprinted from J. T. Dickinson, E. F. Donaldson, D. B. Snyder, Acoustic emission and electron emission during deformation of anodized aluminium, *J. Vac. Sci. Technol.*, 17(1) (1980) 429–432, with the permission of AIP Publishing.

Figure 2.19 shows experimental data from this Dickinson's work [27] on electron emission (EE), acoustic emission (AE), and their ratio EE/AE. Interestingly, there is an initial peak (I) of the EE/AE ratio, at about 2% strain, where cracking of the oxide appears to be noticeable more efficient in generating electrons.

Optical microscopy showed that there were initial cracks formed in the oxide around that initial peak, whereas the cracking occurring later was of primarily crack extension. In the medium portion of the emission curves (II), which correlates with the largest peaks of both EE and AE, the EE/AE ratio is near unit. Figure 2.19 also shows from (II) to (III) the raise of the EE/AE ratio, starting at the 3% strain. The AE curve is seen to drop off faster than the EE curve.

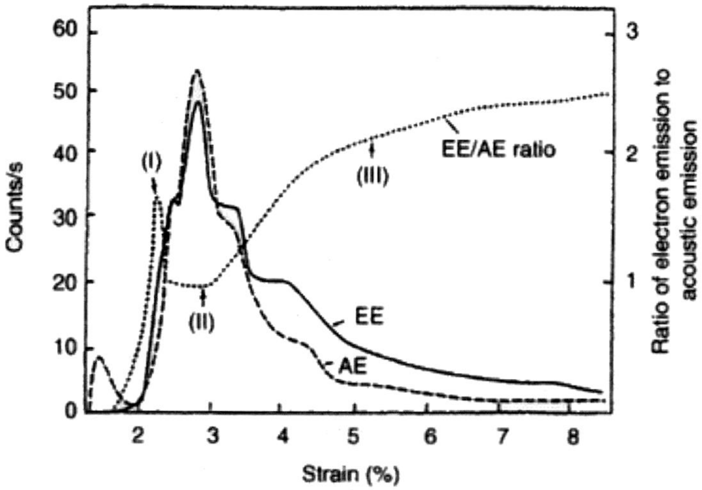

Figure 2.19 Electron emission (EE) and acoustic emission (AE) rates and their ratio (EE/AE) as a function of strain for a sample of anodized aluminum. Reprinted from J. T. Dickinson, E. F. Donaldson, D. B. Snyder, Acoustic emission and electron emission during deformation of anodized aluminium, *J. Vac. Sci. Technol.*, 17(1) (1980) 429–432, with the permission of AIP Publishing.

2.2.1.3 Review of most relevant triboemission research

Extensive research was carried by Nakayama et al. [28–41] on charged-particle and photon triboemission from several material systems of sliding contact. In their early work [28] they reported burst-type Tribo-stimulated exoelectron emission (EEE) from an aluminum disk scratched by a steel blade, and no emission after scratching ceased. For electron measurement they used a ceramic-surface continuous electron multiplier in vacuum of 10^{-3} Pa.

In a series of following papers Nakayama et al. [29–37] reported measurements of currents from a system of a scratching diamond-stylus on a rotating disk of the pertaining material (e.g., metals, ceramics, and semiconductors). These measurements of charges were carried out by "collecting" them in ring-shape metal pieces mounted on the stylus. The voltage between the collectors and the reference emitting surface was amplified and processed, and the output signals were usually measured in arbitrary units. By biasing the collectors +15V and –15V from reference, they collected signals which would, respectively represent negative and positive charges. No simultaneous measurements of both type of charges was performed. In the cases when photons were simultaneously detected [30–32, 34, 35, 37], a photon multiplier tube was employed.

In Nakayama's investigations, both negative and positive charge detected outputs were burst-type. Signal-levels were well above the instrument background (e.g., that obtained during disk rotation without scratching). The experimental setup allowed testing different environments, including low vacuum, for the tribological contacts. Standard atmosphere, reactive and inert gases, and hydrocarbon liquids and gases were investigated for their influence on the charge measurements. It is of note that the Faraday-cup-type charge-detector employed by Nakayama et al. may have been sensitive, in addition to actually emitted particles, to build-up of electrostatic charge on the contact surfaces, and the latter phenomenon may be critically important for insulating materials. Some general features and findings from Nakayama's experimental works [29–37] are summarized as follows: (i) Conditions of severe wear were used for the experiments (i.e., scratching of a rotating disk by a diamond stylus). (ii) Burst-type signals started immediately with scratching and abruptly dropped when scratching cease. (iii) Higher negative charge intensity was measured than that of positive charge for same material. (iv) Increases of negative-charge intensity were observed for increasing electrical resistance of materials [30, 31] in the order: conductor-semiconductor-insulator. (v) Photon emission was also observed and it followed the same trend of the above (iv) [31, 32, 34]. (vi) Coefficient of friction did not correlate with charge intensities [29–34, 36] (vii) Variables that may relate to wear phenomena (i.e., sliding speed [29], normal load [30], hardness [31]) correlated with charge intensities. For insulators, the intensity of both negative and positive charges decreased for increasing hardness. (viii) Conditions that may lower wear under boundary lubrication conditions (i.e., hydrocarbon lubricants [36]) reduced charge intensities. (ix) Increasing charge intensities correlated with oxide-film thickness on metal substrates [29]. (x) Charge intensities from ceramics (e.g., Si_3N_4, Al_2O_3, ZrO_2 and soda lime glass) were affected by different gas species atmospheres (e.g., wet and dry air, O_2, N_2, Ar, and He) at same pressure [32]. (xi) Charge intensities from ceramics reached a maximum for a characteristic value of gas-species pressure (i.e., of He, Ar and O_2) in the range 100 Pa to 10^5 Pa [34]. (xii) Under boundary lubrication with hydrocarbons, charge intensities from ceramics correlated with hydrocarbon chain-length. In this, charge intensity decreased for increasing chain-length [33, 35–37].

More recently, Molina et al. [42–45] employed a high-vacuum (of 10^{-6} Pascal or better) triboemission instrument they developed to measure negative- and positive-charge intensities and emission energy-distribution, using a channel electron multiplier (CEM) detector in the pulse-counting mode. Figure 2.20 displays the main features of that instrument.

Individual charged-particles reaching the CEM were detected as individual counts in acquisition windows as short as 10 ms, from diamond-cone or alumina-ball scratching of ceramic and semiconductor rotating disks. Molina et al. tested material contact pairs of rotating-disks of amorphous alumina, sapphire, Si_3N_4, and Si and Ge that were scratched by fixed pins: either a conical diamond stylus of 90° angle and nominal 0.060-inch tip or a 0.125-inch diameter alumina-ball. Low speeds and loads were employed (2 to 10 rpm for linear speeds of 0.096 to 0.48 cm/s at the circular wear tracks, and 2 to 5 N load for diamond-cone, or 10 N for alumina-ball sliding). The contact was applied for varied sliding periods; typical electron triboemission outputs are shown in Figs. 2.21 to 2.25.

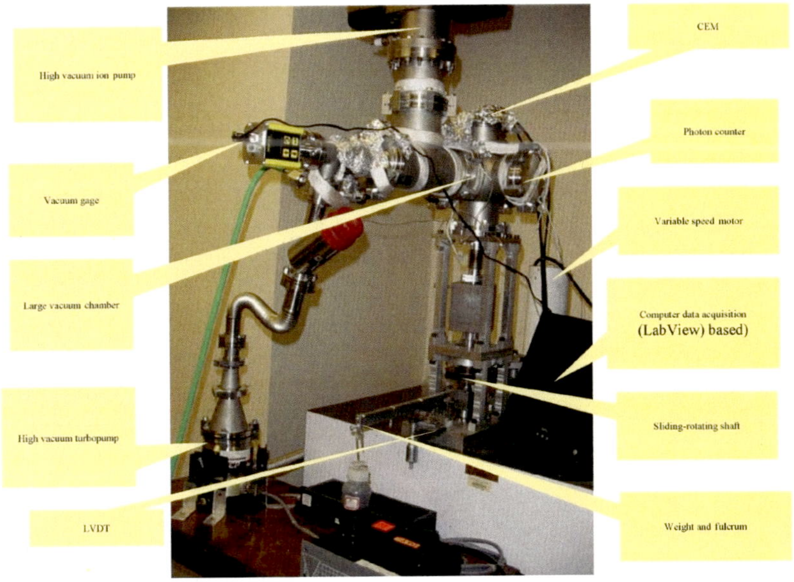

Figure 2.20 Photograph of Molina et al.'s vacuum tribometer as operating in Tribology Laboratory, Georgia Southern University.

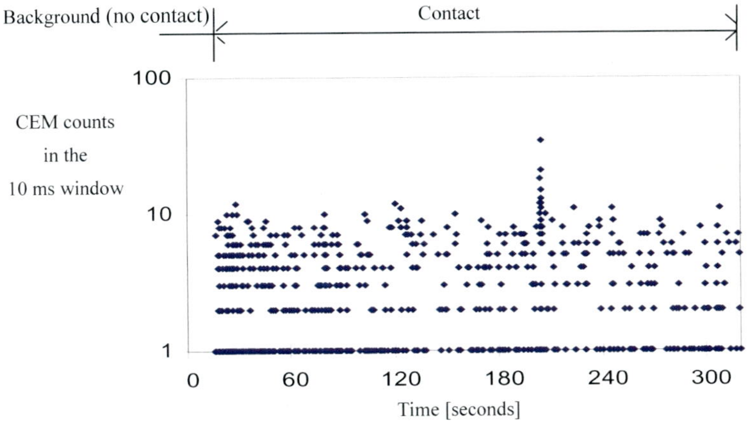

Figure 2.21 Negatively charged triboemission from diamond-cone on alumina sliding contact. Acquisition window: 10 ms. Load: 2 N. Speed: 0.48 cm/s. Reproduced with permission from G. J. Molina, M. J. Furey, C. Kajdas, N. Steika, On the mechanism of electron triboemission from ceramics, *The Annals of University "Dunarea de Jos" of Galati, Romania, VIII (Tribology)* (2005) 17–23.

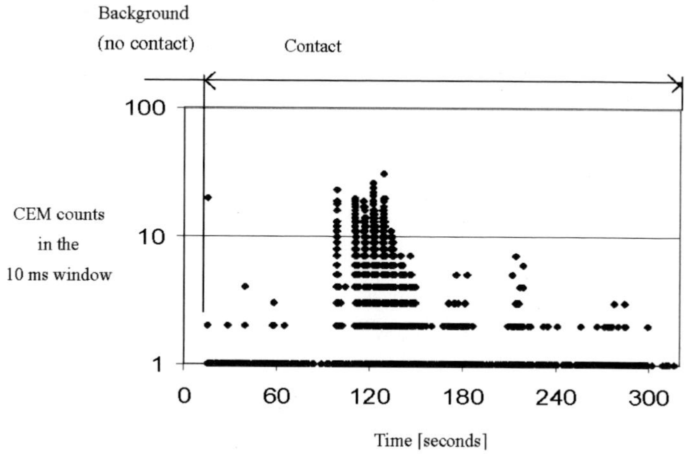

Figure 2.22 Negatively charged triboemission for alumina-ball on alumina disk. Acquisition window: 10 ms. Load: 10 N. Speed: 0.48 cm/s. Reproduced with permission from G. J. Molina, M. J. Furey, C. Kajdas, N. Steika, On the mechanism of electron triboemission from ceramics, *The Annals of University "Dunarea de Jos" of Galati, Romania, VIII (Tribology)* (2005) 17–23.

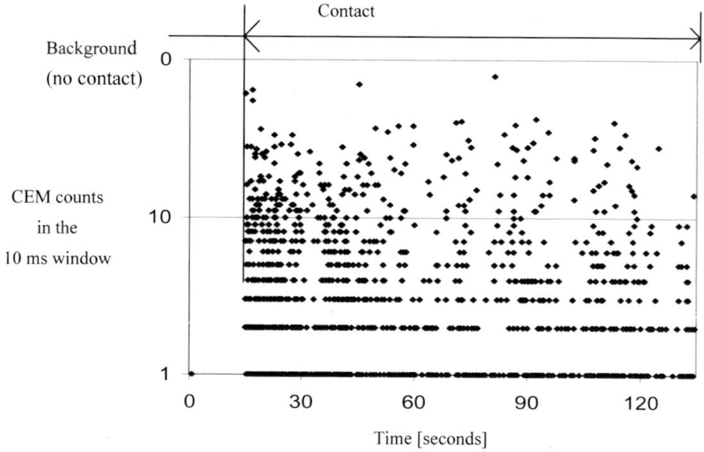

Figure 2.23 Negatively charged triboemission from diamond-cone on Si sliding contact. Acquisition window: 10 ms. Load: 5 N. Speed: 0.1 cm/s. Reproduced with permission from G. J. Molina, D. A. Mazilu, M. J. Furey, C. Kajdas, On triboemission from the sliding contact of Si and Ge, *The Annals of University "Dunarea De Jos" of Galaţi, Romania, VIII (Tribology)* (2004) 22–29.

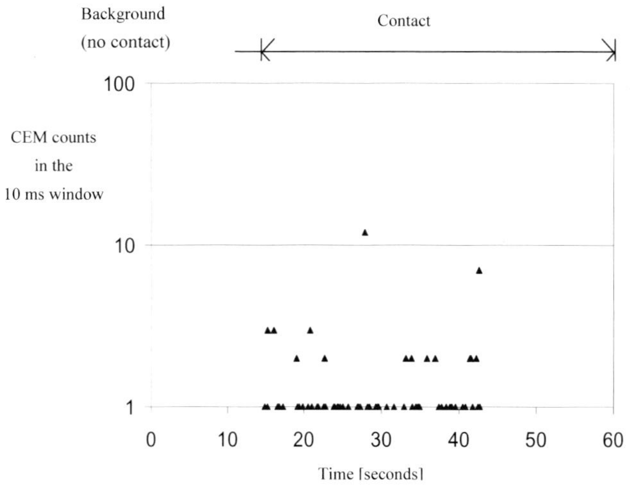

Figure 2.24 Negatively charged triboemission from diamond-cone on Ge sliding contact. Acquisition window: 10 ms. Load: 5 N. Speed: 0.1 cm/s. Reproduced with permission from G. J. Molina, D. A. Mazilu, M. J. Furey, C. Kajdas, On triboemission from the sliding contact of Si and Ge, *The Annals of University "Dunarea De Jos" of Galaţi, Romania, VIII (Tribology)* (2004) 22–29.

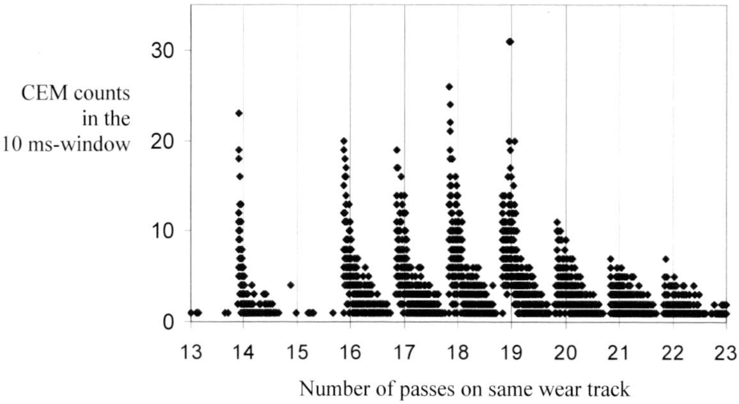

Figure 2.25 Decreasing negatively charged triboemission outputs from diamond-on-alumina repeated sliding contacts on same wear track. Acquisition window: 10 ms. Load: 2 N. Speed: 0.48 cm/s. Reproduced with permission from G. J. Molina, M. J. Furey, C. Kajdas, N. Steika, On the mechanism of electron triboemission from ceramics, *The Annals of University "Dunarea de Jos" of Galati, Romania, VIII (Tribology)* (2005) 17–23.

Figures 2.21 to 2.25 show that the measured negatively charged triboemission outputs are clearly associated with sliding contact. Triboemission rates and statistical significance of these experiment results were discussed by Molina et al. [42–45]. Emission starts at a relatively constant level upon application of sliding contact (in Figs. 2.21, 2.23, and 2.25), but for alumina-ball on alumina-disk (in Fig. 2.22) a considerable delay occurs until the sporadic appearance of large superimposed bursts of triboemission. These delays were consistently observed for the ball-on-flat geometry when compared to scratching by a diamond-cone [45, 46]. Molina et al. also observed that surface modifications and wear for the two compared systems were consistent with the different measured evolutions of the triboemission outputs: while emission and wear occurred upon application of diamond-cone scratching, for the ball-on-disk contacts the onset of wear coincided with the delayed large bursts. Decreasing triboemission also is observed when the same wear track is scratched in repeated passes [45]; this decrease relates to diminishing wear. A conceptual mechanism was postulated by Molina et al. to explain the delays in the occurrence of electron triboemission outputs; the mechanism was based on findings about electron work function reduction from plastic deformation and increased dislocation density which occur during sliding [45].

Triboemission data reported by Nakayama et al. [40] and Molina et al. [42] suggested that the intensity of emission correlates to material hardness. For instance, the found triboemission intensity differences between the two semiconductor materials (i.e., Si and Ge) can be related to their different hardness. Decreasing triboemission is also observed from the semiconductors when the same wear track is scratched in repeated passes, and such decreasing triboemission relate to decreasing semiconductor wear rate. For the cases of diamond-on-alumina and diamond-on-sapphire, retarded-energy spectrometry showed that a large fraction of the triboemitted negative charges were of low-energy (i.e. a large fraction of electrons triboemitted in their experiments was of low energy, in the range of 1–5 eV) [42–45], these ranges of emission energy were later supported by Nakayama et al. measurements [46]. These results are of significant importance to the NIRAM approach, for which it was hypothesized that the energy level of triboelectrons to initiate tribochemical reaction should be 1–4 eV.

Molina et al. did extensive work on frequency analysis of triboemission outputs, including testing of fit to known distributions [47] and interpretation of the mechanism of triboemission as a stochastic multiplicative process with superimposed deterministic features [48] by introducing Shannon spectral entropy as suitable analysis tool for triboemission-output frequency-of-occurrence. For instance, triboemission outputs from diamond-scratching of the semiconductor Si show average emission-rates which are relatively constant when computed for different-length periods (e.g., for different data-acquisition windows); they do not include sporadic but rather repeated large bursts of diminishing intensity. This suggests a characteristics quasi-periodicity for Si triboemission that is not observed for insulators.

A relevant study was recently carried out by Le Rouzic and Reddyhoff [49] of electron triboemission from aluminum oxide surfaces scratched by diamond using a microchannel plate feeding a phosphor screen, to visualize the spatial distribution of emitted electrons in real time. They used the Shannon spectral entropy analysis to study the probability density of the emission, which suggested, as in the conclusions of Molina et al. [48], that partially deterministic mechanisms were responsible for the phenomena. Spatially aligned distributions of electron bursts support the hypothesis that electron emission may arise due to the formation of surface cracks, probably associated to the presence of localized defects as sources. But also sustained after-emission (of length of up to 1 s) was observed from localized defects, and such emission was often reactivated every time the contact repeated over the same defect; that prolonged emission (which fluctuated over time) is orders of magnitude to that typical of crack propagating events, but likely to be the result of charging on the surface. Le Rouzic and Reddyhoff [49] proposed that the measured electron triboemission is due to at least three mechanisms: (i) deformation of the surface in front of the sliding tip, (ii) crack formation on the surface, and (iii) emission from the wear scar after contact has occurred, and most likely from the developed surface charging.

A thorough review of the triboemission phenomena requires consideration to the often simultaneous effects of triboelectrification and of triboluminescence. The recent work of Hiratsuka et al. [50], presented their findings on triboelectrification and triboluminescence of polymers and on possible connections to tribocatalysis [51]; their extensive work [52] in Chiba Institute of Technology, Japan, explored in particular the emission of photons from alumina-on-alumina, and from alumina on typical catalytic metals as sliding contact pairs. Typical results from some of these sliding contact systems, which were generated by chemiluminescence apparatus including a photomultiplier, are shown in Fig. 2.26, in detected arbitrary units (a.u.).

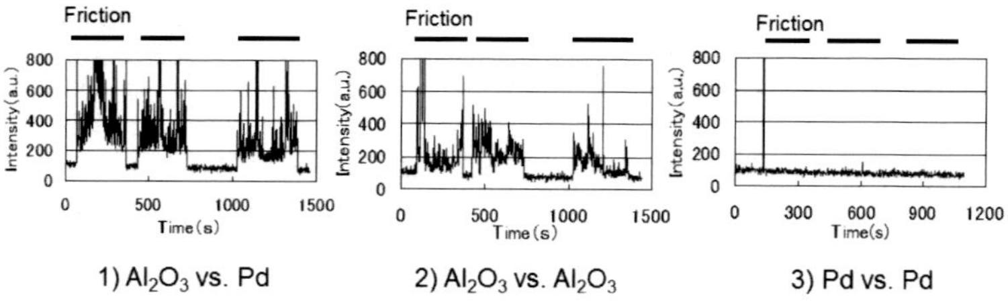

Figure 2.26 Bursts of triboluminescence, in arbitrary units (a.u.), detected from sliding the sliding systems of Al_2O_3/Pd, Al_2O_3/Al_2O_3, and Pd/Pd of (1) Al_2O_3/Pd, (2) Al_2O_3/Al_2O_3, and (3) Pd/Pd. Load of 1.0 N and sliding velocity of 60 mm/s. Reproduced with permission from Tribology and Lubrication Engineering Society. G. J. Molina, R. McDaniel, K. Hiratsuka, C. Kajdas, A preliminary study on tribo-photoemission from sliding surfaces, *Proceedings of the 2010 Society of Tribologist and Lubrication Engineers Annual Meeting and Exhibition*, Las Vegas, NV, May 17–21 (2010).

Friction periods in Fig. 2.26 are shown corresponding to solid lines (bars) in each graph. All the photon counts were obtained for the same experimental conditions, thus the measured intensities were comparable. While bursts of photons were detected from the sliding contact of Al_2O_3/Pd and of Al_2O_3/Al_2O_3, no significant photon emission was detected from Pd/Pd contact system structures; photoemission bursts presented in Fig. 2.26 resemble that of typical bursts of electron triboelectrons in Fig. 2.22 from an alumina ball sliding on alumina disk. More recently Nevshupa [53] observed that both uniform and/or burst-type triboluminescence modes occurred during sliding of a diamond pin along a soda-lime disc in an argon atmosphere in the pressure range between 10^5 and 10^{-2} Pa. He found a progression of combined triboluminescence modes as gas pressure increases: chaotic weak bursts occur for gas pressure below 20 Pa, followed by periodic strong bursts in the pressure range between 20 and 300 Pa, which develops to chaotic bursts with uniform triboluminescence between 300 Pa and 10 kPa, but disappearance of the chaotic bursts occurs above 10 kPa, to remain as uniform triboluminesce only at higher than 10 kPa pressures. Analysis of the output time series generated by the employed photomultiplier-detector in counting mode revealed two frequency-of-occurrence distributions: exponential-like and bell-shaped, that, respectively corresponded to the burst-type and to uniform triboluminesce modes. These results qualitatively match the finds of Molina et al. [47] from their frequency analysis of electron triboemission outputs in vacuum. In Nevshupa's same work [53], measurements of electrostatic potentials suggested that most of the positive charge would be concentrated on the pin while the negative charge was spread over a large surface area on the disc; the fairly symmetric radial-component of the electric field around the pin supports a corona model, where gas-discharge would occur in radial-to-the-pin direction rather than across the gap between the pin and the disc, as it was traditionally assumed.

Recent research regarding the existence of triboplasmas is relevant to the understanding of triboemission. Between 2002 and 2009 Nakayama et al. [54–59] presented their investigations on the emission of UV-photons from the scratching of sapphire and diamond-like carbon, and the production of a micro-triboplasma around and trailing the sliding contacts. More recently [60] they showed that by applying a magnetic field to the sliding contact between diamond and sapphire, the distribution and intensity of the triboplasma can be changed and controlled. In 2013 Nakayama [61] presented work on the effect of normal force on the triboplasma generated under oil lubrication. He concluded that infrared photon emission intensity from the sliding contact increased linearly with the contact force and with the real area of contact produced by it; and that a critical normal force would exist, below which the plasma generation is negligible and above which it would greatly increase. Nakayama hypothesized that air molecules in the oil were excited them by tribocharging and then an electron avalanche process would occur in oil air-bubbles or cavities.

Following sections discuss the triboemission phenomena as part of the boundary friction and tribocatalysis processes; in particular, Molina and Kajdas have discussed the observed triboemission outputs and their connections to tribocatalysis in [62], and they presented a model proposing that the reduction of activation energy in some heterogeneous catalytic reactions can be explained by spatial anisotropy of triboemitted electrons and photons [63].

2.2.1.4 Discussion of triboemission phenomena as part of the boundary friction process

The complexity of the phenomena leading to triboemission, which occur in the boundary lubrication or dry friction regimes, can be illustrated as in Fig. 2.27 [50]. As depicted in this figure, "emission" is the key in a number of tribological interactions. Each particular tribological system produces wear debris (which can be understood as emission of surface material) and tribochemical products (such as degradation or synthesis of molecules) because each encompasses very specific interactions in which triboemission can play a fundamental role. One of the causes for tribochemical reactions is the main theme of this chapter—triboelectron emission. The electron emission is likely triggered by surface cracks associated with wear [27, 64]. In this way, triboemission cycle is established. Other concurrent mechanisms involved in triboemission origin are discussed and reviewed in Section 2.2.5.

The work published by Kajdas et al. in 2002 [65] reviewed phenomena related to the boundary friction regime. The central thrust of that review was on triboemission, on the boundary friction process, and on tribochemistry of the boundary lubrication process. The interactions taking place at the tribological contact follow some specific physical and chemical processes, which in turn control both friction and wear of the involved mating elements. More specifically, forces are transmitted, physical and chemical nature of the materials is changed, and energy is exchanged. All these complex and interacting surface phenomena were presented in 1986 by N. P. Suh in his book on tribophysics [66]. He described the basic mechanisms governing interfacial behavior, illustrating how the basic theories can be used to provide practical solutions to friction and wear issues. More recently the physics and chemistry of tribological wear has been discussed by Kajdas [67].

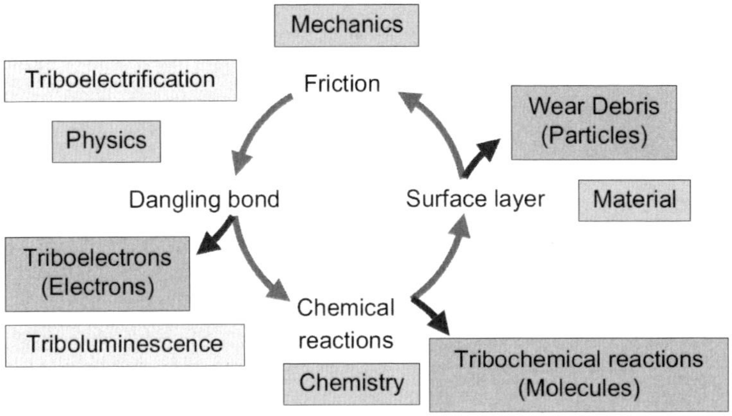

Figure 2.27 Mechanical, physical, chemical, and material interactions, and their outputs in a schematics of general tribology process. Reprinted from *Tribol. Int.*, 55, K. Hiratsuka, K. Hosotani, Effects of friction type and humidity on triboelectrification and triboluminescence among eight kinds of polymers, 87–99, Copyright (2012), with permission from Elsevier.

The most likely energy absorbing mechanisms during abrasive wear and the resulting dissipation processes are depicted in Fig. 1 of ref. [68]. About 90% of the work done during abrasion and in plastically deforming the material is converted to heat, which may subsequently cause local temperature rises and drive phase changes, and it can provide the thermal activation for recovery and recrystallization processes [68].

2.2.2 Conventional Exoelectron Emission

Better understanding of exoelectron emission processes requires some brief discussion of conventional electron emission ones. They include thermionic emission, photoemission and field emission as they, respectively correspond to the different energy sources, thermal, photon excitation, and external electric fields, for ejecting electrons from a solid.

2.2.2.1 Thermionic emission

Thermionic emission, also the energy required for it determined by the work function (ϕ), occurs when the temperature of a material is high enough; such that kT is of the order of the work function ϕ of the material (where T is the absolute temperature and k is the Boltzmann constant). The emission current I is given by

$$I = KT^2 \exp(-\phi/kT),$$

where K is a characteristic constant for the material. The main factor controlling change of the current with temperature is the exponential term. Since the term $\exp(-\phi/kT)$ approximates the order of 10^5 K for typical metals, very high temperatures are needed to obtain significant emission currents.

Thermionic emission of electrons is also known as thermal electron emission. Theory of thermionic emission from clean metal surfaces has been described by Modinos [69]. This theory takes into account the energy band structure of the metal, the inelastic scattering due to electron-electron collisions and the thermal vibration of the atoms. Applying the theory to, for instance, thermionic emission from Cu (100), the thermally emitted current from this crystallographic plane is a function of the applied field, and an almost periodic deviation from the Schottky line is found. The dependence of the measurable quantities on the parameters which enter the theory was analyzed and demonstrated by explicit numerical calculations in [69].

In thermionic emission, the charge carriers relate either to electrons or ions, also known as "thermions." Initially, that is just after emission, the charge is left in the emitting region; it is equal in magnitude and opposite in sign to the total charge emitted. But because of the very low loads and speeds used in experiments that produce electron triboemission, thermionic emission is ruled out as a possible source of such output.

2.2.2.2 Photoemission

Photoemission occurs when an electron absorbs a photon with energy greater than the work function of the material. The photoelectric effect is a quantum phenomenon where the maximum kinetic energy of the emitted electrons is ($h\nu - \phi$). More detailed information on field emission can be found in [70]. Figure 2.28 presents the typical energy-band diagrams for photoemission from a clean metal surface, and from an insulator or semiconductor [71], where E_F is the Fermi level separating the highest-energy filled electron states from a continuum of higher-lying unfilled states.

Because the photoemission effect is extensively documented and described in basic book references, this chapter does not deal with more details on this phenomenon.

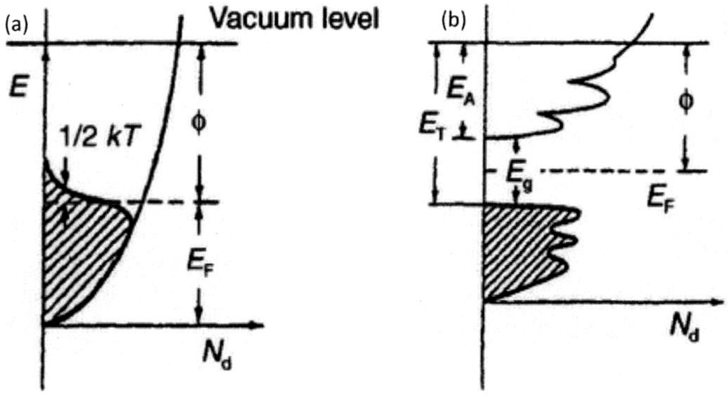

Figure 2.28 Density of occupied and empty states in (a) a metal and (b) a semiconductor or insulator, where: E_F = Fermi level; ϕ = work function; E_A = electron affinity; E_T = photoelectric threshold; E_g = fundamental gap for metal: $E_T = \phi$. *Photoemission in Solids I, Topics in Applied Physics,* Introduction, vol. 26, 1978, 1–104, M. Cardona, L. Ley, with permission of Springer.

2.2.3 Exoelectron Emission and Their Importance in Tribochemistry and Tribocatalysis

2.2.3.1 Historical background on the "Kramer effect"

In the years 1950–1952, Kramer studied emission of low-energy electrons (also known then as slow electrons), from metal oxides after mechanical treatment at room temperature [72, 73]. The treatments encompassed grinding, stretching, twisting and similar mechanical actions. His studies also included that of non-metals after they were exposed to photon and electron radiations. Kramer called the emission observed in his experiments "exoelectron-emission" (EEE). The EEE process can be described as a phenomenon evolving low-energy electrons from mechanically treated solid surfaces; the intensity of such emitted electrons was later shown that can be enhanced by heating (called tribo-stimulated emission), irradiation with ultraviolet light (called photon-stimulated emission), and an electric field (called field-stimulated emission).

The review work of Oster et al. [74] noted that works of Tanaka in the years 1935–1940 reported emission of slow electrons (of energy below 1 eV), which was initiated by preliminary exposure of oxide metals to radiation at room temperature. Another very early discovery relates to the Russell effect. According to the Grunberg and Wright review [78], Russell discovered in 1897 [79] that freshly abraded metal surfaces produce an image on photographic plates and suggested that hydrogen peroxide is responsible for this action. Further work was performed in the first decade of the 20th century [80, 81]. A more recent paper was published in 1939 [82] and it confirmed that hydrogen peroxide was produced when fresh surfaces of certain metals were exposed to the action of water and oxygen. For example, when Al, Mg, Ni, or zinc was cut under water, measurable quantities of hydrogen peroxide were generated.

The work of Oster et al. [74] earlier developed a classification of the relaxation electron emission processes. On the basis of this classification they examined and proposed experimental criteria for distinguishing different types of EEE mechanisms. Papers [80, 81] suggested and applied criteria to thermally stimulated exoelectron emission from magnesia (MgO) and calcium sulfate ($CaSO_4$); adequate confirmation was achieved by the Auger mechanism, for which MgO peaked at 350 K, and $CaSO_4$ for 450 K. It is of note, however, that for calcium sulfate, the EEE excitation was provided by both the simple trapping of the primary electron and the electron of the electron-hole pairs [81].

2.2.3.2 Electron attachment reactions

It is well known that low-energy electrons are very reactive in that they are effectively captured (or attached) by many molecules, which then undergo rapid decomposition [1]. For instance, the work of Bald [82] presented the decomposition of propionyl chloride triggered by slow electrons; propionyl chloride, also named propionic acid chloride (PAC), reacts easily with water, amines and alcohols (PAC belongs to the group with the general formula R–C(O)X, where X is a halogen atom and R may be aliphatic, hydrogen, aromatic, or any other organic rest).

Low energy electron attachment to PAC was studied by means of a crossed beam experiment [82] in which two distinct resonances, located near 0 and 0.7 eV, had been observed: the first resonance was attributed to a temporary occupation of the lowest unoccupied molecular orbital (LUMO). It mainly decays into chlorine, but also the fragments $(M{-}CH_2)^-$, $(COCl)^-$ and $(M{-}Cl)^-$ were observed [82]. Additionally the non-decomposed parent molecular anion M^- is observed. The second resonance exclusively leads to Cl^- generation.

The above-discussed experiments performed by Bald et al. were the first study of the PAC interaction with low energy electrons, ranging from 0 to 3 eV. That paper discusses the origin of the found resonances, their decay channels, and their possible role in the PAC degradation or decomposition, along with compounds related to the earth's atmosphere, as discussed by the work of Graedel and Crutzen [83].

Important to this discussion, electron-attachment reactions to molecules in the gas phase have been studied in the last three decades, but experiments in clusters have been performed only recently. In his very detailed and specific paper [1], Illenberger focused on reactions in weakly bound clusters induced by low-energy electron capture. Particular emphasis was placed on the question of how basic quantities of such processes (i.e., attachment energy, evolution of the negatively charged compound, and energy distribution of the products ultimately formed) could behave if proceeding from an isolated molecule which undergoes a single collision to a molecular aggregate [1].

Negatively charged clusters are of particular interest because they are considered to provide models for excess electrons in liquids [84, 85]. Studies on electron-attachment reactions in homogeneous and heterogeneous clusters allow insight into fundamental properties of molecular ions, including electron-transfer processes. At this point, it is convenient to briefly recall some important facts on negative ions: The electron affinity in the relation of a neutral particle (M) with its negative ion (M^-) is the basic quantity. The adiabatic electron affinity of a particle might be defined as the energy difference between the neutral and the anion in their respective ground states. By convention, the electron affinity of M is considered positive if the ground state of M^- lies below that of M, and negative, if M^- lies above the neutral molecule M [84, 85]. A positive value for the electron affinity indicates the existence of a stable anion in which the extra electron exists in a (thermodynamically) bound state.

In molecules which are characterized by considerable geometry changes between the anion and the neutral, a difference between the EA (adiabatic electron affinity) of M and the VDE (vertical detachment energy) of M. The VDE is the number which is experimentally obtained in a Franck-Condon transition in photo-detachment from molecular anions, in complete analogy to adiabatic and vertical ionization energies in photo-ionization [1]. In systems where the geometry change is not too large, optical photo-detachment does directly allow measure the adiabatic electron affinity [86]; in the gas phase under single collision conditions a stable molecular anion may be produced by electron transfer from neutrals or anions according to either of the following processes:

$$A + M \rightarrow A^+ + M^-$$

$$A^- + M \rightarrow A + M^-$$

Because the binding energy of the extra electron to M is usually less than the ionization energy of A, the first reaction is typically endothermic and can only occur if A and M contain sufficient energy, referred to as translational/internal energy [86]. Further details along with references until year 1992 are described and discussed in [1]. Electron attachment mass spectroscopy (EA-MS) allows determination of the types of reactive intermediates formed from organic compounds [87]. Under conditions of EA-MS, negative ion spectra can be obtained.

2.2.3.3 Significance of EEEs in tribochemistry

In Section 2.1 we define tribochemistry as the process proceeding between an environment and friction surfaces, and/or under boundary lubrication conditions, being controlled either by emission or transfer of electrons. We consider the tribochemistry process as a subfield of mechanochemistry, because it is activated by the mechanical energy. In the background of Section 2.1, several works by Carey-Lea [88, 89] are described. His works were greatly interested in the relations of energy to the chemical changes. He was the first mechanochemist/tribochemists because his experiments related to the first systematic investigations on the chemical effects of the mechanical action.

To avoid repeating text of Section 2.1, here we only underline that both mechanochemistry and tribochemistry can be well compared with term order, as in the names of chemical physics and physical chemistry. In the former term, chemistry is first and physics goes second. In tribochemistry, friction (tribos) is the first. By and large, further details on the importance of exoelectrons for tribochemistry can be found in Section 2.1 and related references.

2.2.3.4 Importance of EEs in catalysis and tribocatalysis

Section 3.1 presents all detailed information on tribocatalysis, along with a novel approach to its mechanism based on EEs. In this section we focus on specific validity of electron emission processes in catalysis, starting with pioneered experiments demonstrating that thermally stimulated EEE occurred in the same range of temperatures as that typical for heterogeneous catalysis [4]. Other researchers [9] also observed emission of negative charges (electrons and ions) during catalytic decomposition of methanol and formic acid. Hoenig and Tamjidi [8] found that the EEE level correlated to the rate of reaction for catalytic oxidation of carbon oxide, hydrogen and ammonia over a heated platinum wire. It is worthwhile noting that Krylova et al. [9] suggested that the measured thermal emission curves characterized the energy spectrum of the surface trapping levels for electrons participating in the catalysis process. The emitted electrons can also be applied to monitor oxidation rate. For instance, reference [8] suggested that the electrons from the catalyzed reaction may be used to monitor the rate of oxidation of carbon monoxide and methane over palladium catalysts.

It had been previously shown that the rate of EEE-decay from catalysts surfaces correlates with oxygen adsorption, which is known to be required for heterogeneous catalysis [90]. At the same time, the paper by Tomishko [91] considers and provides experimental proof for low-energy plasma, which is formed in the catalysis zone. It has been stressed that plasmas may play an important role in the mechanisms of catalytic processes. He found that a catalytically active surface is being charged under reaction conditions, and that it can exchange charges with the reactants. An earlier research by Grinevich et al. [92] demonstrated that during heterogeneous catalytic oxidation processes the catalyst surface becomes positively charged by emitting negatively charged particles into the gas phase.

2.2.3.5 EEE as the background for the NIRAM (negative ion radical action mechanism) approach

Publications [93, 94] review details of the NIRAM approach and of its role in tribochemistry. The latter review work [94] also presents its connection with tribocatalysis. The NIRAM approach comprises the following major steps:

(a) Low-energy electron emission and generation of positively charged spots; (b) Interaction of the emitted electrons with the lubricant molecules producing negative ions and radicals on tops of the rubbing surfaces; (c) Reaction of negative ions with positively charged sites of friction surfaces, producing organometallic/inorganic film, protecting the rubbing surfaces from wear; (d) Cleaving chemical bonds of organometallic compounds, resulting in producing inorganic films; (e) Eventual destruction of protective layer caused by wear, followed by electron emission and subsequent formation of a new protective film.

The possibility that chemical reactions in solids can be initiated by mechanical deformation have been considered since early last century, although at that time little was known about the equivalence of different forms of energy and whether these were convertible one to the other [95]. The enhanced surface activity was expressed in 1974 as follows [96]:

Enhanced reactivity = Exoelectron + Elevated temperature + High pressure

The mechanical action at solid surfaces tends to promote chemical reactions and produce surface chemistry that may be entirely different to that observed in static studies. Under boundary friction conditions, the exposed surface is extremely reactive due to mechanical activity. For instance, hydrocarbons can chemisorb and even chemically react with the metal surface [97]. In other words, friction initiates and accelerates chemical reactions that otherwise would take place at much higher temperatures or would not initiate. These phenomena make up a special branch of chemistry dealing with chemical and physico-chemical changes of solids due to the influence of mechanical energy [98].

It might be assumed that the first work pertinent to tribochemistry was performed by Carey-Lea [88, 89], who showed that shearing action was more effective than simple hydrostatic pressure and, also that the halogens produced by grinding of silver halides were able to react with moisture to produce the corresponding acid. Twenty years later, Parker [99] demonstrated more conclusively the efficacy of mechanical shearing in bringing about chemical reaction. Another work clearly related to tribochemistry was performed in early forties by Shaw [100]. Ten years later Grunberg [101] succeeded in demonstrating that metal cutting process under water can generate hydrogen peroxide and, accounted for the results in terms of the Kramer effect.

Goldblatt [102] explained the lubricating properties of polynuclear aromatics by assuming that, as a first step in the chemical processes, radical anions of the aromatic hydrocarbons are generated at the freshly abraded surface. He assumed that one of the first processes taking place under boundary lubrication conditions is the electron transfer. Rosenfeld [103] reported hydrogen formation in some lubricating oils during sliding wear experiments; they concluded that hydrogen formation is the result of a chemical reaction which occurs between the lubricating oils and the fresh metal surfaces.

To understand the chemical behavior of lubricant components under boundary lubrication conditions works of Kajdas et al. [104–106] proposed a concept of negative-ion-radical reactive intermediates formation for some components. The concept is based upon the ionization mechanism of lubricant compounds caused by the action of electrons of low energy (of 1 to 4 eV). Electrons of such energy are spontaneously emitted from most fresh surfaces formed during friction. The principal thesis of the model is that lubricant components form anions which are then chemisorbed on the positively charged areas of rubbing surfaces. The formation of negative ions and their decomposition process were simulated by electron attachment mass spectrography (EA MS). This type of apparatus uses electrons of an energy range similar to those of exoelectrons [87].

2.2.3.6 Importance of the tribopolymerization process in combination with the NIRAM approach

Tribopolymerization models, as a general mechanism of boundary lubrication combined with as an antiwear action mechanism, were presented and discussed by Furey and Kajdas 28 years ago [107]. To evaluate the validity of the anionic-radical mechanism, two metal systems were investigated [108, 109], a hard steel ball on a softer steel plate and a hard ball on an aluminum plate. Both metals emit exoelectrons under tribological conditions, but aluminum was found to produce more exoelectrons than steel did [110]. The obtained results clearly support the negative ion-radical concept, showing that addition-type tribopolymerization can be initiated by EEE. For example, with aluminum on steel, the addition of 1% styrene to hexadecane reduced the wear volume of the disc by over 65%. In the steel-on-aluminum system, conclusive evidence of polystyrene was found via FTIR microspectrometry on the discs lubricated with styrene-containing solutions.

It has also been found [111] that laurylmethacrylate, diallyl phthalate and vinyl acetate reduced alumina wear in a pin-on-disc test by 65%, 80% and 57%, respectively. However, the anti-wear (AW) effect of vinyl octadecyl ether under the same friction conditions was marginal. It was assumed that the difference can be explained in terms of polymerization mechanisms. Lauryl methacrylate, diallyl phthalate, and vinyl acetate can polymerize according to anionic or radical mechanisms, whereas vinyl octadecyl ether polymerizes only by cationic mechanism. These results are in line with what the negative ion-radical action mechanism predicts: vinyl octadecyl ether, polymerizing only by a cationic mechanism, cannot be effective as an antiwear additive for alumina lubrication under the friction conditions described in references [111, 112].

2.2.4 Mechanisms of Triboemission

The detailed physics of triboemission is still unknown, but is clear that, as exoelectrons, triboelectrons are emitted for external excitation lower than the electron work function of the material, customarily referred as work function (WF); they are the low-energy triboelectrons measured by Molina et al., and Nakayama et al. Electron triboemission is a case of the more general phenomena of electron exoemission, which is a suitable probe of electronic excitation: the experimental work of Nienhaus et al. [113] showed that exoelectron emission is just a fraction of the total electronic excitation on the surface, the majority being detectable as internal chemically generated hot carriers (chemicurrents), for energy levels smaller than the reaction enthalpy. They estimated that for the oxidation of Mg surfaces approximately 10% of the chemical reaction energy is dissipated by non-adiabatic electronic excitation and it is produced as exoemission.

Several surface mechanisms have been postulated to explain ejection as electron triboemission of a part of the mechanically-involved charge: microfracture and surface-charge effects [113], gas-discharge tribomicroplasma [114] and mechanical modification of work function (WF). In the latter, Zharin et al. [115] used a non-contacting Kelvin probe to measure variation of the WF during tribological damage, and Bhushan et al. [116] to observe changes of Kelvin-probe potential to detect wear precursors at ultra-low loads for Au, Al, Si and alumina. More recently Li et al. [117] correlated elastic deformation to decreasing WF for Cu, Al and steel during sliding, and reported a stable WF value from the onset of plasticity, which was measured at 0.3 to 0.4 eV lower than those of the unworn surfaces. They also presented a model by which increases in dislocation density and surface roughness led to lower WF for Cu [118, 119]. Molina et al. observed evolution of the electron triboemission outputs can correspond to the sequence of (a) WF reduction and low-rate emission during plastic deformation and (b) emission for stable lower WF from the onset of wear and during the continuous sliding and contact [45].

The absence of post-contact electron emission for semiconductors can be related to the fact that only negligible surface charge exists for semiconductors under vacuum [120] and therefore no triboemission would be produced. However, the proposed mechanisms do not explain triboemission features that are likely related to wear evolution, as the decreasing triboemission rates when sliding for repeated passes on the same wear track, or the non-deterministic features of the outputs [48].

2.2.5 Conclusions

The emission of electrons, ions, neutral particles, photons, and acoustic emission under conditions of tribological contact and damage is called triboemission. Research works on charged-particle triboemission were mainly carried out by Molina et al. [42–45, 47, 48], Nakayama et al. [28–41, 54–60], and Dickinson et al. [27, 63]; the recent works of Hiratsuka et al. [50–52] focused on the emission of photons from similar material systems and sliding conditions. They produced extensive results on triboemitted electrons from the scratching of ceramics (i.e., alumina, sapphire and silicon nitride) and semiconductors (i.e., Si and Ge) in vacuum and several atmospheres; these charged-emissions are characteristically burst-type for low-energy electrons (e.g., up to 10 eV). The possibility of thermionic emission was ruled out as the origin of this triboemission because of the very low loads and velocities employed in experiments. Several surface mechanisms have been postulated to explain ejection as electron triboemission of a part of the mechanically-involved charge, but the detailed physics of triboemission is still unknown, while it is apparent that triboelectrons are emitted for external excitation lower than the electron work function of the scratched material.

Recent research by Nakayama et al. [58–61] proved the existence of micro-triboplasmas, which are relevant to the understanding of triboemission and to its connections to tribochemistry and tribocatalysis. This chapter discussed evidence on the specific validity of electron emission processes in catalysis, as experiments demonstrating that EEE occurred in the same range of temperatures as that typical for heterogeneous catalysis, that catalytically active surfaces are being charged under reaction conditions, and that they can exchange charges with the reactants, as well as the existence of low-energy plasmas [91] which are formed in the catalysis zone. This chapter also discussed how a specifics boundary lubrication process, the tribopolymerization, can be explained by means of low-energy triboemitted electrons and the NIRAM approach.

References

1. E. Illenberger, Electron-attachment reactions in molecular clusters, *Chem. Rev.*, 92 (1992) 1589–1609.
2. W. D. von Voss, F. R. Brotzen, Electron emission from plastically strained aluminum, *J. Appl. Phys.*, 30(11) (1959) 1639–1645.
3. J. A. Ramsey, Exoelectron emission from abraded metal surfaces at high and ultrahigh vacuums, *J. Appl. Phys.*, 37 (1966) 452–453.
4. N. Sato, M. Seo, Chemically stimulated exo-emission from silver catalyst, *Nature*, 216 (Oct) (1967) 361–362. DOI:1.1038/216361a0.
5. N. Sato, M. Seo, Chemically stimulated EEE from silver catalyst during partial oxidation of ethylene, *J. Catal.*, 24 (1972) 224–232.
6. G. J. Molina, M. J. Furey, A. L. Ritter, C. Kajdas, Importance of low-energy triboelectrons for the initiation of tribochemical reactions, *Tribologia i Praktyka (Poland)*, 1(187) (2003) 103–122.
7. A. Kulczycki, C. Kajdas, A new attempt to better understand Arrhenius equation and its activation energy, in: *Tribology in Engineering* (H. Pihtili, ed.), ISBN 978-953-51-1126-9, May (2012), chapter 4, InTech Open.
8. S. A. Hoenig, F. Tamjidi, EEE during heterogeneous catalysis (the effect of external electric potentials), *J. Catal.*, 28 (1973), 200–208.
9. V. Krylova, N. I. Konyushkina, I. A. Rodina, V. I. Svitov, Exoelectron emission accompanying catalytic reactions in adsorbed layers, *J. Catal.*, 60(1) (1979) 8–14.

10. S. A. Hoenig, M. G. Utter, Observations of exoelectron emission associated with heterogeneous catalysis, *J. Catal.*, 47 (1977) 210–213.
11. J. Chrzanowski, H. Teterycz, R. Klimkiewicz, Exoelectron emission from tin oxide materials, *J. Electrochem. Soc.*, 150(6) (2003) H140–H146.
12. M. M. Tomishko, Plasma in catalytic processes, *High Energy Chem.*, 37(5) (2003) 333–340.
13. V. I. Grinevich, N. V. Kolobova, V. V. Kostrov, Influence of catalysts on sulfur dioxide and carbon monoxide oxidation in a barrier-plasma discharge, *High Energy Chem.*, 31(6) (1997) 400–441.
14. U. Landman, B. Yoon, C. Zhang, U. Heiz, M. Arenz, Factors in gold nanocatalysis: Oxidation of CO in the non-scalable size regime, *Top. Catal.*, 44(1–2) (2007) 145–158.
15. B. Yoon, H. Hakkinen, U. Landman, A. S. Worz, J. M. Antonietti, S. Abbet, K. Judai, U. Heiz, Charging effects on bonding and catalyzed oxidation of CO on Au_8 Clusters on MgO, *Science*, 307, 21 (2005) 403–407.
16. T. M. Bernhardt, R. N. Barnett, U. Landman, S. M. Lang, Methane activation and catalytic ethylene formation on free Au_2, *Angew. Chem. Int. Ed.*, 49/5 (2010) 980–983.
17. H. Hakkinen, U. Landman, Gas-phase catalytic oxidation of CO by Au_2, *J. Am. Chem. Soc.*, 123 (2001) 9704–9705.
18. C. Zhang, B. Yoon, U. Landman, Predicted oxidation of CO catalyzed by Au nanoclusters on a thin defect-free MgO film supported on a Mo(100) surface, *J. Am. Chem. Soc.*, 129 (2007) 2228–2229.
19. M. Amft, N. V. Skorodumova, Catalytic activity of small MgO-supported Au clusters towards CO oxidation: A density functional study, *Phys. Rev. B*, 81 (2010) 195443.
20. S. M. Lang, T. M. Bernhardt, R. N. Barnett, B. Yoon, U. Landman, hydrogen-promoted oxygen activation by free gold cluster cations, *J. Am. Chem. Soc.*, 131 (2009) 8939–8951.
21. L. M. Molina, B. Hammer, Theoretical study of CO oxidation on Au nanoparticles supported by MgO(100), *Phys. Rev. B*, 69/15 (2004) 155424.
22. X. Wu1, L. Senapati, S. K. Nayak, A. Selloni, M. Hajaligol, A density functional study of carbon monoxide adsorption on small cationic, neutral, and anionic gold clusters, *J. Chem. Phys.*, 117 (2002) 4010–4015.
23. L. M. Molina, B. Hammer, Some recent theoretical advances in the understanding of the catalytic activity of Au, *Appl. Catal. A Gen.*, 291(21) (2005) 21–31.

24. J. C. Fierro-Gonzalez, B. C. Gates, Evidence of active species in CO oxidation catalyzed by highly dispersed supported gold, *Catal. Today*, 122(3–4) (2007) 201–210.

25. L. D. Socaciu, J. Hagen, J. Le Roux, D. Popolan, T. M. Bernhardt, L. Woste, S. Vajda, Strongly cluster size dependent reaction behavior of CO with O_2 on free silver cluster anions, *J. Chem. Phys.*, 120(5) (2004) 2078.

26. K. Nakayama, Tribophysical phenomena and tribochemical reaction, *Jpn. J. Tribol.*, 42(9) (1997) 712–717.

27. J. T. Dickinson, E. F. Donaldson, D. B. Snyder, Acoustic emission and electron emission during deformation of anodized aluminium, *J. Vac. Sci. Technol.*, 17(1) (1980) 429–432.

28. K. Nakayama, T. Fujiwara, H. Hashimoto, Exoelectron measurement in a scanning electron microscope, *J. Phys. E*, 17 (1984) 1199–1203.

29. K. Nakayama, H. Hashimoto, Y. Fukuda, Triboemission from aluminum oxide film in atmosphere, *Proceedings of Japan International Tribology Conference*, Nagoya, Japan (1990) 1141–1146.

30. K. Nakayama, H. Hashimoto, Triboemission from various materials in atmosphere, *Wear*, 147 (1991) 335–343.

31. K. Nakayama, H. Hashimoto, T. Susuki, Triboemission of charged particles and photons from solid surfaces during frictional damage, *J. Phys. D*, 25 (1992) 303–308.

32. K. Nakayama, H. Hashimoto, Triboemission of charged particles and photons from wearing ceramic surfaces in various gases, *Tribol. Trans.*, 35(4) (1992) 643–650.

33. K. Nakayama, Triboemission of charged particles from various solids under boundary lubrication conditions, *Wear*, 178 (1994) 61–67.

34. K. Nakayama, H. Hashimoto, Effect of surrounding gas pressure on triboemission of charged particles and photons from wearing ceramics surfaces, *Tribol. Trans.*, 38(1) (1995) 35–42.

35. K. Nakayama, H. Hashimoto, Triboemission of charged particles and photons from wearing ceramic surfaces in various hydrocarbon gases, *Wear*, 185 (1995) 183–188.

36. K. Nakayama, H. Hashimoto, Triboemission of electrons and ions during wear of solids under boundary lubrication with saturated hydrocarbon liquids, *Tribol. Trans.*, 38(3) (1995) 541–548.

37. K. Nakayama, H. Hashimoto, Triboemission, tribochemical reactions, and friction and wear in ceramics under various n-butane gas pressures, *Tribol. Int.*, 29(5) (1996) 385–393.

38. K. Nakayama, H. Ikeda, Triboemission characteristic of electrons during wear of amorphous carbon and hydrogenated amorphous carbon films in a dry air atmosphere, *Wear*, 198 (1996) 71–76.
39. K. Nakayama, H. Hashimoto, Friction, wear, and triboelectron emission of hydrogenated amorphous carbon films, *Tribol. Trans.*, 40(3) (1997) 507–513.
40. K. Nakayama, B. Bou-Said, H. Ikeda, Triboelectromagnetic phenomena of hydrogenated carbon films–tribo-electrons, -ions, -photons, and -charging, *Trans. ASME: J. Tribol.*, 119 (1997) 764–767.
41. K. Nakayama, *Triboemission and Triboplasma Generation with DLC Films in Tribology of Diamond-Like Carbon Films: Fundamentals and Applications* (C. Donnet, A. Erdemir, eds.), Springer (2008) 291–310.
42. G. J. Molina, M. J. Furey, A. L. Ritter, C. Kajdas, Triboemission from alumina, single crystal sapphire, and aluminum, *Wear*, 249(3–4) (2001) 214–219.
43. G. J. Molina, D. A. Mazilu, M. J. Furey, C. Kajdas, On triboemission from the sliding contact of Si and Ge, *The Annals of University "Dunarea De Jos" of Galați, Romania, VIII (Tribology)* (2004) 22–29.
44. G. J. Molina, M. J. Furey, A. Ritter, C. Kajdas, The role of triboemission in ceramic tribology: A review, *Tribol. Finnish J. Tribol.*, 21(4) (2003) 3–11.
45. G. J. Molina, M. J. Furey, C. Kajdas, N. Steika, On the mechanism of electron triboemission from ceramics, *The Annals of University "Dunarea de Jos" of Galati, Romania, VIII* (*Tribology*) (2005) 17–23.
46. K. Nakayama, T. Fujimoto, The energy of electrons emitted from wearing solid surfaces, *Tribol. Lett.*, 17(1) (2004) 75–81.
47. G. J. Molina, M. J. Furey, A. Ritter, C. Kajdas, Frequency analysis and modeling of charged-particle triboemission from ceramics, *Wear*, 255(1–6) (2003) 686–694.
48. G. J. Molina, M. J. Furey, C. Kajdas, A deterministic-chaos study of electron triboemission outputs, *ASME Trans. J. Tribol.*, 129(3) (2007) 679–683.
49. J. Le Rouzic, T. Reddyhoff, Spatially resolved triboemission measurements, *Tribol. Lett.*, 55 (2014) 245–252.
50. K. Hiratsuka, K. Hosotani, Effects of friction type and humidity on triboelectrification and triboluminescence among eight kinds of polymers, *Tribol. Int.*, 55 (2012) 87–99.
51. K. Hiratsuka, C. Kajdas, M. Yoshida, Tribo-catalysis in the synthesis reaction of carbon dioxide, *Tribol. Trans.*, 47(1) (2004) 86–93.

52. G. J. Molina, R. McDaniel, K. Hiratsuka, C. Kajdas, A preliminary study on tribo-photoemission from sliding surfaces, *Proceedings of the 2010 Society of Tribologist and Lubrication Engineers Annual Meeting and Exhibition*, Las Vegas, NV, May 17–21 (2010).

53. A. Nevshupa, Effect of gas pressure on the triboluminescence and contact electrification under mutual sliding of insulating materials, *J. Phys. D Appl. Phys.*, 46 (2013) 185501.

54. K. Nakayama, The plasma generated and photons emitted in an oil-lubricated sliding contact, *J. Phys. D. Appl. Phys.*, 40 (2007) 1103–1107.

55. K. Nakayama, T. Fujimoto, The energy of electrons emitted from wearing solid surfaces, *Tribol. Lett.*, 17 (2004) 75–81.

56. K. Nakayama, R. A. Nevshupa, Characteristics and pattern of plasma generated at sliding contact, *Trans. ASME J. Tribol.*, 125 (2003) 780–787.

57. K. Nakayama, R. A. Nevshupa, Effect of dry air pressure on characteristics and pattern of tribomicroplasma, *Vacuum*, 74 (2004) 11–17.

58. K. Nakayama, Triboemission of electrons, ions, and photons from diamondlike carbon films and generation of tribomicroplasma, *Surf. Coatings Technol.*, 188–189 (2004) 599–604.

59. K. Nakayama, Contact geometry and distribution of plasma generated in the vicinity of sliding contact, *Japan J. Appl. Phys.*, 46 (2007) 6007–6014.

60. K. Nakayama, Effect of magnetic field on the plasma generated during a sliding contact, *J. Phys. Conf. Ser.*, 301 (2011) 012069, *13th International Conference on Electrostatics*, IOP Publishing, doi:10.1088/1742-6596/301/1/012069.

61. K. Nakayama, Effect of normal force on the triboplasma generation under oil lubrication, *Tribol. Lett.*, 53 (2014), 449–456.

62. G. J. Molina, C. Kajdas, Triboemission phenomena and possible connections to tribocatalysis, *Proceedings of the European Conference on Tribology, EcoTrib 2011-Viennano 2011* (F. Franek, W. J. Bartz, A. Pauschitz, J. Vizintin, E. Ciulli, R. Crockett, eds.), Vienna, Austria, June 7–9 (2011) 779–784.

63. C. K. Kajdas, A. Kulczycki, K. J. Kurzydłowski, G. J. Molina, Activation energy of tribochemical and heterogeneous catalytic reactions, *Mater. Sci. Poland*, 28(2) (2010) 523–533.

64. M. Kim, S. C. Langford, J. T. Dickinson, Electron and photon emission accompanying the abrasion of MgO with diamond, *Tribol. Lett.*, 1 (1995) 147–157.

65. C. Kajdas, M. J. Furey, A. L. Ritter, G. J. Molina, Triboemission as a basic part of the boundary friction regime, *Lubrication Sci.*, 14(2) (2002) 223–254.

66. N. P. Suh, *Tribophysics*, Prentice-Hall, New Jersey, USA (1986).

67. C. Kajdas, Physics and chemistry of tribological wear. *10th International Colloquium on Tribology, Technische Akademie Esslingen, Ostfildern,* Germany, January 9–11 (1996).

68. M. A. Moore, Energy dissipation in abrasive wear, *ASTM International Conference on Wear of Materials*, Dearborn, MI, USA (K. C. Ludema, W. A. Glaeser, S. K. Rhee, eds.) (1979) 636–638.

69. A. Modinos, Theory of thermionic emission, *Surf. Sci.*, 115(3) (1982) 469–500.

70. J. S. Escher, NEA semiconductor photoemitters, in: *Semiconductors and Semimetals* (R. K. Willardson, A. C. Beer, eds.), vol. 15, chapter 3, Academic Press, New York (1981).

71. M. Cardona, L. Ley, Introduction, in: *Photoemission in Solids I, Topics in Applied Physics* (M. Cardona, L. Ley, eds.), vol. 26, Springer-Verlag, Berlin (1978) 1–104.

72. J. Kramer, *Der metallische Zustand*, Vandenhoeck u. Ruprecht, Goettingen (1950).

73. J. Kramer, Oberflächenusuchung an metallen und nichtmetallen mit exo- und photoelectronen, *Z. Phys.*, 133 (1952) 629–646.

74. L. Oster, V. Yaskolko, J. Haddar, Classification of exoelectron emission mechanisms, *Phys. Stat. Sol. (a)*, 174 (1999) 431–439.

75. L. Grunberg, K. H. Wright, Russell effect on evaporated metal films, *Nature*, 170 (1952) 456–457; doi:10.1038/170456b0.

76. W. J. Russell, On the action exerted by certain metals and other substances on a photographic plate, *Proc. R. Soc. Lon.*, 61 (1897) 424–438.

77. W. J. Russell, Action of plants on a photographic plate in the dark, *Proc. R. Soc. Lon. B*, 78 (1906) 385–390.

78. W. J. Russell, On the action of wood on a photographic plate in the dark, *Proc. R. Soc. Lon.*, 74 (1904) 131–134.

79. J. R. Churchill, The formation of hydrogen peroxide during corrosion reactions, *Trans. Electrochem. Soc.*, 76(1) (1939) 341–357.

80. A. R. Krasnaya, L. N. Oster, V. Yaskolko, Exoelectron emission from calcium sulphate excited by slow electrons, *Phys. Stat. Sol.* (*a*), 85 (1984) K145.

81. A. R. Krasnaya, L. N. Oster, V. Yaskolko, Exoelectron spectroscopy of dielectrics under photon and low-energy electron excitation, *Phys. Stat. Sol.* (*a*), 97 (1986) K153.

82. I. Bald, E. Illenberger, O. Ingolfsson, Decomposition of propionyl chloride triggered by slow electrons, *Chem. Phys. Lett.*, 442 (2007) 270–274.

83. T. E. Graedel, P. J. Crutzen, *Atmospheric Change: An Earth System Perspective*, W. H. Freeman, New York (1993).

84. P. S. Drzaic, J. Marks, J. I. Brauman, Gas phase ion chemistry; M. T. Browers (editor) vol. III, Academic Press, New York, NY, USA (1984) 168–213.

85. D. M. Wetzel, J. I. Brauman, Electron photodetachment spectroscopy of trapped negative ions, *Chem. Rev.*, 87(3) (1987) 607–622.

86. K. Lacmann, Collisional ionization, in: *Potential Energy Surfaces* (K. P. Lawley, ed.), Wiley, Publishers, New York, NY, USA (1980).

87. M. von Ardenne, K. Steinfelder, R. Tuemmler, *Elektronenanlagerungs-Massenspektrographie organischer Substanzen*, Springer Verlage, Berlin, Germany (1971).

88. M. Carey-Lea, Disruption of the silver haloid molecule by mechanical force, *Philos. Mag.*, 34 (1892) 46–50.

89. M. Carey-Lea, On endothermic reactions effected by mechanical force (First Part), *Philos. Mag.*, 36 (1893) 351–354.

90. J. Chrzanowski, H. Teterycz, R. Klimkiewicz, Exoelectron emission from tin oxide materials, *J. Electrochem. Soc.*, 150(6) (2003), H140–H146.

91. M. M. Tomishko, Plasma in catalytic processes, *High Energy Chem.*, 37(5) (2003) 333–340.

92. V. I. Grinevich, N. V. Kolobova, V. V. Kostrov, Influence of catalysts on sulfur dioxide and carbon monoxide oxidation in a barrier-plasma discharge, *High Energy Chem.*, 31(6) (1997) 400–441.

93. C. Kajdas, Importance of anionic reactive intermediates for lubricant component reactions with friction surfaces, *Lubrication Sci.*, 6 (1994) 203–228.

94. C. Kajdas, Importance of the triboemission process for tribochemical reaction, *Tribol. Int.*, 38 (2005) 337–353.

95. P. G. Fox, Review on mechanically initiated reactions in solids, *J. Mater. Sci.*, 10 (1975) 340–360.

96. C. N. Rowe, W. R. Murphy, in: *Proceedings of the Tribology Workshop* (F. F. Ling, ed.), National Science Foundation, Washington, D. C. (1974) 327.
97. D. H. Buckley, *Surface Effects in Adhesion Friction, Wear, and Lubrication*, Elsevier, Amsterdam (1981).
98. G. Heinicke, *Tribochemistry*, Carl Hansen Verlag, Munich, Germany (1984).
99. L. H. Parker, Reactions between solid substances, *J. Chem. Soc.*, 113 (1918) 396–409.
100. M. Shaw, The chemico-physical role of the cutting fluid, *Metal Prog.*, 42 (1942) 85–88.
101. L. Grunberg, The formation of hydrogen peroxide on fresh metal surfaces, *Proc. Phys. Soc.*, 66B (1953) 153–161.
102. I. L. Goldblatt, Model for lubrication behavior of polynuclear aromatics, *Ind. Eng. Chem. Prod. Res. Dev.*, 10 (1971) 270–278.
103. L. Rosenfeld, Gas release during wear: I. The phenomenon and preparations for identification, *Wear*, 40(2) (1976) 165–177.
104. C. Kajdas, On a negative-ion concept of EP reaction of organo-sulfur compounds, *ASLE Trans.*, 28 (1983) 21–30
105. C. Kajdas, About a negative-ion concept of the antiwear and antiseizure action of hydrocarbons during friction, *Wear*, 101 (1985) 1–12.
106. C. Kajdas, About an ionic-radical concept of the lubrication mechanism of alcohols, *Wear*, 116 (1987) 167–180.
107. M. Furey, C. Kajdas, Models of tribopolymerization as an antiwear mechanism, *Proc. Jap. Int. Trib. Conf. Nagoya Japan*, 2 (1990) 1089.
108. P. A. Marin-LaFleche, A study of tribopolymerization under fretting contact conditions, MSc. Thesis, Deptment of Mechanical Engineering, Virginia Polytechnic Institute and State University, Blacksburg, VA, USA, May (1990).
109. C. Kajdas, P. M. Marin-Lafleche, M. J. Furey, J. W. Hellgeth, T. C. Ward, A study of tribopolymerisation under fretting contact conditions, *Lubrication Sci.*, 6(1) (1993) 51–89.
110. M. Connely, E. Rabinowicz, Detecting wear and migration of solid film lubricants using simultaneous exoelectron emission, *ASLE Trans.*, 26 (1983) 139–143.
111. M. J. Furey, C. Kajdas, R. Kempinski, B. Tripathy, Tribopolymerization and behavior of oxygen-containing monomers in reduction of ceramic wear, *Proceedings of the 6th International Congress on Tribology, EUROTRIB'93*, Budapest, Hungary (1993).

112. M. J. Furey, C. Kajdas, R. Kempinski, B. Tripathy, Action mechanism of selected vinyl monomers under boundary lubrication of alumina-on-alumina system, *Lubrication Sci.*, 10 (1997) 3–25.

113. H. Nienhaus, S. Glass, Probing the distribution of hot charge carriers generated in Mg surfaces by oxidation, *Surf. Sci.*, 600 (2006) 4285–4289.

114. K. Nakayama, R. Nevshupa, Plasma generation in a gap around a sliding contact, *J. Phys. D Appl. Phys.*, 35 (2002) L53–L56, PII: S0022-3727(02)33009-2.

115. E. S. Zanoria, K. Hamall, S. Daniluk, A. Zharin, The non-vibrating Kelvin probe and its application for monitoring surface wear, *J. Test. Eval.*, 25(2) (1997) 233–238.

116. B. Bhushan, A. V. Goldade, Kelvin probe microscopy measurements at surface potential change under wear at low loads, *Wear*, 244(1) (2000) 104–117.

117. W. Li, D. Y. Li, Exploring the application of the Kelvin method in studying the history prior to wear and the onset of wear, *Wear*, 253 (2002) 746–751.

118. W. Li, D. Y. Li, A study on the kinetic response of the electron work function to wear, *Wear*, 255 (2003) 333–340.

119. W. Li, D. Y. Li, Effect of dislocation on electron work function of metal surface, *Mater. Sci. Technol.*, 18 (2002) 1057–1060.

120. Y. S. Zharkikh, V. V. Pianitsky, O. V. Tretiak, Effect of the weak form of adsorption on the Si surface charge, *Appl. Surf. Sci.*, 134 (1998) 263–266.

2.3 Effect of Fluid Flow on Oxidation of Lubricants

Akira Sasaki

Maintek Consultant, Yokohama, Japan

akirassk@ps-yokohama.com

2.3.1 Review of the Past Studies on Static Electricity

Static electrification is an old subject. It is known that when a low conductive liquid like hydrocarbon flows in a tube or porous materials, electric charges appear in the liquid and the opposite charges will appear somewhere nearby simultaneously and that when it flows into a storage tank, electric charges will accumulate in the liquid and cause spark discharges. However, it is said that little attention was paid to it because of poor repeatability and that the state of the knowledge of it was little more advanced in 1940 than it was 50 years ago, according to Henry [1]. Swann [2] made investigation and reported that explosions were reported in the middle 1920s in the dry cleaning industry, where benzol was used in washing machines, which rotated at 300–400 rpm. Winter [3] mentioned that the generation of electric charges in hydrocarbon fuels during transfer is recognized as an explosion hazard. He studied the generation of electric charges in hydrocarbon fuel during fueling to aircraft. According to him, the conductivity of aviation fuels is within the range one to 5 picomho/m. When a charged fuel enters an aircraft tank at a fast rate, the mixing time for the fuel within the tank may be high compared with the relaxation time. Therefore, the charge tends to be distributed throughout the fuel and sets up a corresponding electric field in the vapor space. He calculated that the discharges of electricity will occur in the range of 3000 kV/m in air at atmospheric conditions. Van der Meer and White [4] reported that three oil tankers had explosion in 1969. Mackeown and Wouk [5] studied the magnitude of electric charge, when gasoline is handled. They tried to determine the magnitude of electrical charges that could be produced in the handling of gasoline. They measured charges produced when gasoline was pumped from a filling rack to a tank truck with 3 inch fill pipe in the range of filling rates, from 5 to 800 gallons per minute. They reported that the current produced was directly proportional to the rate of flow and that the high current was 300×10^{-8} coulomb in 1 min and the voltage was 3000 V, when the tank truck was completely insulated. They stated that no spark could occur between two materials, no matter how short the gap was, unless a potential of approximately 300 V existed across the gap. They calculated that when an ordinary passenger car having a capacitance to ground of 750 micromicrofarads filled 15 gallons of gasoline at a service station at a filling rate of 12 gallons per minute, the total current would be 225×10^{-9} coulomb and the voltage 300 V. Klinkenberg [6] also reported explosions at refineries due to possible spark discharges of static electricity and proposed to use antistatic additives of polar organic substances (like alcohols, amines, mercaptans, acids, esters) in fuels to prevent explosion by static electricity. According to him, virgin products that have conductivities less than 5 picomho/m can produce very strong electric field and oil products that have a conductivity of the order of 1000 picomho/m are considered safe and in no case the conductivity should be reduced to below the value of 50 picomho/m for refinery handling. The recommended dosage is 2 kg/1000 m^3 of product. The important factor for antistatic additive is that it should not decompose during storage and not be removed by any means.

According to Canadian Centre for Occupational Health and Safety [7], the explosions, which were investigated by them in association with static electricity, occur during the handling of gasoline, benzol, jet fuels, oil products in refineries, and oil in oil tankers. The flash points of gasoline is as low as −40°C, that of benzol about 6°C and that of jet fuel or kerosene about 38°C. Refineries handle various kinds of liquid from gasoline to lubricants, and crude oils in oil tankers contain various kinds of volatile substance. Therefore, when vapor of liquids having low flash points forms a burnable mixture with air and spark discharges of static electricity ignite the gas mixture, explosion will occur. On the other hand, the flash points of lubricating oils such as turbine oils, hydraulic oils, gear oils, etc. are as high as 200–250°C. They themselves have neither flammable nor explosive gases but spark discharges of static electricity in lubricating oils crack oil molecules to generate hydrogen gas and low-molecular-weight hydrocarbon gases such as methane, ethane, ethylene, etc. Therefore, it is important to know the mechanism of generation of electric charges in hydrocarbon liquids.

Many researchers studied the mechanism of generation of electric charges in low conductive liquids like hydrocarbons. There are two lines of studies: One is charge generation when liquids flow tubes or pipes and the other when they flow porous materials like filter media. Keller and Hoelscher [8], Gravis and Koszman [9], Goodgellow and Graydon [10], and Walmsley [11] studied generation mechanism of electric charge when hydrocarbon liquids flow through pipes or tubes. Keller and Hoelscher paid attention to increase hazards from fire and explosion in hospital and refineries, where new synthetic materials are used. They measured development of static electricity by using the Faraday cage method while hydrocarbon (heptane) passes from a reservoir tank through a poly vinyl chloride tube into an insulated tank. They found that a plot of charge generation rate vs. Reynolds number on logarithmic coordinates resulted in a single straight line relationship for all tube length within the limits of accuracy of the measurement techniques. They also confirmed that the zeta potential, which is defined as the potential at the surface of shear between the bound layer of ions and those carried along with the flowing fluid, was flow dependent, because increased flow rates produced larger shear forces at the double layer. They found that the electrification per unit weight of heptane was a linear function of the flow rate and independent of tubing length from 6 to 18 feet and that pure liquids did not electrify. Their findings indicated that the presence of a high molecular weight polar organic molecule in the liquid or one to be leached from the plastic tubing is essential to the development of significant charge [8]. Gravis and Koszman reviewed the past theories and found that the theories based on the classical double layer are in consistent with the observation at the points of (1) no dependence of charge generation rate on tube length and (2) difficulty in explaining the phenomena that the generation rate goes through a maximum and then decrease with increasing conductivity, as the theories explain that the double-layer zeta potential increase with ion concentration and thus with conductivity. They worked to establish the theory of charge generation in low conductive liquids like hydrocarbons [9]. Goodfellow and Graydon also measured using a small diameter stainless steel tubes charging current on low conductive fluids with small quantity of additives and found that charging currents increase as the conductivity increases and that the tube current polarity is dependent on the chemical nature of the polar species (additives). They also found that charging current magnitudes are independent of the nature of the polar additives but on solution conductivity and fluid velocity [10]. Walmsley extended the Carruthers and Marsh model for potentials generated by liquid flow through plastic pipes to cover pipes that are surrounded by an earthed conductive media, because the model by Carruthers and Marsh is constructed on the condition that pipes are earthed at both ends but without earth contact along the wall, although many pipes are in contact with a conductive, earthed out layer or medium. They confirmed that their extended model was in good agreement with the result of their experiment, which was conducted with a pipe covered with conductive layer [11].

Gavis and Wagner [12], Lauer and Antal [13], Leonard and Carhart [14], and Huber and Sonin [15, 16] studied the mechanism of charge generation when hydrocarbon liquids flow through porous media or filters. Gavis and Wagner studied electric charge generation during flow of hydrocarbons through microporous media and found that the generation of electric charge is a function of the liquid relaxation time, its superficial velocity, and the nominal pore diameter of the filter media [12]. Lauer and Antal confirmed that the streaming currents are proportional to the flow velocity and that the magnitude of the current with small pore filter is larger than that with large pore filter [13]. Leonard and Carhart examined generation of electrostatic charge when hydrocarbon fuels flow through a fiber glass filter element by adding additives as a function of conductivity of fuel. They found that the quantity of electrostatic charges generated is dependent on the conductivity of the liquid and the flow velocity through the filter and that as the conductivity of the fluid is increased, the charging tendency increases, passes through a maximum, and then decrease. They also found that the sign of the charge on the fuel is dependent on the nature of the surface of the filter material. They divided the flow of fuel from a storage tank into two streams each containing different filter materials, one causing the fuel to acquire a positive charge and the other a negative charge, and by carefully controlling the flow through each filter, it was possible to deliver fuel to the receiving tank with virtually no net charge on it. The results suggest a possible method of reducing the charge level on fuel in a practical system [14]. Huber and Sonin worked for proposing a theory for the charging mechanism in filtration [15, 16]. Their studies were conducted with various types of Millipore filters and a Velcon fuel filter with Shell ASA-3 antistatic additives. As Millipore filters are membrane type filters, they cut a Velcom cartridge filter and used the filter sheet to test. The test data with Millipore filter had a good agreement with their theory but the test data with the Velcom filter did not have good agreement [16]. Miaoulis et al. [17, 18] theoretically and experimentally studied generation of electric charge in hydrocarbon by two kinds of wire net screen of copper and stainless steel and confirmed that the sign of the charge in liquid is dependent on the nature of the filter surface.

There are two kinds of theory about how electric charge is generated when hydrocarbon liquids flow through tubes and porous materials. One is the theory that electric charge can be generated by friction and the other the theory based on electrical double layers.

Ernsberger [19] denied the double-layer theory that preferential adsorption of small amount of ionized impurity form Helmholtz double-layer at the tube wall and that the charge separation rate is independent of tube length, stating that they found by their experiment that the charge-separation rate increases toward a limiting asymptote as the tube length increases and that an electronic current can be collected continuously and independently from a metal tube carrying a continuous flow of hydrocarbon fuel. He insisted that electrostatic charge is generated by frictional contact to solid and non-conducting fluids on the assumption that pure hydrocarbons are not charged by contact with clean metal. However, MacDonald [20], Klinkenberg [6], Gravis [9], Goodfellow [10], Lauer [13], Gibbongs [21], Schenck [22], Felici [23], Vazquez-Garcia [24] discuss their studies on the basis of the electrical double layer theory. Therefore, it is reasonable to accept the theory that static electricity is generated in the flowing oil by separation of the charges of electrical double layers.

2.3.2 Experimental for Investigating the Mechanism of Oil Electrification

Sasaki et al. [25] measured potential of static electricity generating while an insulated metallic pin was sliding on an oil-lubricated metallic disc using a pin-on-disc machine, in order to investigate whether static electricity was generated by friction. The schematic diagram of test device is shown in Fig. 2.29. The disc is in an oil bath and rotates. The ball on the pin contacts the disc. The pin is insulated by a polytetrafluoroethylene (PTFE) holder and connected to the potentiometer and AD converter assembly through a diode that allows the electricity accumulated on the pin and ball assembly only to go to the potentiometer.

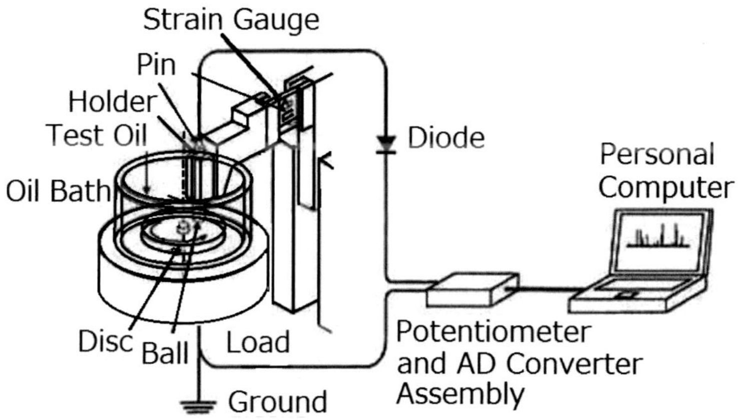

Figure 2.29 Schematic diagram of test device. Reprinted from A. Sasaki, et al., Measurement of the potential of static electricity generated by the friction of oil lubricated metal on metal, *Tribo. Trans.*, 45(1) (2002) 55–60, by permission of Taylor & Francis Ltd.

The lubricants that were used for the experimental were (1) PAO with the viscosity of 28.8, 150, 320, 420 mm^2/s, (2) a group II base oil with the viscosity of 32 mm^2/s, and (3) a grease with a base oil of the viscosity of 30 mm^2/s. The viscosity of PAO was adjusted by blending two base stocks of 28.8 and 420 mm^2/s. The friction of high viscosity lubricant is high. The viscosity of grease is high with thickener. If static electricity is generated by friction, the potentials of static electricity generated by the tests with grease and high viscosity PAO shall be higher than those tested with low viscosity lubricants like 28.8 mm^2/s PAO and 32 mm^2/s group II oil (P-150). The experimental results were shown in Figs. 2.30 to 2.32.

Figure 2.30 shows that the potential of the generated electricity is higher with lower viscosity oil than with higher viscosity ones. The potential of the generated electricity is measured by the electrometer in accordance with the following equation. As the test condition is constant, C is constant. Therefore, the quantity of the electric charge is proportional to the potential of the electric charge.

$$Q = CV, \tag{2.6}$$

where Q is the quantity of electric charge (Coulomb), C the capacitance of the test device (F), and V the potential (volt).

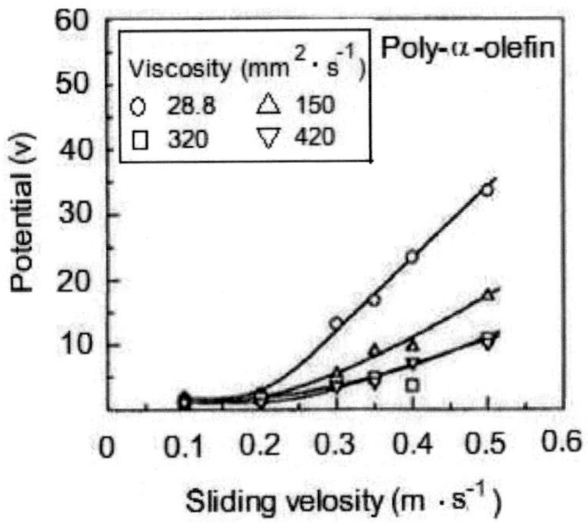

Figure 2.30 The relation between the oil viscosities and the generated potentials. Reprinted from A. Sasaki, et al., Measurement of the potential of static electricity generated by the friction of oil lubricated metal on metal, *Tribo. Trans.*, 45(1) (2002) 55–60, by permission of Taylor & Francis Ltd.

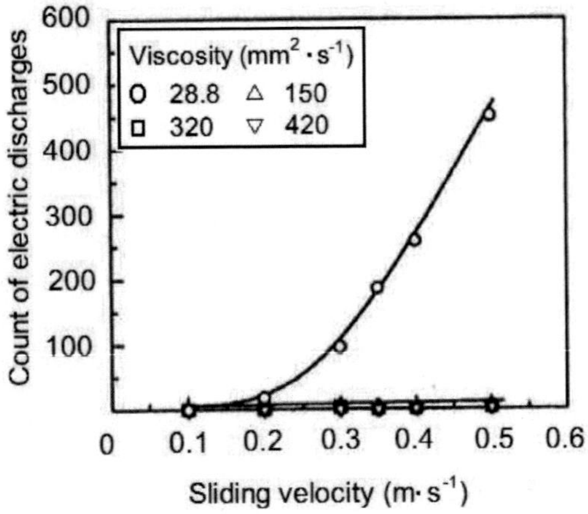

Figure 2.31 Numbers of spark discharges and oil viscosities. Reprinted from A. Sasaki, et al., Measurement of the potential of static electricity generated by the friction of oil lubricated metal on metal, *Tribo. Trans.*, 45(1) (2002) 55–60, by permission of Taylor & Francis Ltd.

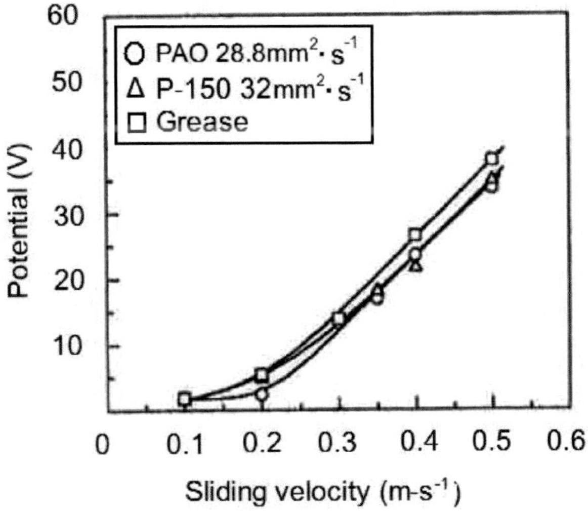

Figure 2.32 The potentials of three different lubricants with similar viscosity. Reprinted from A. Sasaki, et al., Measurement of the potential of static electricity generated by the friction of oil lubricated metal on metal, *Tribo. Trans.*, 45(1) (2002) 55–60, by permission of Taylor & Francis Ltd.

Figure 2.31 also shows that the count of electric discharges tested with PAO of 28.8 mm^2/s viscosity is substantially larger than those tested with PAO of 150, 320, and 420 mm^2/s viscosity, because the PAO with low viscosity generates larger quantity of electricity than those with high viscosity PAO. The results show that electricity is not generated by friction of lubricants and metals, because if electricity is generated by friction of lubricant and metal, the potential of electric charge must be higher with high viscosity lubricants than with low viscosity ones, as the friction of metal with high viscosity lubricants is larger than that with low viscosity lubricants.

The nominal viscosity of grease, which is semi-solid, is substantially higher than that of liquid type lubricants. However, Fig. 2.32 shows that the potentials of electricity, which was generated by the tests using three kinds of lubricant of (1) PAO with the viscosity of 28.8 mm^2/s, (2) P-150 base oil (group II oil) with the viscosity of 32 mm^2/s and (3) grease with base oil with 30 mm^2/s, are almost similar. Grease is made of thickener and liquid lubricant with additives. The thickener works as a kind of oil tank of micron size. When grease lubricates, the liquid in grease oozes out of the thickener to lubricate. Therefore, grease lubricates not by thickener but by the liquid lubricants in the grease. The data of Fig. 2.32 suggests that the liquid lubricants with the viscosity of 30 mm^2/s ooze out of the thickener and the flow of it generates electricity by the mechanism of the electrical double layer.

2.3.3 Electric Discharges and Decomposition of Oil Molecules

It is known that oil deterioration is initialized by formation of free radicals due to thermal decomposition of hydrocarbons. Rice et al. [26] investigated the thermal decomposition of paraffin hydrocarbons based on the assumption that the unimolecular decomposition consists of the primary rupture of a C-C bond with the formation of free radicals. They calculated that the energy of activation to be 67,000 cal per mole (280.465 kJ per mole) and found that organic compounds decompose into free radicals when heated in the range 800–1000°C. There are some other sources to decompose hydrocarbons into free radicals. Bubula et al. [27] caused electric discharges in three types of hydrocarbon liquids (hexadecane representing paraffin, cis-decaline representing naphthalene, and α-methylnaphthalene representing aromatic) to produce gas bubbles by breakdown of liquids using high voltage of 18 kV and measured the emitted light to know electron energy using a photomultiplier. They detected a light emission of energies in excess of 10 eV, which is the luminous ring around the high voltage electrode and in the UV range of the electromagnetic spectrum at the first stage. At the second stage they detected the smallest discharges of short streamer, which is originated by photoionization either of the molecules of the liquid or of the gas absorbed on the electrode or dissolved in the liquid and can be measured by an electrical method with a sensibility of about 0.2 pC. At the third stage, they observed a gas bubble produced by thermal ionization, which indicated decomposition of the hydrocarbon, due to the high temperature of the spark channel. They demonstrated that a much larger number of hydrogen and low hydrocarbon gaseous products were generated by spark discharges in hexadecane than in the other liquids. Bårmann et al. [28] caused sparks in a dielectric liquid (transformer oil) and measured the spectra of streamer filaments at the electric breakdown at the voltage between +60 and −60 kV with the electrode gap of 8 mm. The gas chromatographic measurement showed hydrogen molecules are the most abundant stable species (80%) and the rest low-molecular-weight hydrocarbons such as methane, ethane and ethylene, in addition to C_2 and C_3 radicals. The breakdown arc temperature of the streamer was estimated to be slightly above 10000 K 300 ns. According to Murooka [29], corona discharges can be caused even at 1 V and with current of 1×10^{-5} A. The temperature of corona discharge is said that 10000 K or higher. The density of oil molecules is much higher than that of gas. When spark discharge of static electricity occurs at a sharp edge in oil, the stream of electrons will be concentrated at the sharp edge. Then concentrated electrons will attack oil molecules and break molecular bonding to lead to oil oxidation.

2.3.4 Measurement of Potential of Filter and Oil

2.3.4.1 Experimental apparatus and test conditions

Sasaki et al. [30, 31] conducted an experiment to demonstrate that mechanical filters generate static electricity. Schematic diagram of the test apparatus is shown in Fig. 2.33.

For measurement of potential of oil, a combined unit of a Faraday cage and an electrometer was used as show in Fig. 2.34. As the electrometer is detachable, it was used for measuring potential of filter housing. As confirmed by the experimental in the section 1–2 for investigating the mechanism of oil electrification, generation of electricity is higher with low viscosity oil than with large viscosity oil. Therefore, the oil used for this experimental was naphthenic base transformer oil with the viscosity of 7.1 mm^2/s at 40°C. The test conditions were as follows:

(a) Room temperature: 25°C ± 1°C;
(b) Oil temperature: 35°C ± 2°C in the test oil tank;
(c) Measurement of oil potential: Faraday cage and electrometer set was used, as shown in Fig. 2.34;

(d) The volume of the inner cup of Faraday cage was 0.5 L; Capacitance (*C*) of the Faraday cage is 27.0 pF and potential will be measured on accordance with the equation of 2.6;

(e) The potentials of the electrified filter housing were measured by the detached electrometer at the distance of 50 mm from the filter housing.

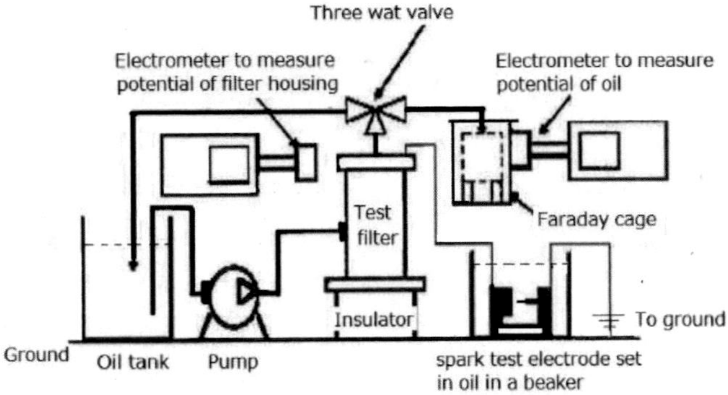

Figure 2.33 Schematic diagram of filter electrification test apparatus. Reproduced with permission from Society of Tribologists and Lubrication Engineers. A. Sasaki, T. Yamamoto, Generation of static electricity during oil filtration, *Lubrication Eng.*, 55(9) (1999) 14–21.

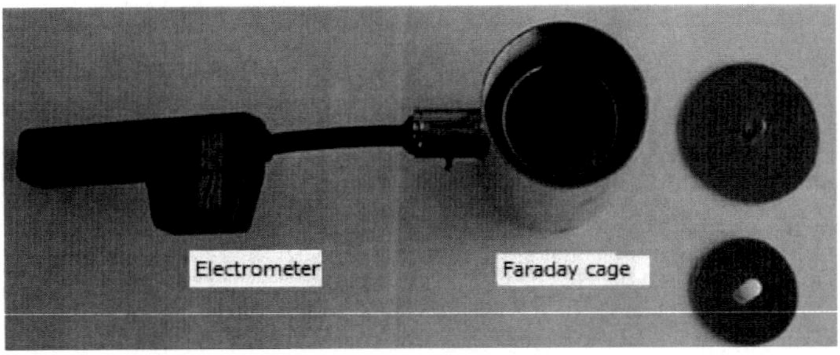

Figure 2.34 A combined unit of Faraday cage and electrometer to measure potential of oil and filter. Reproduced with permission from Society of Tribologists and Lubrication Engineers. A. Sasaki, T. Yamamoto, Generation of static electricity during oil filtration, *Lubrication Eng.*, 55(9) (1999) 14–21.

A commercially available pleated absolute 3 micron filter that was initially used is shown in Fig. 2.35.

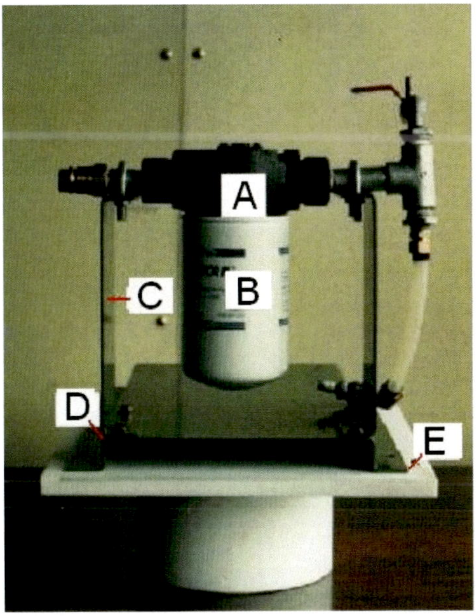

Figure 2.35 A commercially available pleated filter on a test stand.

According to the maker's catalog of the pleated filter, the flow velocity of the oil with viscosity of 32 mm^2/s is about 60 L/min when the pressure drop of the filter assembly is about 98 kPa (1 kgf/cm^2). However, the preliminary test that was performed with the flow velocity in the range of 2.6 to 7.4 L/min caused spillover of oil out of the Faraday cage and oil spillover contaminated the sensor of the electrometer. Therefore, a polypropylene nominal 2 μm depth filter with 0.15 L in volume was specially made for measurement of potential of oil, as shown in Fig. 2.36. The depth filter has a vinyl hose of 6 mm in the inside diameter and of 350 mm in length.

As the test apparatus consists of (a) a depth filter placed on an insulated plate, (b) an electrode set submerged in oil in a beaker, (c) a pump and motor (grounded) and (d) an oil tank (grounded), as shown in Fig. 2.36. The electrode set has two electrodes: One is a flat electrode, which is connected to a filter housing by a high voltage electric cable, and the other a needle electrode, which is connected to the ground by a low voltage electric cable. As soon as oil returned from the filter to the oil tank, spark discharges occurred. Spark discharge phenomena of static electricity at the flow rate of 4.0 L/min using the test apparatus of Fig. 2.36 were recorded on a video tape. His experimental video is uploaded on YouTube ("Auto Oxidation in Turbine Oil" www.youtube.com/watch?v=sj16rWrj6dk) by someone under the name of RCCNOW.

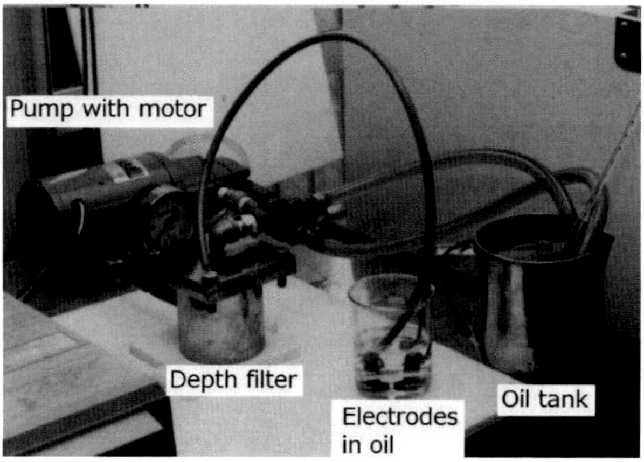

Figure 2.36 Spark discharge test apparatus with depth filter. Reproduced with permission from Society of Tribologists and Lubrication Engineers. A. Sasaki, T. Yamamoto, Free radicals and oil auto-oxidation due to spark discharges of static electricity, *Lubrication Eng.*, 55(9) (1999) 24–27.

2.3.4.2 Measurement of potentials

Before starting experimental of spark discharges using this filter, a preliminary test was conducted without filter element in the filter housing. The measured potential is shown in Fig. 2.37. The potential of oil, which flowed an empty filter housing at 3.3 L/min, was almost 0 V and that at 4.0 L/min was 8 V, which is negligible comparing with the potential of static electricity of kV level. Therefore, the potential of oil passing through filter media was measured at the flow rate in the range of 0.75 to 4.0 L/min.

The potentials of 0.1 and 0.2 L oil taken in a Faraday cage were measured independently. The test results show that the potential is proportional to the volume of oil in Faraday cage, as shown in Fig. 2.38. Therefore, the potential of oil was measured by taking 0.2 L oil in Faraday cage.

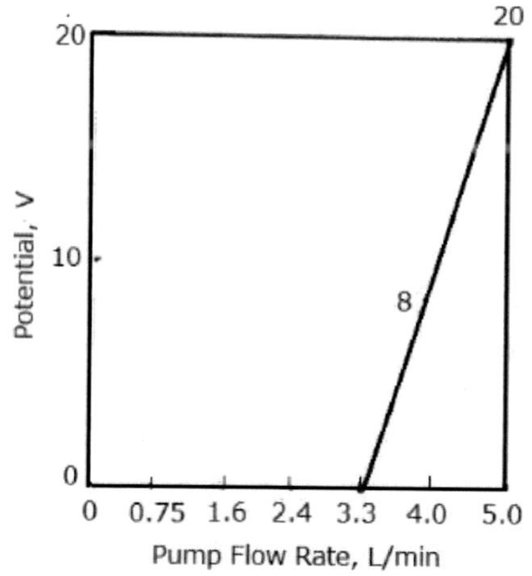

Figure 2.37 The potential of oil when the oil flowed an empty filter housing using the test apparatus shown in Fig. 2.36. Reproduced with permission from Society of Tribologists and Lubrication Engineers. A. Sasaki, T. Yamamoto, Generation of static electricity during oil filtration, *Lubrication Eng.*, 55(9) (1999) 14–21.

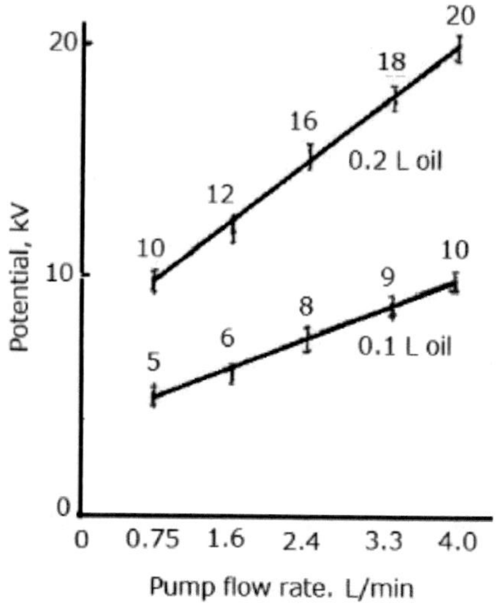

Figure 2.38 Potentials of 0.1 and 0.2 L oil taken in Faraday cage. Reproduced with permission from Society of Tribologists and Lubrication Engineers. A. Sasaki, T. Yamamoto, Generation of static electricity during oil filtration, *Lubrication Eng.*, 55(9) (1999) 14–21.

All machines that use hydraulic and lubricating oils have filters on the oil flow line to protect important components of machines from oil contamination. Filter elements have micron size pores. When the machine starts to be operated, lubricating oil flows through filter elements on the lubrication system with high velocity. Then electricity will be generated. As electricity has polarity, electric charge will be separated into two, when oil flows out of filter elements. One will go on oil and the other on filter housing. Therefore, potential of oil and that of filter housing must be measured independently. The potential of oil was measured when the filter housing is grounded and when it is insulated. The results are shown in Fig. 2.39. The potential of oil was apparently higher when the filter housing was grounded than when it was insulated.

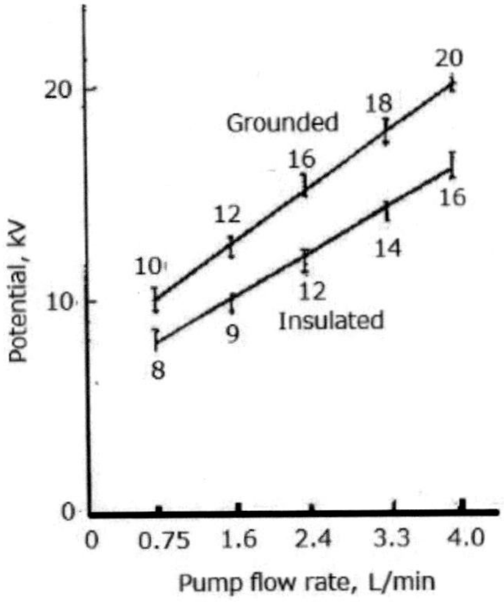

Figure 2.39 Potentials of oil when the filter housing is grounded and insulated. Reproduced with permission from Society of Tribologists and Lubrication Engineers. A. Sasaki, T. Yamamoto, Generation of static electricity during oil filtration, *Lubrication Eng.*, 55(9) (1999) 14–21.

All machines are grounded and the oil having high potential will go to the oil tank. As confirmed in Fig. 2.38, the potential becomes high in proportion to the volume of oil. Oil tanks have large volume of oil. For instance, the volume of an oil tank of a steam turbine having the output of about 1000 MW is about 100 kL.

Therefore, the potential of oil will be very high. However, the potential that reaches near the saturated level will be released by spark discharges at a sharp edge of some metal in oil tank and/or by relaxation. A photo of spark discharge in an oil tank is shown in Fig. 2.40.

Figure 2.40 An example of a spark discharge in oil tank.

The spark discharges in oil tank were recorded on video tape by the author and his colleagues. The spark video is uploaded on YouTube ("Sparks in Oil Tank" www.youtube.com/watch?v=_5cTtbq2X_Q uploaded by the name of RCCNOW; "Electrostatic Discharge in Oil Tank" www.youtube.com/watch?v=UpOJ8aRTWMg uploaded by the name of GlobalIndustrial219; ""Revised Sparks in Oil Tank" www.youtube.com/watch?v=qmsfsWNalEl by the name of Filter Systems UK).

The potential of filter housing was measured at the flange of the filter housing when the filter housing is grounded and when it is insulated. When the filter housing was grounded, the potential of filter housing was zero regardless of the oil flow rate, but when it was insulated, it increased to the saturated level quickly or gradually depending on the oil flow rate, as shown in Fig. 2.41.

The electricity of the oil was "positive" and that of filter housing was "negative" in the case of the depth type filter element made of polypropylene. The results of measurement suggest that the filter housing is a kind of electrical condenser. When the filter housing is grounded, the negative charge immediately goes to the ground and the condenser is empty. Therefore, the potential of the filter housing is zero and we can touch it. As the condenser is empty by connecting to the ground, negative charge can be unlimitedly supplied to the condenser and separation of the electric charge increases.

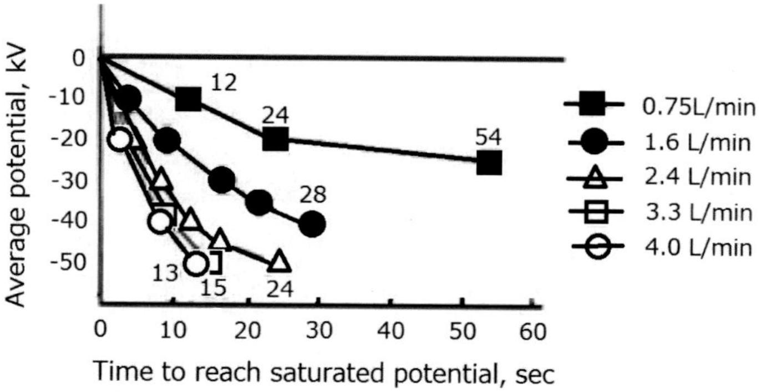

Figure 2.41 Potentials of filter housing when it was insulated. Reproduced with permission from Society of Tribologists and Lubrication Engineers. A. Sasaki, T. Yamamoto, Generation of static electricity during oil filtration, *Lubrication Eng.*, 55(9) (1999) 14–21.

When the filter housing is insulated electric charge supplied to the condenser will be accumulated. If the condenser become full, it will not accept additional supply of electric charge. However, a part of the accumulated electric charge will be lost into the atmosphere or by relaxation. Then the condenser can accept supply of the electric charge equivalent to the lost one. The electric charge that was separated but not accepted by condenser will be offset with the opposite charge in the oil. That is why the potentials of oil with the grounded filter housing are higher than those with the insulated one as shown in Fig. 2.39.

The potential of the housing of the pleated filter, which is shown in Fig. 2.35, was measured by changing the flow rate in the range from 2.6 to 7.4 L/min. The shape of the measuring points of A, B, C, D and E are shown in Fig. 2.42.

The potentials at the point A, B, and C, which had a large radius and was almost flat, were in the range of 0.05 to 0.3 kV but those at the point D and E, which had a small radius or a sharp edge showed large potentials of 5 to 10 kV at 7.4 L/min. The potential of the test looks not so high comparing with those of the depth filter of Fig. 2.41. However, the pleated filter is designed to flow oil with the viscosity of 32 mm^2/s at 60 L/min. If the filter is used with the designed flow rate, the potential of the generated electricity may be higher than the data of Fig. 2.43.

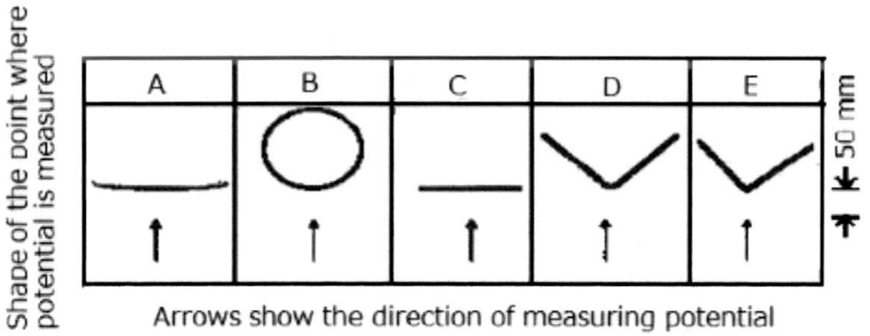

Figure 2.42 The shape of the point A, B, C, D, and E at which the electrometer pointed for measurement of the potential of the housing of the pleated filter. Reproduced with permission from Society of Tribologists and Lubrication Engineers. A. Sasaki, T. Yamamoto, Generation of static electricity during oil filtration, *Lubrication Eng.*, 55(9) (1999) 14–21.

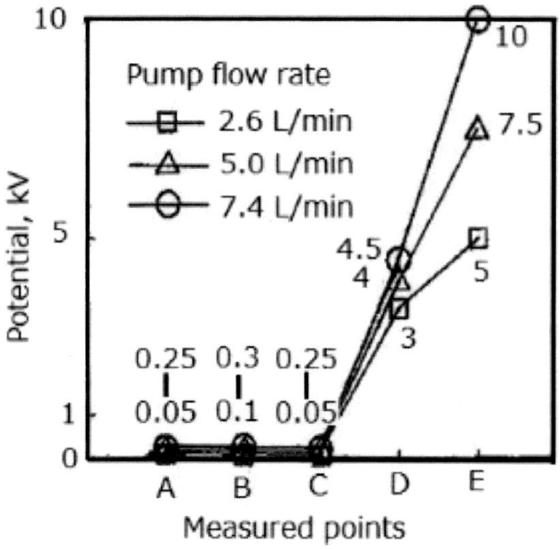

Figure 2.43 Measured potentials at five points of A to E on the pleated filter housing and the filter stand placed on an insulating pedestal. Reproduced with permission from Society of Tribologists and Lubrication Engineers. A. Sasaki, T. Yamamoto, Generation of static electricity during oil filtration, *Lubrication Eng.*, 55(9) (1999) 14–21.

The data of Fig. 2.43 also indicates that the higher the pump flow rate, the shorter the time to reach the saturated potential. While the potentials of the filter housing were measured, discharging noises were heard periodically. This suggests that the electric charge higher than –50 kV is discharged into the atmosphere. The test results indicate a possibility on a working machine that the accumulated electric charge on the dielectric filter element will be discharged with a spark at the nearest sharp point of metal in a filter installed on a grounded machine.

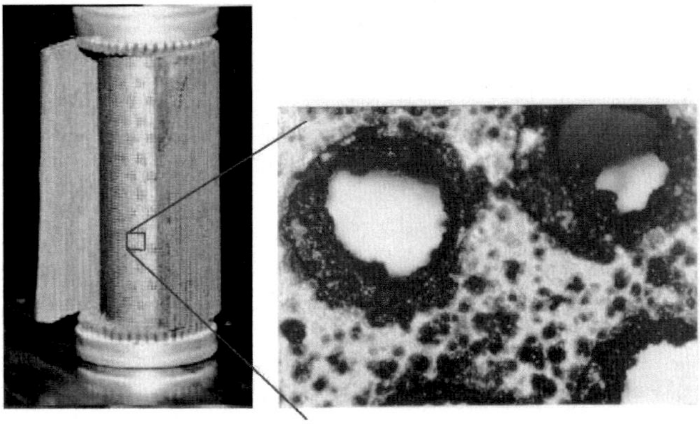

The surface of a center core of a pleated filter

A microscopic photo of sparked edges of holes

Figure 2.44 A center core of a pleated filter that has many sparks at the edges of punched holes. Reproduced with permission from Society of Tribologists and Lubrication Engineers. A. Sasaki, T. Yamamoto, Generation of static electricity during oil filtration, *Lubrication Eng.*, 55(9) (1999) 14–21.

On this assumption, a used pleated filter was examined by cutting the pleated part as shown in Fig. 2.43. The photo of Fig. 2.44 shows that the metal is melt by sparks at the edges of punched holes on the center core of a pleated filter and that the above assumption is correct.

2.3.4.3 Is kV level of electric charge harmful?

Kaufman and Boyd [32] demonstrated by applying voltage in the range of 2–20 V to plain bearings that electric current increases wear rate and that the larger the current with the same voltage the larger the pitting damage. They also showed that pitting damage was caused even when the voltage was as low as 2 V with 0.9 A.

Komatsuzaki et al. [33] investigated electrical wear with the applied voltages in the range of 0 to 4.2 V and demonstrated that deterioration of lubricating grease were accelerated by wear particles, which were produced by electrical wear. Chiou et al. [34] studied the effect of supply voltage, supply current, and oil film thickness on the characteristics of electrical pitting for the lubricated surface of steel pair by using a static electrical pitting tester with high precision under the influence of shaft voltage of AC electric field. They confirmed the study of Erdman et al. [35] that the bearing breakdown voltage threshold is 0.4 V for an oil film of 0.2 μm and that the interface voltage greater than 0.4 V might develop harmful bearing current at an oil film of 0.2 μm for the mineral oil. These studies suggest that bearing surface will be damaged by spark discharges with the electricity having the potentials lower than 4 V and that lubricants are deteriorated by sparks.

2.3.5 Summary

It has been known that static electricity will be generated by flow of oil. Problems by spark discharges of static electricity have been studied for many years. Oils that are dielectric and insulating fluids generate static electricity when they pass through oil filter media that have numerous fine pores. The quantity of electric charge on oil passing through a grounded filter was larger than that of an ungrounded filter. All machines and filters are grounded. Therefore, the potential of oils circulating the lubricating system must be high in general. While oil having electric charge is circulating on the lubricating system, it will sometimes come near sharp corners. There it will lose some of electric charge with sparks. The circulating oil comes back to oil tank with remaining electric charge. As the volume of oil is large in oil tank, the quantity of electric charge on oil becomes large, although electricity has relaxation time. Therefore, spark discharges occur in oil tank, too. Thus, spark discharges crack oil molecules and produce free radicals to lead to auto-oxidation of oil.

References

1. P. S. H. Henry, Survey of generation and dissipation of static electricity, *Brit. J. Appl. Phys.*, Suppl. 2 (1953) S6–S11.
2. H. W. Swann, Survey of harmful static electrification, *Brit. J. Appl. Phys.*, Suppl. 2 (1953) S68–S71.
3. E. F. Winter, The electrostatic problem in aircraft fuelling, *J. Roy. Aeron. Soc.*, 66 July (1962) 429–446.
4. D. Van der Meer, J. W. White, Electrostatic charge generation during tank washing, Project Summary and Introduction, *J. Electrostatics*, 1 (1975) 265–271.
5. S. S. Mackeown, V. Wouk, Electrical charges produced by flowing gasoline, *Ind. Eng. Chem.*, 34(6) (1942) 659–663.
6. A. Klinkenberg, B. V. Poulston, Antistatic additives in the petroleum industry, *J. Inst. Petro.*, 44(419) (1958) 379–393.
7. Canadian Centre for Occupational Health and Safety, Flammable & combustible liquids–hazards, http://www.ccohs.ca/oshanswers/chemicals/flamable/flam.html.
8. H. N. Keller, H. E. Hoelscher, Development of static charges in a nonconducting system, *Ind. Engr. Chem.*, 49(9) (1957) 1433–1438.
9. J. Gavis, I. Koszman, Development of charge in low conductivity liquids flowing past surfaces: A theory of the phenomenon in tubes, *J. Coll. Sci.*, 16 (1961) 375–391.
10. H. D. Goodfellow, W. F. Graydon, Electrostatic charging current characteristics for different fluid systems, *Chem. Engr. Sci.*, 23 (1968) 1267–1281.
11. H. L. Walmsley, The electrostatic fields and potentials generated by the flow of liquid through plastic pipes, *J. Electrostatics*, 38 (1996) 249–266.
12. J. Gavis, J. P. Wagner, Electric charge generation during flow of hydrocarbons through microporous media, *Chem. Engr. Sci.*, 23 (1968) 381-391.
13. J. L. Lauer, P. G. Antal, Electric charge generation during nonuniform flow of hydrocarbon through porous insulators, *J. Colloid Interface Sci.*, 32(3) (1970) 407–423.
14. J. T. Leonard, H. W. Carhart, Effect of conductivity on charge generation in hydrocarbon fuel flowing through fiber glass filters, *J. Colloid Interface Sci.*, 32(3) (1970) 383–394.

15. P. W. Huber, A. A. Sonin, Theory for electric charging in liquid hydrocarbon filtration, *J. Colloid Interface Sci.*, 61(1) (1977) 109–125.

16. P. W. Huber, A. A. Sonin, Electric charging in liquid hydrocarbon filtration: A comparison of theory and experiments, *J. Colloid Interface Sci.*, 61(1) (1977) 126–145.

17. I. N. Miaoulis, B. Abedian, M. Darnahal, Theory for electric charging in flow of low conductivity liquids through screens, *J. Electrostatics*, 25 (1990) 287–294.

18. I. N. Miaoulis, B. Abedian, M. Darnahal, Electric charging in flow of low-conductivity liquids through screens: A comparison of theory and experiments, *J. Electrostatics*, 25 (1990) 295–307.

19. F. M. Ernsberger, Mechanism of frictional electrification of dielectric liquids, *J. Appl. Phys.*, 27 (1956) 418–419.

20. J. R. MacDonald, Static space-charge effects in the diffuse double layer, *J. Chem. Phys.*, 22(8) (1954) 1317–1322.

21. J. C. Gibbings, G. S. Saluja, The electrostatic boundary-layer in stationary liquids, *J. Electrostatics*, 3 (1977) 335–370.

22. J. F. Schenck, A transistor method for measuring changes in double layer potentials, *J. Colloid and Interface Sci.*, 61(3) (1977) 569–576.

23. N. J. Felici, J. P. Gosse, A. Solofomboahangy, Liquid flow electrification and zeta-potential in hydrocarbons, *J. Electrostatics*, 12 (1982) 369–375.

24. J. Vazquez-Garcia, J. Rivenc, A. Agneray, T. Paillat, G. Touchard, A critical approach to measure streaming current: Case of fuels flowing through conducive and insulating polymer pipes, *IEEE Trans. Ind. Appl.*, 41(5) (2005) 1335–1342.

25. A. Sasaki, et al., Measurement of the potential of static electricity generated by the friction of oil lubricated metal on metal, *Tribo. Trans.*, 45(1) (2002) 55–60.

26. F. O. Rice, W. R. Johnston, B. L. Evering, The thermal decomposition of organic compounds from the standpoint of free radicals, II. Experimental evidence of the decomposition of organic compounds into free radicals, *J. Am. Chem. Soc.*, 54 (1932) 3529–3543.

27. E. Bubula, J. Golonski, J. H. Calderwood, S. Zoledziowski, Decomposition of hydrocarbon liquids by partial discharges, *J. Electrostatics*, 12 (1982) 565–572.

28. P. Bårmann, S. Kroell, A. Sunesson, Spectroscopic measurements of streamer filaments in electric breakdown in a dielectric liquid, *J. Phys. D: Appl. Phys.*, 29 (1990) 1188–1196.

29. Y. Murooka, *Corona Phenomena*, New Corona Series No. 7, Corona Publishing Co., Ltd. Tokyo, Japan (1989).
30. A. Sasaki, T. Yamamoto, Generation of static electricity during oil filtration, *Lubrication Eng.*, 55(9) (1999) 14–21.
31. A. Sasaki, T. Yamamoto, Free radicals and oil auto-oxidation due to spark discharges of static electricity, *Lubrication Eng.*, 55(9) (1999) 24–27.
32. H. N. Kaufman, J. Boyd, The conduction of current in bearings, *ASLE Trans.*, 2 (1959) 67–77.
33. S. Komatsuzaki, T. Uematsu, F. Nakano, Bearing damage by electrical wear and its effect on deterioration of lubricating greases, *Lubrication Eng.*, 45(1) (1987) 25–30.
34. Y.-C. Chiou, R.-T. Lee, C.-M. Lin, Formation criterion and mechanism of electrical pitting on the lubricated surface under AC electric field, *Wear*, 236 (1999) 62–72.
35. J. Erdman, R. J. Kerkman, D. Schlegel, G. Skibinski, Effect of PWM inverters on AC motor bearing currents and shaft voltages, OEEE APEC Conference Dallas, TX March (1995).

2.4 Silicon Nitride Tribochemistry: A Paradigm for Ceramic Nitrides

Roberto C. Dante[a,b] and Czesław Kajdas[c]

[a]*Department of Rheology and Complex Materials,*
Instituto de Investigaciones en Materiales,
Universidad Nacional Autónoma de México, Mexico
[b]*Laboratorio de Materiales Avanzados ETSIIAA,*
University of Valladolid, Spain
[c]*Institute of Chemistry in Płock,*
Warsaw University of Technology, Poland

khiratsuka@it-chiba.ac.jp; rcdante@yahoo.com

2.4.1 Introduction

2.4.1.1 General information on tribology of silicon nitride (Si_3N_4) and the aim of this chapter

Among a broad group of engineering ceramic materials that combine low density with very good mechanical properties in a wide temperature ranges, nitrides are of particular importance.

Tribological component materials demanding high reliability in the use are presently available and might be applied in various severe wear, thermal mechanical applications. The most recent work [1] provides a better understanding of the Si_3N_4 tribology and demonstrates its wide variety engineering applications. It has also been emphasized that the tribological properties of silicon nitride depend on tribochemical reactions, which modify both surface composition and morphology, e.g., in air or humid conditions, hydroxylated silicon oxide is the main product of tribochemical wear, which flattens surface decreasing stresses on asperities and reducing wear rate and friction level. Boundary lubrication with *n*-alcohols is discussed in terms of tribochemical wear, which leads to the formation of silicon alkoxides and polysiloxanes.

This section is specifically focused on Si_3N_4 tribochemistry as a paradigm for ceramic nitrides and aims at providing better understanding of some nitrides' tribology, especially of Si_3N_4, AlN, and TiN, by accounting for the mechanisms of their real tribochemistry. Our working model, based on the triboemission phenomenon is here considered and discussed in detail. We demonstrate that during friction of a silicon nitride material, positively charged silicon sites and nitrogen free radicals are produced. Basing on our earlier works, especially [1], detailed tribochemistry of ammonia generation from the milled silicon nitride is developed. NIRAM-HSAB derived tribochemistry of silicon carbide and silica is compared with the tribochemistry approach of these ceramics. Thus, tribochemistry of silicon nitride has been taken into account as the paradigm for ceramic nitrides.

2.4.1.2 Selected properties of nitrogen-based ceramics toward tribochemistry

Ceramic materials are very stable chemically, but under friction conditions they become quite reactive. Actually, the friction and wear behavior of selected ceramics might be more sensitive to the environment than some metals. The difference in the nature of inter-atomic bonding between metals and ceramics differentiate their chemical and physical properties. Covalent bonds in ceramics are strong bonds due to shared electrons between atoms of similar electronegativity. Metallic bonds are moderate in strength. By and large, silicon nitride is the most significant engineering ceramic material; other ceramics are also important, particularly SiAlONs. "Sialons" are phases in the Si–Al–O–N and related systems and are comparable in variety and diversity with the mineral silicates. They are being explored for their thermal, mechanical, chemical, and electrical properties. These types of nitrogen containing ceramics were reviewed in detail some 40 years ago [2].

Boundary lubrication of silicon nitride is described in detail in [3]; the work reviewing/discussing reactivity and bond strength of Si_3N_4 suggests that its reactivity with water is consistent with wear test results on the effect of humidity/water on wear. The proposed reaction process is based on the regular thermochemical reaction.

Taking into account the present trend toward the incorporation of engineering ceramic materials in tribological applications, more comprehensive understanding of the ceramics' thermal chemistry and tribochemistry is needed. Most often papers consider tribochemistry of silicon nitride with H_2O in terms of the thermochemical process. Paper [3] describes tribochemistry along with microfracture in the friction and wear of Si_3N_4. Testing Si_3N_4 with water, work [4] found that after an initial high wear rate, water generated a strong lubrication effect with silicon nitride. The following chemical reaction had been postulated:

$$Si_3N_4 + 6H_2O \rightarrow 3SiO_2 + 4NH_3 \tag{2.7}$$

Well-sophisticated wear testing research described in [3] was performed with controlled water concentrations, and demonstrated that the lubricating effect of alcohol is not due merely to the effect of dissolved water; see also [4]. Some direct surface chemical reaction involving the hydroxyl functional group of organic compounds was suggested. Papers [5–7] discuss and describe a mechanistic model of ammonia formation from silicon nitride in the presence of water.

Work [8] presents results clearly showing that ammonia originates from the mechanical grinding of any kind of silicon nitride, irrespective of the preparation method adopted. The likely reaction is mechanically activated hydrolysis by water vapor, i.e., a reaction caused by direct contact during the milling process between disturbed Si–N bonds and water, yielding silica and ammonia. Mechanically activated process of hydrolysis is a complex phenomenon. Therefore, the importance of comprehensive understanding of the ceramic surface physics and chemistry cannot be understated. Mechanical hydrolysis of Si_3N_4 is discussed in the book on Surface Modifications and Mechanisms [9]. Silicon nitride is an electrical insulator and can outperform the commonly used materials with long life and very reliable low maintenance operation. Other nitrides will be considered only in a general way and their tribochemistry will be compared with the detailed one of Si_3N_4, particularly of AlN and TiN [10–13].

2.4.2 Detailed Characteristics of Other Nitrides

2.4.2.1 Properties of nitrogen-based engineering ceramics

In the würzite phase AlN is a wide-bandgap semiconductor material with possible applications in opto-electronics. The attention to microelectronics and microelectromechanical systems MEMS was mainly due to a high thermal conductivity (up to 285 $WK^{-1}m^{-1}$ for single crystals) coupled with properties of an electrical insulator [11]. For example, aluminum nitride is used for silicon substrates for its linear expansion coefficient similar to that of silicon. However, AlN tribological characteristics are still scarcely studied.

Industry today requires lighter materials and as the use of low weight components of aluminum, magnesium or titanium alloys increases, there is an increased demand on the tooling industry to machine and form these materials to high surface finish and narrow geometrical tolerances. Tool materials are generally much harder than the light metals and consequently, tool wear due to abrasion or erosion is very limited.

2.4.2.2 Applications of ceramics in tribology

Major ceramics used in tribology include Si_3N_4, SiC, Al_2O_3, and ZrO_2. Ceramic coatings are also widely applied. This is because ceramics can improve the surface characteristics of metals while keeping the metal body and in many cases can be the right balance among mechanical properties, density, low wear, and costs.

Surface nitrides are very common, especially TiN, which combines a low density bulk with a hard surface, as those based on carbon such as diamond (diamond like carbon, DLC) as hard disk coat to prevent wear and carbon nitrides. The latter is a new class of nitride that can compete with DLC. The most uses of engineering ceramics are related to the wear improvement of tools, automotive (internal combustion engines, ICE) and aerospace parts (e.g., SiC, TiC) and for grinding polishing and cutting (e.g., diamond, Al_2O_3), however, forecasting the tribological performance of ceramics is an extremely complicated problem due to the very specific wear mechanisms of ceramics [14].

The value of the coefficient of friction for ceramic-ceramic contact lies typically in the range of 0.25–0.8. These are similar to those of metallic couple in air in the presence of oxide films. The ionic (e.g., Al_2O_3) or covalent bonds (e.g., SiC, Si_3N_4) of ceramics lead to much less ductile materials. In spite of a reputation of chemical inertness, the surfaces of most ceramics are susceptible to tribochemical reactions, which lead to the formation of surface films and thus modify their frictional behavior.

Non-oxide ceramics, e.g. silicon nitride, silicon carbide, titanium nitride, and titanium carbide, in air commonly form oxide films on their sliding surfaces. Oxide ceramics will react with water.

$$SiO_2 + 2H_2O \rightarrow +Si(OH)_4 \tag{2.8}$$

Non-oxide ceramics can react with water too (see Eq. 2.7). Friction of ceramics is very sensitive to the environment. Another important factor in ceramics is fracture; widespread brittle fractures can occur in the contact zone, often intergranular in polycrystalline ceramics, but under more severe conditions, may be transgranular. Friction of ceramics is largely affected by tribochemical reactions and fracture extents, which dissipate energy. The transition from mild wear to severe wear is usually associated to the transition from tribochemical wear to the fracture (either intergranular or transgranular) dominated wear.

2.4.2.3 Examples of silicon nitride practical applications

Silicon nitride is appreciated especially for its excellent resistance to thermal shock, fracture toughness and wear resistance. It is used for bearings, rotors, valve seats, pistons and piston heads, helicopter gearboxes, etc. High hardness, fracture toughness, and wear resistance. Sialon, which is basically a silicon nitride containing small amounts of sintering agents, is used in machining the alloys that cannot be machined by the aluminum oxide-base tools, for internal combustion engine parts, for example the rings, injectors, bearings, turbo-superchargers, rotors, cam followers, valve-sleeve guides, and so on [15].

Si_3N_4 is a leading candidate for gas turbines and other high temperature engineering applications. It exhibits polymorphism and three phases are known: the two long-established hexagonal polymorphs α and β, which are built up on corner sharing SiN_4 tetrahedra, and a cubic spinel in which two silicon atoms each coordinate six nitrogen atoms octahedrally, and one silicon atom coordinates four nitrogen atoms tetrahedrally [16]. The cubic spinel phase (c) can only be synthesized under high pressures and temperatures and has a hardness of 35 GPa [15]. The spontaneous thermally activated transformation is from c to α to β, being β the most stable phase. The β phase is more stable in the whole range between 0–4200 K according to theoretical calculations [16], and α phase is formed below 1227°C (1500 K) only for kinetic reasons, this accounts for the lack of α to β transition. However, the cubic (c) and α phases are more compact and harder than β phase. The most tribological tests have been carried out with the common and stable β phase.

Si_3N_4 is difficult to produce as bulk material because at over 1850°C (2123 K), it gets decomposed before melting. The most used processes to obtain bulk materials are hot pressing, isostatically hot pressing, reaction bonded silicon nitride, sintered reaction bonded silicon. Sintering of silicon nitride is carried out with sintering agents such as Al_2O_3, MgO, Y_2O_3, and Al in quantities usually between 2 and 10 wt%; for instance, sialon corresponds to a silicon nitride doped with Al_2O_3 and Y_2O_3 to bind silicon nitride particles [17]. The sintering temperature is usually between 1500–1800°C (1773–2073 K), and pressure between 20 and 100 MPa. During sintering process α to β transition is thermally activated. Recently, new techniques have been used such as spark plasma sintering (SPS), called also pulsed electric current sintering (PECS) leading to materials with high strength and fine-grained silicon nitrides.

Hardness of silicon nitride has values between 16 and 20 GPa, fracture toughness K_{IC} has values in the range of 3 and 6 MPa $m^{1/2}$, while density is around 3 g cm^{-3} [18–21]. Papers [4, 22–27] demonstrate specific importance for tribology and tribochemistry of silicon nitride as well as for other ceramic materials. Work [28] concerns generation and reactivity of silica surface radicals when inhaled; it combines health problems with sophisticated chemistry.

All the mentioned research papers and works and discussed in more detail in the following sections. Unfortunately, many tribology papers do not report the silicon nitride characteristics in detail and the great variety of silicon nitride-based materials makes difficult the quantitative and comparative analysis of the collected data, which can have especially a qualitative use to identify trends and general behaviors.

2.4.3 Tribochemistry of Silicon Nitride

2.4.3.1 Silicon oxide and ammonia formation

The surroundings are very important for the ceramics because they determine their tribological characteristics. Tribometric tests show that the coefficient of friction of silicon nitride as well as wear depends markedly on the surrounding atmosphere and sliding velocity indicating that surface tribochemical changes are fundamental. Two kinds of tribochemical phenomena can be distinguished: reactions with atmospheric molecules (e.g., O_2 and H_2O), and the reactions with adsorbed molecules (alcohols, hydrocarbons, lubricants, etc.). Basically both reactions imply the breaking of a bond between silicon and nitrogen, according to the following equation:

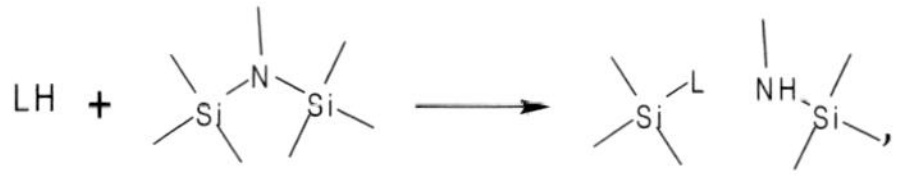

(2.9)

where L can be either a polymer chain or OH in case of water. The affect of water is decreasing both the coefficient of friction and wear, leading to smooth surfaces. This behavior characterizes silicon ceramics and oxides. Details are discussed in [1] with presenting the dependence of both the coefficient of friction and wear on relative humidity in pin-on disk tests [23] along with formation and reactivity of surface radicals [28].

The surface oxygen was early detected by energy dispersion X-ray spectroscopy (EDX) with SEM microscopy, and the silicon bonded to oxygen was found by XPS analysis [28]. All these results are in accordance with the surface silicon oxidation. The reaction of Eq. 2.10 can progress up to silicon oxide (or silanes, silicon alkoxides, etc.) and ammonia removal:

(2.10)

which is a classic example of tribochemical wear: i.e., removal of mass is caused by the immediate formation of new more stable bonds. The formation of ammonia and the straight correlation between wear loss and the produced ammonia amount is an indication that in certain conditions tribochemical wear mode is dominating as shown in Fig. 2.45 for pin-on-plate tests [24].

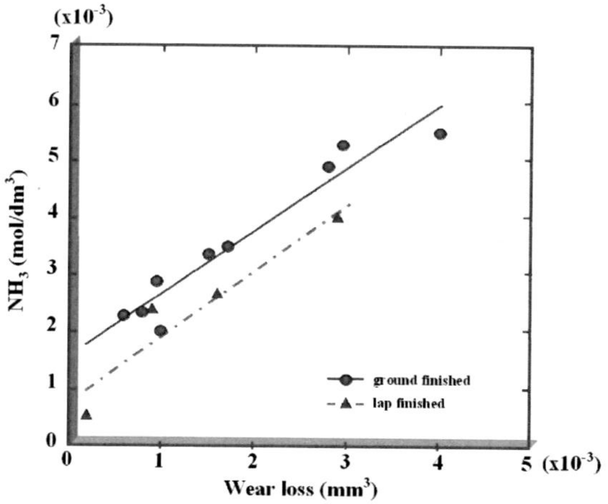

Figure 2.45 The amount of ammonia formed as a function of worn volume. Materials: β-Si_3N_4. Pin-on-plate tribometer. Speeds between 0.13 and 0.38 m/s (frequency: 1.6 and 4.8 rounds per sec, respectively), load of 11.4 N at room temperature. Reprinted from *Wear*, 205, T. Saito, Y. Imada, F. Honda, An analytical observation of the tribochemical reaction of silicon nitride sliding with low friction in aqueous solutions, 153–159, Copyright (1997), with permission from Elsevier.

2.4.3.2 Tribochemistry of Si–OH in the presence of salt solution

In NaCl solution, the Si–OH bond can be decomposed by the Na^+ substitution reaction, as shown below in Eq. 2.11.

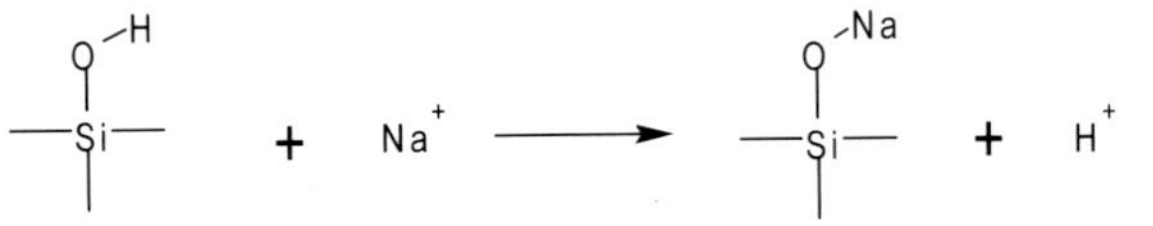

(2.11)

This reaction is supported by the experimental result, demonstrating that the wear debris is composed of Si-O-Na bonds on a silica surface [24]. The H^+ ion formed as the result of the above reaction could be partially neutralized by the NH_3 formed as an intermediate step. As a result, the pH value of the solution is practically unchanged. This reaction consumes the Si–OH on the interfaces, and produces soluble Na-silicate, which is dispersible into the aqueous solution.

A sufficient amount of soluble Na-silicate is difficult to retain on the slid surface with a thickness beyond the surface roughness, and it is considered unable to produce a smooth surface and accordingly to lower the friction coefficient on the sliding surfaces as observed in distilled water or humid conditions.

In this case, worn materials does not form ether debris or layers, and consequently do not interact with the contact surface in an appreciable way to affect friction. This suggested a method to investigate tribochemical wear separately from the other types of wear (adhesive and abrasive) by forming soluble and gaseous by-products [1].

2.4.3.3 Si_3N_4 wear diagram with five (I through V) distinct regions

The wear transition diagram of Si_3N_4 in self-mated sliding, made by Dong and Jahanmir [18] for silicon nitride, is depicted in Fig. 2.46. Tests were carried out with a ball-on-flat tribometer with a relative humidity in the range of 40–70%. Each region is characterized by specific dominant mechanism; I tribochemical reaction, II selective oxidation, III formation of crystalline precipitates, IV oxidation of silicon nitride, V microfracture.

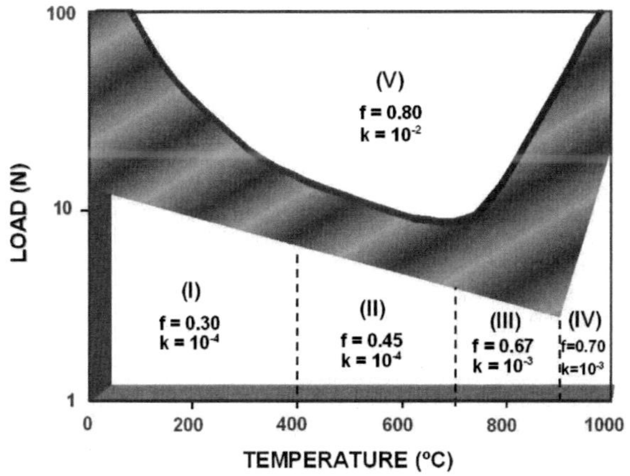

Figure 2.46 Wear transition diagram for silicon nitride, showing five distinct regions plus one transition zone. The friction coefficient, *f*, and wear coefficient, *K*, for each region are indicated on the figure. Reprinted from *Wear*, 165, X. Dong, S. Jahanmir, Wear transition diagram for silicon nitride, 169–180, Copyright (1993), with permission from Elsevier.

The average coefficient of friction, *f*, and wear coefficient, *K*, for each region are indicated in Fig. 2.46. Region I is characterized by formation of hydroxylated silicon oxide roll-like debris (Fig. 2.47). These formed during sliding may be rolled between the sliding surfaces. In region II selective oxidation occurs; this is similar to region I but, in this case, some impurities are preferentially oxidized (e.g., W from milling balls). In region III crystalline precipitates are formed and worn such as particles of Mg silicate. Mg comes from the additives used to sinter silicon nitride particles. In this phase, the binding between grains starts to fail.

Region IV is characterized by the massive oxidation of silicon nitride accompanied by large flat films of oxide removed mainly by plastic deformation. The silicon oxide rolls are observed also in this case (see Fig. 2.48a–d) [4].

In region V no rolls are observed and the wear track has the appearance of a typical fracture surface of silicon nitride, indicating a possible intergranular fracture (see Fig. 2.49a–b). This observation suggests that microfracture is the predominant mode of wear in region V.

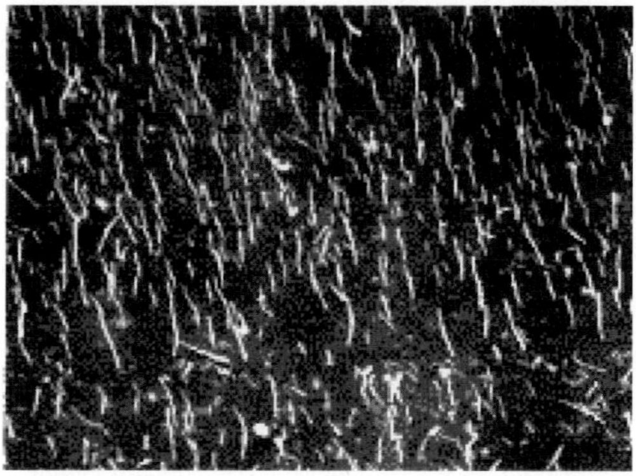

Figure 2.47 Hydroxylated silicon oxide rolls scanning electron micrographs of the wear track on the flat specimen tested at 23°C (296 K) under a load of 4.9 N in the wear track. Reprinted from *Wear*, 165, X. Dong, S. Jahanmir, Wear transition diagram for silicon nitride, 169–180, Copyright (1993), with permission from Elsevier.

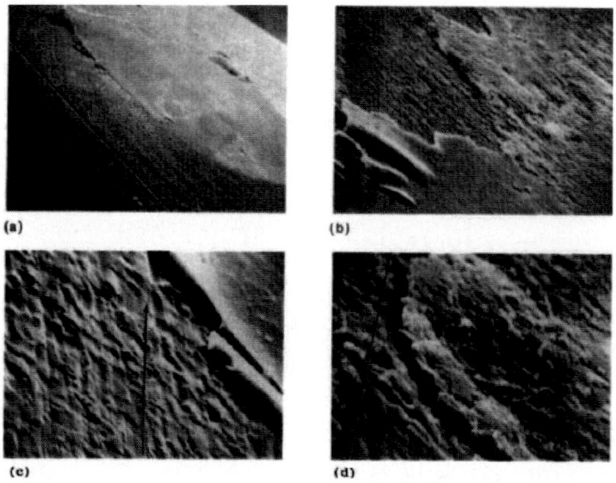

Figure 2.48 SEM micrographs of an Si_3N_4 pin after sliding against on a Si_3N_4 plate in 98% humid argon (room temperature; load, 10 N; speed 1 mm s^{-1}; sliding distance, 3 m): (a) overview; (b) detail of the surface showing evidence of the film; (c) fracture surface of the film; (d) fine wear particles on the film, including rolls formed from the surface film. The arrows show the sliding direction of (a) the plate and (b)–(d) the counterface and represent a length of 10 μm. Reprinted from *Wear*, 105, T. E. Fischer, H. Tomizawa, Interaction of tribochemistry and microfracture in the friction and wear of silicon nitride, 29–45, Copyright (1985), with permission from Elsevier.

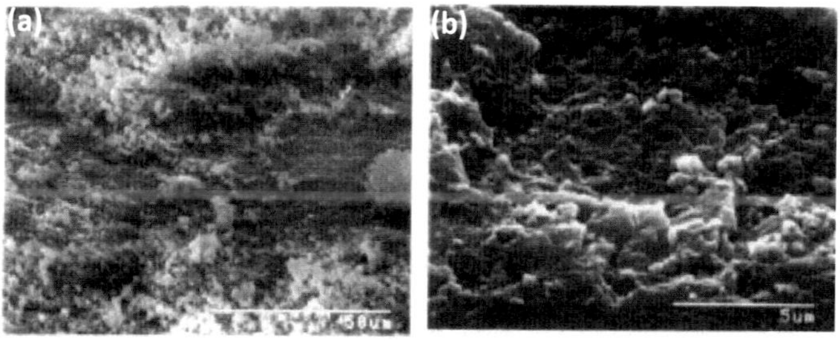

Figure 2.49 Scanning electron micrographs of the wear track on the flat specimen tested at 600°C under a load of 49 N. The micrograph in (b) was taken after ultrasonic cleaning to show the wear surface. Reprinted from *Wear*, 165, X. Dong, S. Jahanmir, Wear transition diagram for silicon nitride, 169–180, Copyright (1993), with permission from Elsevier.

2.4.3.4 Tribochemical wear and mechanism [1]

The absence of water favors wear abrasive and fracture modes leading to higher wear rates and rough surfaces as shown by the pin-on-plate experiments of Fischer et al. [4]. Figure 2.50 shows the rough surface of a silicon nitride in a test carried out in dry Ar, which is similar to that of Region V of severe wear in air, and to be compared with the flat surface formed by the hydroxylated film of Fig. 2.50.

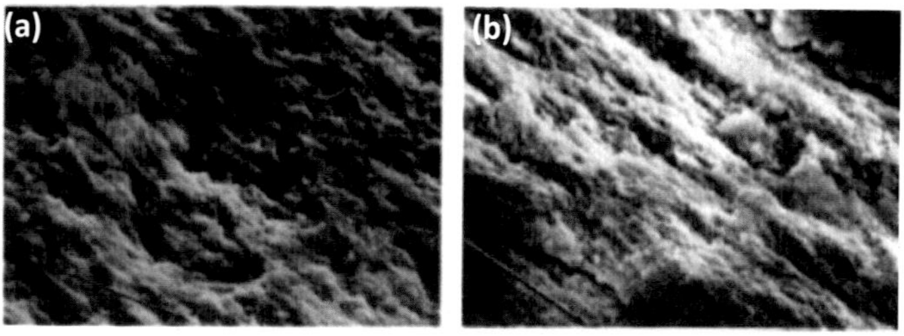

Figure 2.50 SEM micrographs of the wear scar on an Si_3N_4 pin after sliding 3 m in dry argon at room temperature: (a) fracture surface; (b) friction surface with cracks that will lead to removal of the lamella and exposure of the surface. The arrows show the sliding direction of the counterface and represent a length of 10 μm. Reprinted from *Wear*, 105, T. E. Fischer, H. Tomizawa, Interaction of tribochemistry and microfracture in the friction and wear of silicon nitride, 29–45, Copyright (1985), with permission from Elsevier.

The friction track of a test performed at 23°C and the applied load of 2 N, corresponding to Region I conditions, exhibits a transitory phase in which the coefficient of friction quickly decreases from the initial value of 0.6 to 0.4 and then continue to decrease much slowly (Fig. 2.51a). On the contrary, in Region V, at 600°C (873 K) and an applied load of 49 N, the coefficient of friction promptly reaches the steady state value of 0.8 (Fig. 2.51b). The transitory phase of Region I can be attributed to the hydroxylated silicon oxide formation, as the film grows the coefficient of friction is more stabilized following the reaction passes of Eqs. 2.7 and 2.8. The temperature of 600°C (873 K) is much above the dehydroxylation temperature (above 673 K), so that the surface silicon oxide is completely dehydroxylated, and leading to a steady and high coefficient of friction. The reverse of Eq. 2.8 corresponds to the dehydroxylation of the silicon oxide film.

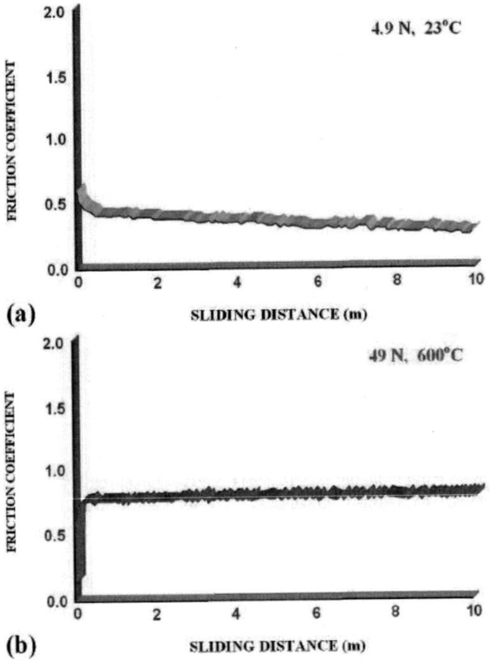

Figure 2.51 Tracks of the coefficient of friction for silicon nitride in pin-on-plate tests in Region I (a) and Region V (b). Reprinted from *Wear*, 165, X. Dong, S. Jahanmir, Wear transition diagram for silicon nitride, 169–180, Copyright (1993), with permission from Elsevier.

$$Si(OH)_4 \rightarrow SiO_2 + 2H_2O \tag{2.12}$$

This indicates that the hydroxylated silicon oxide film has an action that is more wear protective than only silicon oxide.

2.4.4 Tribochemical Wear and Mechanisms

2.4.4.1 Wear rate vs. friction coefficient of Si_3N_4, according to [1, 25]

Muratov et al. [25] used oxidant and etching solutions of CrO_3, KOH, and $KMnO_4$ in a ball-on-disk tribometer. They studied the effects of both chemical (temperature and concentration of reagent) and tribological parameters (load and sliding speed) on the kinetics of the reaction, i.e., the rate of material removal. In CrO_3 and in KOH, the removal rate increases with temperature; the apparent activation energy is 20 kJ mol^{-1} in CrO_3 and 22 kJmol^{-1} in KOH. In water, the apparent activation energy is close to zero. In thermal reactions, the corrosion of silicon nitride is between 80–100 kJ mol^{-1}.

The material removal is temperature independent, in $KMnO_4$, its rate decreases with increasing temperature. $KMnO_4$ decomposes during friction and the brown particles found eventually form a protective layer. These changes are accompanied by parallel variations in the coefficient of friction. In water and CrO_3 solutions, a strong dependence of friction and material removal rate was observed with the load change. With the changes in temperature, concentration, and loading, it was found that the reaction rate (in mm^3 N^{-1}m^{-1}) is linear with the coefficient of friction and null below a threshold value of the friction coefficient, indicating that the reaction is activated by friction (see Fig. 2.52). For example, the threshold is $\mu_{th} < 0.26$ for CrO_3.

In general the tribochemical wear rate constant K can be expressed in the following way:

$$K = K_0\,(T, W, C)\,(\mu - \mu_{th}), \tag{2.13}$$

where K_0 is a constant, which depends on the variables temperature T, load W and concentration C, and μ_{th} is the threshold coefficient of friction above which tribochemical wear is activated. The velocity dependence is complicated by the phenomena of mixed lubrication. The proposed model concerning the experiments at different velocity is based on the observation that the coefficient of friction is due to two contributions:

$$\mu = f\mu_{HD} + (1-f)\mu_C, \tag{2.14}$$

where f is the load fraction carried by the liquid phase, μ_{HD} the coefficient of friction due to viscous drag, μ_C is the contribution of the contact asperities.

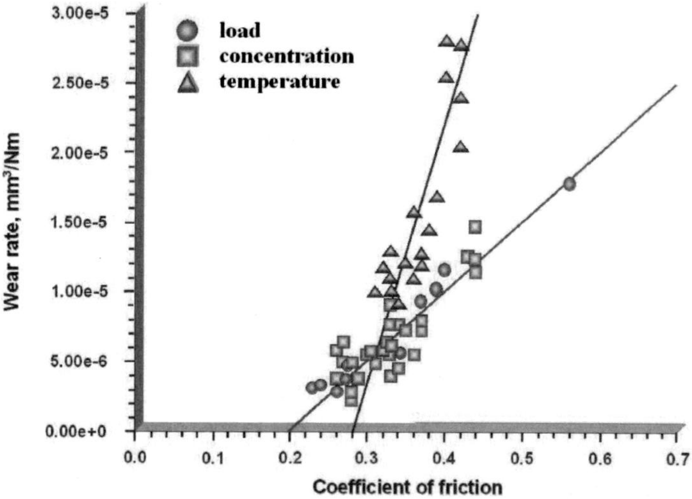

Figure 2.52 Universal linear relationship between the tribochemical reaction rate of Si_3N_4 and coefficient of friction in 3 wt% CrO_3. Constant values: Load = 4 N, v = 20 mm s^{-1} (T or Concentration changing), v = 6.8 mm s^{-1} changing load. Reprinted from *Tribol. Int.*, 31, V. A. Muratov, T. Luangvaranunt, T. E. Fischer, The tribochemistry of silicon nitride: Effects of friction, temperature and sliding velocity, 601–611, Copyright (1998), with permission from Elsevier.

The wear rate constant K acquires the following form:

$$K = K_C(1-f) = K_C\frac{\mu-\mu_{HD}}{\mu_C-\mu_{HD}}, \tag{2.15}$$

where K_C is the rate constant due to the contact asperities, and μ_{HD} is in this case a threshold value that depends on velocity. It is noteworthy to mention that at velocity tending to 0, also f tends to be null. With increasing velocity the tribochemical wear decreases as the coefficient of friction does. This depends on the growth rate of the film, above a certain velocity (below 10^{-3} m s^{-1} for the CrO_3 solution at 3%) the removal is faster than growth and this accounts for the absence of residues and the lower wear rates.

In all cases, the lack of solid wear particles has verified the tribochemical nature of the material removal. In their experimental conditions, flash temperature (velocity was below 0.12 ms^{-1} and the load was 2 N in the velocity-dependence measurements) cannot account for these low activation energies. This was a step forward toward a different look on tribochemical reaction activation not only based on flash temperature [26].

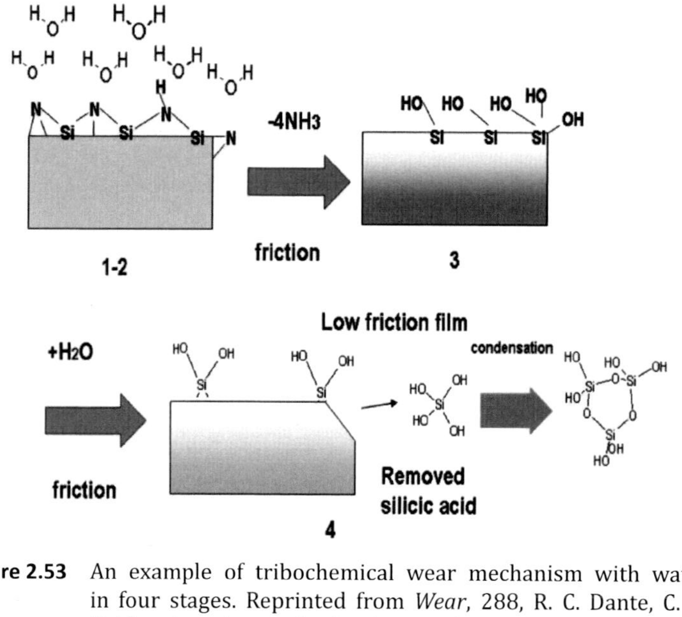

Figure 2.53 An example of tribochemical wear mechanism with water in four stages. Reprinted from *Wear*, 288, R. C. Dante, C. K. Kajdas, A review and a fundamental theory of silicon nitride tribochemistry, 27–38, Copyright (2012), with permission from Elsevier.

Work [25] also demonstrates the idea that a combination of different factors can be the cause of the enhancement of the reaction rate (tribochemical wear rate) such as the bond stretching that lowers the gap between HOMO (Highest Occupied Molecular Orbital) and LUMO (Lowest Unoccupied Molecular Orbital), vibrational excitation pulses, and the dissipated heat. Additionally, it emphasizes that friction activates the reaction and this implies also that some sites are produced by friction work.

The mechanism of tribochemical reactions involves several steps [1]:

(1–2) Formation of active sites/adsorption layer;
(3) Formation of products by tribochemical reaction;
(4) Removal of the products (e.g., light molecules, larger molecules, particles).

Figure 2.53 shows how this mechanism works for water, with the chemical formation of silicic acid and hydroxylated silicon oxide condensation products. The determination of the nature of active sites is difficult matter since they react during sliding.

2.4.4.2 Reactive sites in silicon ceramics

Back some 20–30 years, several works of research groups [8, 27–29], were focused on pathogenicity of ceramics, and were carried out as a response to the many questions opened by asbestos, opening the path to a better understanding of the reactivity of the wrongly supposed inert ceramics. A significant contribution was given by the group of Fubini [8, 27–30]. These papers emphasize the presence of active sites as source of interferences with macrophage metabolism stimulating fibrogenic factors. Electron paramagnetic resonance (EPR), which only detects unpaired electrons such as those of dangling bonds, allowed these researchers to distinguish the surface reactive species from the others that are EPR inactive. They detected many active sites by EPR in ground materials and underlined the different reactivity due to friction (mechanical) work. The found radical active sites are summarized in Table 2.1 [28], and they are divided into two groups. The first one implies homolytic cleavage of Si bond, and the second one relates to the heterolytic bond cleavage. Some radicals are ionic, and in specific cases it was possible to isolate an ionic couple such as for superoxide radical obtained in air and H_2O_2 presence.

Table 2.1 Functionalities found in silicon ceramic materials active to EPR [28]

Functionalities of the silica surface		
Functionalities		**Origin**
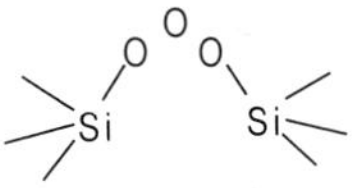 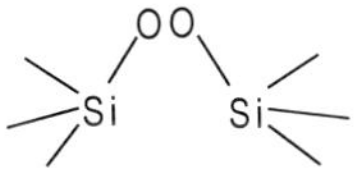	Peroxide bridges	Mechanical grinding in presence of O_2
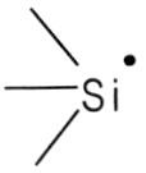	E′ center	Mechanical grinding-homolytic cleavage
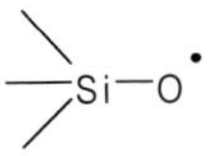	Radical	Mechanical grinding-homolytic cleavage
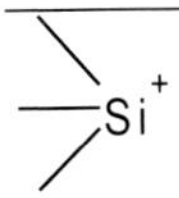 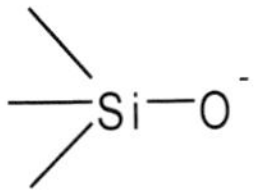	Surface charges	Mechanical grinding-heterolytic cleavage
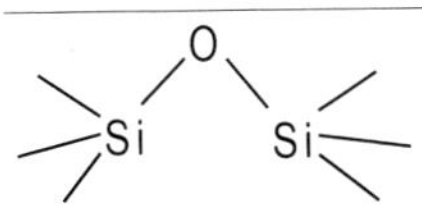	Siloxane bridges	**Regular** Elimination of H_2O from two adjacent silanols by thermal treatment
		Distorted Surface reconstruction after mechanical cleavage
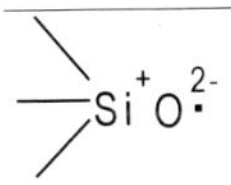	Superoxide radical	Mechanical grinding in presence of O_2 or H_2O_2
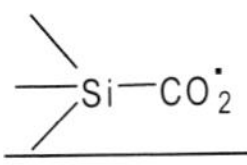	Radical	Mechanical grinding in presence of CO_2

2.4.4.3 Tribochemistry of Si_3N_4 with water

Volante et al. observed that a sample α-Si_3N_4, ground in air, produced large amounts of ammonia. Samples ground in an inert atmosphere (Ar) adsorb more water (data from micro-calorimetry with volumetric apparatus) and with higher release of heat than the material as is, and this reactivity cannot be ascribed only to the small surface area increasing due to grinding [8]. The adsorption curves of ground and "as is" samples, both previously outgassed at 400°C (673 K), are presented in Fig. 2.54.

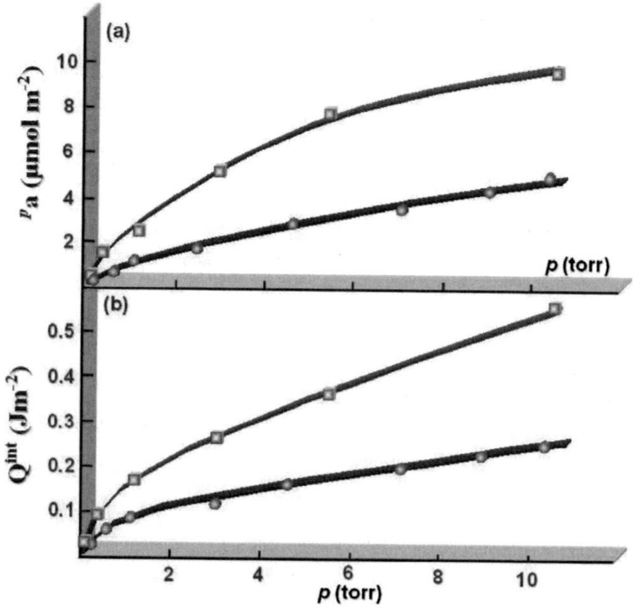

Figure 2.54 (a) Volumetric isotherms, (b) Calorimetric isotherms; (●) sample as is, (■) sample ground in inert atmosphere. Silicon nitride (BET surface area = 19 $m^2\ g^{-1}$) outgassing temperature = 400°C (673 K). *Mater. Sci. Lett.*, Reactivity induced by grinding in silicon nitride, 8, 1989, 1076–1078, M. Volante, B. Fubini, E. Giamello, V. Bolis, with permission of Springer.

Based on these results, the average heat released equals 53 kJ mol^{-1} at 1.067 kPa and is higher than the water adsorption heat of 44 kJ mol^{-1} at the same pressure. The higher heat value could be due to a mix of reactions leading to several gaseous products such as ammonia, nitrogen and hydrogen formation. However, only a fraction of the total water adsorbed molecules reacted.

$$\frac{n_{NH_3}}{n_{H_2O\ adsorbed}} = 0.11. \tag{2.16}$$

According to Eq. 2.7, the ratio between the ammonia moles and the reacted water moles is

$$\frac{n_{NH_3}}{n_{H_2O\ reacted}} = 0.66, \tag{2.17}$$

and if the only product was ammonia, dividing Eq. 2.16 by Eq. 2.17 the fraction of reacted water in respect to total adsorbed water is found

$$\frac{n_{H_2O\ reacted}}{n_{H_2O\,adsorbed}} = 0.17. \tag{2.18}$$

The enthalpy of ammonia formation per water molecule is −106 kJ mol^{-1}, and the total heat due to both reaction and adsorption is

$$\begin{aligned} -\Delta H &= x_{H_2O reacted}\, 106\ \text{kJ mol}^{-1} + (1 - x_{H_2O})_{ads}\, 44\,\text{kJ mol}^{-1} \\ &= 0.17 \times 106\ \text{kJ mol}^{-1} + 0.83 \times 44\,\text{kJ mol}^{-1} = 54.5\,\text{kJ mol}^{-1} \end{aligned} \tag{2.19}$$

The value of 54.5 kJ mol^{-1} is not far from the 53 kJ mol^{-1} found experimentally. This means that the major product is ammonia. However, the ammonia amount seemed to be lower than in the case of grinding in air. Active sites are created during grinding but the formed sites are only partially preserved by the inert atmosphere, probably a surface recombination before water adsorption decreases their number. Silicon nitride outgassed (dehydrated) at 800°C (1073 K) shows a water adsorption behavior similar to that of the material ground in inert gas; however, the average heat released is different (77 kJ mol^{-1}). There is no significant ammonia evolution from the 800°C (1073 K) treated sample. This suggests that in spite of quantitative similarities, the active sites and mechanism are quite different in tribochemical reactions [30].

The finding of several active sites such as radicals and ions suggests that many tribochemical reactions can be related to triboemitted electrons. According to the NIRAM theory, developed by Kajdas, low energy triboemitted electrons with 1–4 eV (262–1048 kJ mol^{-1}) can initiate tribochemical reactions [31]. This theory was used to explain lubricant (LH) reactions with the contact surface (S), and can explain also the reactivity of silicon nitride toward water to form ammonia. The NIRAM–HSAB mechanism is illustrated in Fig. 2.55 [31]. The surface contact site S, activated by friction work, forms a cation S^+ and a low energy electron e^-, which reacts with the lubricant molecule LH (L stays for a lubricant ligand, and H for the reactive hydrogen) producing the radical-ion $LH^{\bullet -}$, which is split into the hydrogen radical $H^\bullet$ and the anion L^- (NIRAM). Finally, the base L^- is recombined with the acid S^+ (HSAB) to give the new species LS.

$$S \rightarrow S^+ + e^-$$

$$LH + e^- \rightarrow LH^{\cdot -} \rightarrow L^- + H^\cdot$$

NIRAM

$$S^+ + L^- \rightarrow SL$$

HSAB

Figure 2.55 NIRAM-HSAB mechanism, S—surface site, LH—lubricant molecule, e^-—electron emitted under boundary lubrication sliding conditions. Reproduced with permission from C. K. Kajdas, Tribochemistry of selected ceramic materials, *Solid Sate Phenomena*, 113 (2006) 339–347.

2.4.4.4 Detailed tribochemistry mechanism of Si_3N_4 with water

The Si_3N_4 tribochemistry is a paradigm for tribochemistry of other ceramics, and especially silicon-based ceramics. General approach to the silicon nitride tribochemistry was presented and discussed in previous sections of this chapter. Figure 2.56 presents the initiation process of Si_3N_4 reaction with water. For simplicity, in Fig. 2.56 reactions of Si–OH species are omitted. Since each electron emitted generates one hydrogen free radical, this process is also deleted in that reaction mechanism.

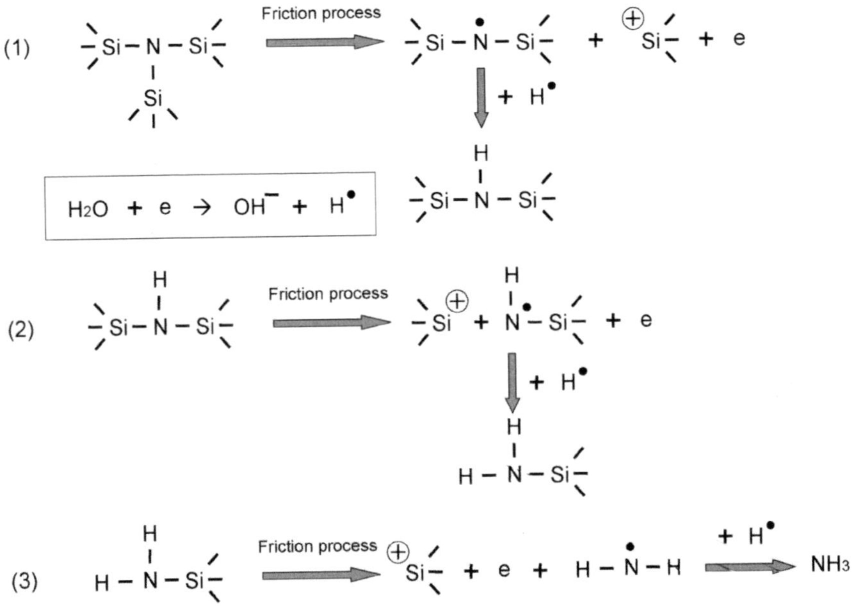

Figure 2.56 Schematic of further tribochemical reaction steps of Si_3N_4 with water. Reproduced with permission from C. K. Kajdas, Tribochemistry of selected ceramic materials, *Solid Sate Phenomena*, 113 (2006) 339–347.

The mechanism clearly shows that the consumed molecule of water does not produce any gas. Further steps of the reaction chain should produce ammonia. Taking into account that not all hydrogen radicals ($H^{\bullet}$) recombine with only one nitrogen atom, even 5–6 water molecules might be chemisorbed to generate only one ammonia molecule. Figure 2.57 extends the first reaction step of silicon nitride with water.

Following the Si_3N_4 tribochemistry paradigm, the same mechanism concept should be due to silicon carbide. Figure 2.58

depicts the first step of silicon carbide reaction mechanism with H_2O. It leads to methane (CH_4) generation.

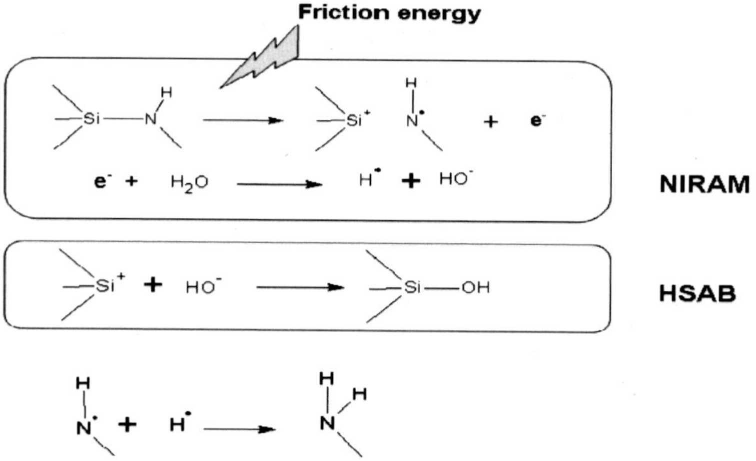

Figure 2.57 Tribochemical reaction of water with silicon nitride based on NIRAM-HSAB theory. Reprinted from *Wear*, 288, R. C. Dante, C. K. Kajdas, A review and a fundamental theory of silicon nitride tribochemistry, 27–38, Copyright (2012), with permission from Elsevier.

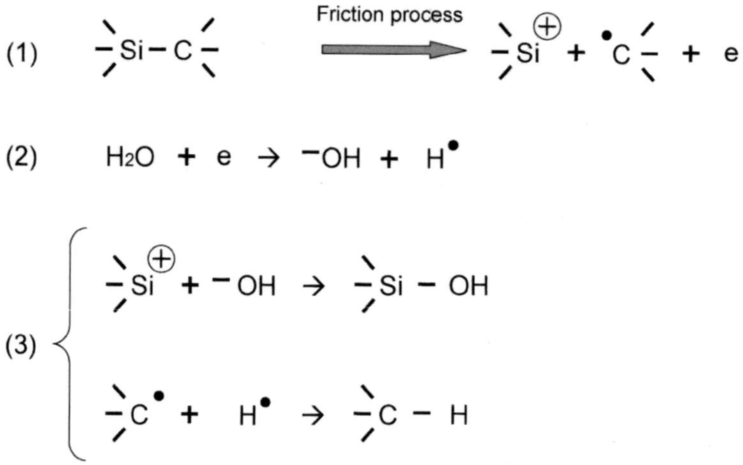

Figure 2.58 Initiation stage of the SiC reaction mechanism with H_2O. Reproduced with permission from C. K. Kajdas, Tribochemistry of selected ceramic materials, *Solid Sate Phenomena*, 113 (2006) 339–347.

2.4.4.5 Tribochemical model of Si/H_2O system, according to [7]

Investigation on tribochemical reaction of H_2O and oxygen on silicon surface [7] demonstrate that Si forms Si–H bonds whose concentration increased with humidity. Similarly, the –OH concentration in wear particles increases with higher humidity. Accordingly, it was concluded that O_2 is much less reactive than H_2O. The tribochemical model of Si and H_2O was suggested and described as follows.

Mechanical force breaks Si–Si bonds near the surface with or without a water molecule.

An OH of water molecules reacts with one of the silicon dangling bonds to generate Si–OH and H reacts to the other to produce Si–H.

Oxygen is less reactive than water since surface dangling bonds cannot be terminated by oxygen.

This approach to silicon tribochemistry and the model for silicon carbide are presented in Figs. 9 and 11 of ref. [7], respectively.

Bearing in mind the Si_3N_4 tribochemistry paradigm we can also consider in similar way tribochemistry of silica (SiO_2) which is presented in Fig. 2.59.

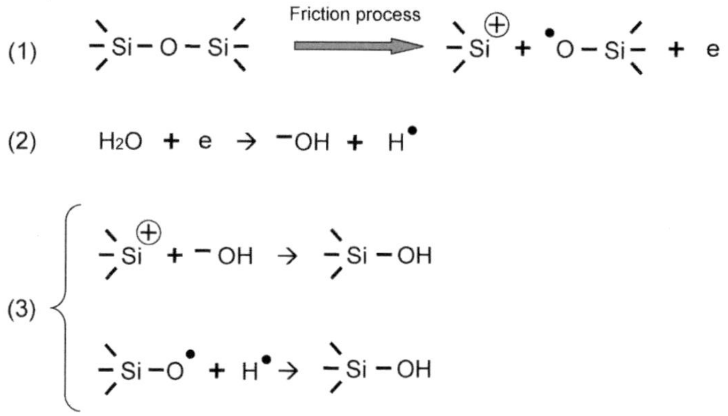

Figure 2.59 NIRAM-HSAB model for how H_2O reacts with silica.

Besides NIRAM theory, the higher rates of tribochemical reactions are thought to be due to different mechanisms that involve excited radical-ions forming intermediates, in accordance to achievements of irreversible processes thermodynamics. The overall lowering of the apparent activation energy is due to the stabilization of an intermediate species having a Gibbs free energy lower than that of the thermal transition state [33]. Electrons emitted in boundary lubrication act as catalysts. Figure 2.60 shows the irreversible processes thermodynamic interpretation of the NIRAM theory by means of the curves of Gibbs free energy as function of the reaction coordinate [33].

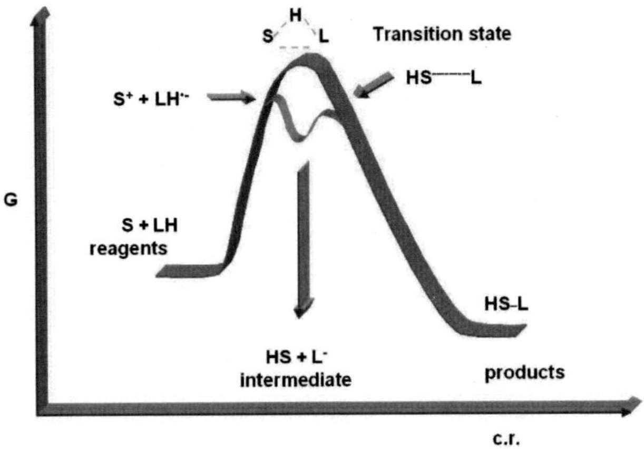

Figure 2.60 Gibbs free energy *G* as function of the coordinate of reaction (c.r.). S is a surface site, LH a lubricant molecule. Reprinted from *Wear*, 288, R. C. Dante, C. K. Kajdas, A review and a fundamental theory of silicon nitride tribochemistry, 27–38, Copyright (2012), with permission from Elsevier.

In the intermediates electron and proton are definitely exchanged between the two reagents S and L. It is noteworthy point out how this mechanism is based on the exchange of the smallest (and subatomic) particles involved in chemistry. Also, the reaction of water with silicon can be interpreted through NIRAM theory.

2.4.5 Tribochemistry of Straight Alcohols

2.4.5.1 Effect of *n*-alcohols on silicon nitride wear

Purified *n*-alcohols effectively reduce friction and wear of silicon nitride even at lower sliding velocity; as the carbon number increased wear decreased. Y. Hibi and Y. Enomoto reported the effect of n-alcohols as function of the carbon number with a ball-on-block machine, and found that wear dramatically decreases with $n > 5$ (Fig. 2.61) having water ($n = 0$) the highest wear rate value [34].

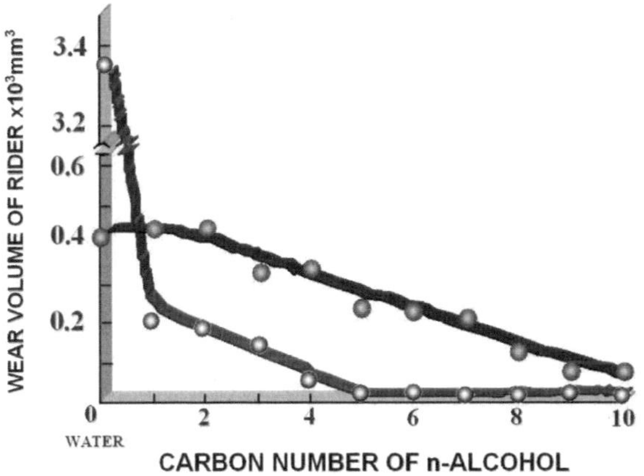

Figure 2.61 Tests with a ball-on-block machine in a glove box. The load applied was 9.8 N and the sliding speed was 2.4 mm s^{-1}. Humidity less than 10%. The material tested was alumina–yttria–titania-toughened hot-pressed Si_3N_4 manufactured by Toshiba Co. Hot pressed Si_3N_4 (green circles), and SiC (red circles) in the same test conditions. Reprinted from *Wear*, 133, Y. Hibi, Y. Enomoto, Tribochemical wear of silicon nitride in water, n-alcohols and their mixtures, 133–145, Copyright (1989), with permission from Elsevier.

2.4.5.2 Application of the NIRAM-HSAB approach to account for Si_3N_4/alcohol tribochemistry, based on [1]

Gates and Hsu carried out a series of experiments by ball-on-flat in air that emphasized the role of the film of silicon alkoxides and polysiloxanes-polyalkoxides in n-alcohol lubrication [3, 6]. They proved that the formation of silicon alkoxides is obtained under friction, and detected silicon alkoxides by "static" secondary ion mass spectrometry (SIMS) in the residue. This high vacuum microanalysis technique used heavy ions (in that case Ga) to bombard the surface of the sample. Surface material is kicked off from the surface and is subsequently analyzed with a time of flight mass spectrometry. Tribochemical wear flattened the surface as observed by other researchers [23–25]. For instance, the mass spectra of a sample tested with decanol (C10) presents several fragment peaks with low mass/charge ratio, while silicon and Ga peaks have m/z = 28 and 69, respectively, and are overlapped to the peaks of the hydrocarbon ions. Other small peaks at high m/z ratios (285–718) belong to silicon alkoxides and polymers, as depicted in Fig. 2.62.

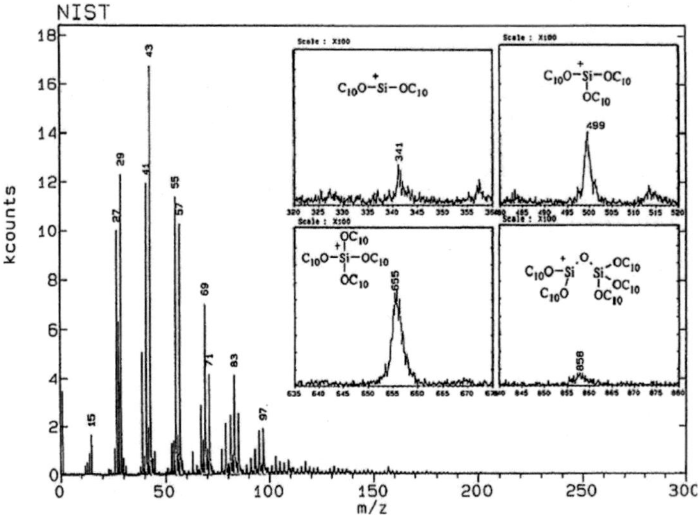

Figure 2.62 SIMS analysis of residue from neat *n*-decanol lubricated hot pressed Si_3N_4 in a ball-on-flat [3]. Reprinted courtesy of the National Institute of Standards and Technology, U.S. Department of Commerce. Not copyrightable in the United States.

The "solid" residue is composed by polyalkoxides with siloxane backbones and forms an effective lubricant film. As inferred by gel permeation chromatography, graphite furnace atomic absorption (GPC-GFAA) analysis, the molecular weights of over 1000. Experiments carried out by Hibi and Enomoto showed that soluble (in alcohols) silicon alkoxides form polyoxysilanes that further condensate and precipitate (alkoxide) silica gel in milling mechanochemical tests with ethanol (C_2) [34].

The semi-solid residue, composed by silicon alkoxide polymers, is that really decreases friction forming a protective film. Reaction and removal of silicon may take place on the tip of asperities, reducing stress concentration. The smooth regions of the wear scar may be indication of tribochemical polishing. The formation of higher molecular weight, higher viscosity products will result in thicker elastohydrodynamic (EHD) films.

The semi-solid residue can be simulated by alkoxide polymers obtained by thermal reactions starting from a tetra-silicon alkoxide. The step that seems to be strictly activated only by friction is that leads to the tetra-silicon alkoxide as shown in Fig. 2.63 with the sketch of the reaction pathway toward polyoxysiloxane.

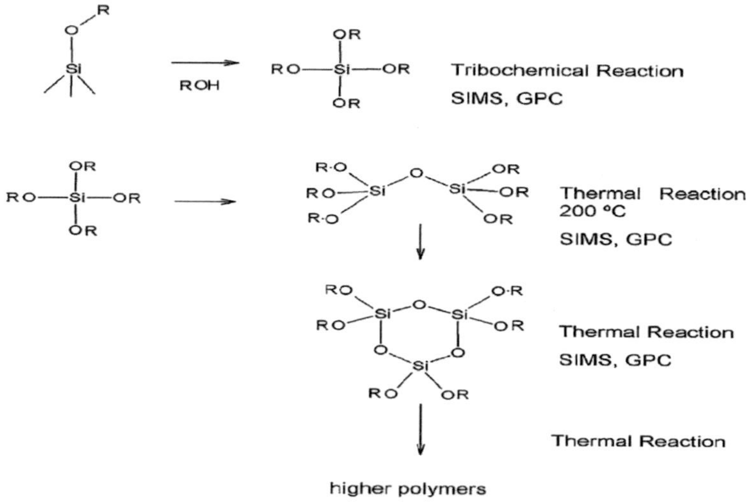

Figure 2.63 Reaction pathway toward the formation of polyoxysiloxane under friction. The techniques used to detect the products are indicated. The tribochemical reaction is that of tetra-silicon alkoxide formation [3]. Reprinted courtesy of the National Institute of Standards and Technology, U.S. Department of Commerce. Not copyrightable in the United States.

Applying the NIRAM–HSAB approach, the formation of such silicon compounds from silicon nitride lubricated by alcohols can be represented as in Fig. 2.64. It should be noted that the intermediate is the anion RO^-, which is formed also during polymerization of the silicon alkoxides, thus further increasing the rate of the tribochemical reaction.

Room-temperature ionic liquids (ILs) exhibit a medium with different properties in comparison with polar and non-polar solvents and are currently investigated in many fields including tribology as boundary lubricants. For instance, an extensive recent review paper [35] shows a good deal of works demonstrating the importance of ILs in silicon nitride lubrication. Since ILs molecules are self-organized on the surface, they clearly follow up and/or extend the real tribochemical action mechanism.

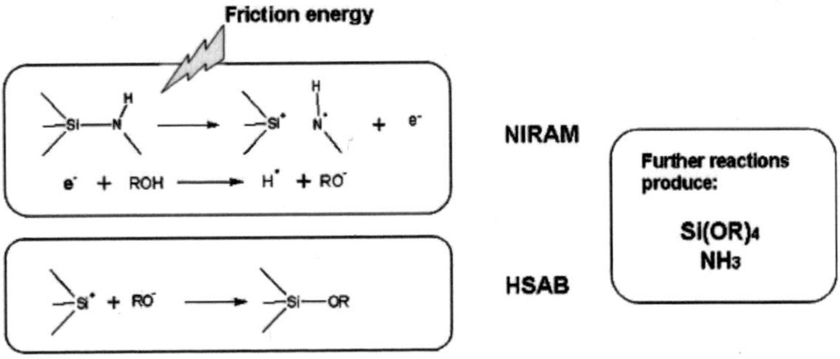

Figure 2.64 Interpretation of the tribochemical reaction of alcohols (ROH) with silicon nitride based on NIRAM–HSAB theory. Reprinted from *Wear*, 288, R. C. Dante, C. K. Kajdas, A review and a fundamental theory of silicon nitride tribochemistry, 27–38, Copyright (2012), with permission from Elsevier.

2.4.6 Tribology of Other Ceramics and Composites

2.4.6.1 Tribology of aluminum nitride

The data available on tribology of aluminum nitride compounds are very heterogeneous due to both the great variety of mating sliding surfaces and tribometric tests. Many reports are dedicated to ternary compounds because they usually exhibit higher hardness. However, there are series of comparative tests that allowed us to underline some common characteristics. For example, TiN, TiAlN are compared with TiB_2 evidencing the lower friction coefficient of TiB_2 with all the tested counterparts (Al, Ti–Al–V alloy and inconel), and this was attributed to the higher chemical resistance of this compound [7]. In coatings deposited by a lateral rotating arc process on high speed steel (HSS), the ratio of Al/Ti is determinant to decrease abrasive wear in a slurry of silicon carbide with a micro scale abrasion tester, as reported by Ding et al. [36]. The maximum hardness was achieved for Al/Ti = 1.22 as well the lower abrasive wear. This indicates that aluminum increases the tribological properties of pure TiN. In fretting experiments against an alumina ball, a dependence of the coefficient of friction on the relative humidity (RH) was found for TiAlN (either multilayer or monolayer) [37]. The decreasing of both the coefficient of friction and wear is attributed to tribochemical reactions. The maxima of both coefficient of friction and wear are positioned at the 30% of RH. A similar behavior was found also for TiN. Ball on disc experiments against steel in ambient humidity shows a transient behavior during which the coefficient of friction grows up to reach a steady state. This behavior is typical of a process of surface mating during which also wear debris usually oxidized acts as a third body [38].

Aluminum nitride was synthesized in 1877; however, potential applications were foreseen only the last three decades. For example, aluminum nitride is used for silicon substrates for its linear expansion coefficient similar to that of silicon. AlN tribochemical characteristics are not widely studied. Most reports are focused on coatings composed of ternary or quaternary systems with nitrogen, and often in combination with titanium for tool applications without a deep tribochemical approach as for silicon nitride.

Tool materials are generally much harder than the light metals and consequently, tool wear due to abrasion or erosion is very limited. However, the tendency for most light metals to adhere to the surface of tools in sliding contact is strong. Therefore, adhesive wear is a major problem in tribological contacts involving light metals. Since the dimension and surface finish of the product is replicated by the tool surface, it is essential that the tool, e.g., a cutting edge or an extrusion die, has a clean surface. For these applications, self-mating surfaces are not common as in silicon nitride and thus the tribological results are much less homogeneous [7].

2.4.6.2 Tribology of TiN and related coatings

Since the introduction of commercial coatings prepared by physical vapor deposition (PVD) at the beginning of the 1980s, their industrial application has grown rapidly. The conventional TiN coating, which is used for machining tools in particular, has dominated the market and is still the most frequently used. This situation seems to be changing. A great number of laboratory tests and field tests have shown examples where other PVD coatings clearly outperform TiN [11]. The best documented and acknowledged examples are those related to TiCN-coated milling tools and TiAlN-coated tools for cast iron. However, there is a class of coatings based on ternary and quaternary compounds with general formula A–Al–N–Z, where usually A ≡ Ti or Cr, and Z can be C or a transition metal depending on the applications. The main role of Al is passivating, since it forms oxides preferentially. The techniques to prepare such coatings are very wide from PVD, to magnetron sputtering, vacuum arc system, pulsed laser deposition [12], etc. Coefficient of friction of various coatings as a function of relative humidity in fretting wear tests sliding against alumina ball is detailed in [40] at load: 1 N, *f*: 10 Hz, D: 100 mm, fretting cycles: 3500.

Speculations on the higher coefficient of friction due to Al_2O_3 formation is reported by Shum et al. [39]. It is thought that the lower coefficient of friction of TiN is due titanium hydroxide-oxide or ortho-titanic acid (TiO_2-xH_2O or $Ti(OH)_2O$), as reported for pin-on-disc self-mated experiments [40]. This compound is not stable and depends on the relative humidity and test temperature. The result of the reaction of the thin TiO_2 layers with the water vapor may be a lubricating and wear-reducing layer. In these experiments, the ternary compound TiAlN has a coefficient of friction in humid conditions higher than that in dry conditions. In general, the ternary compound TiAlN has a higher coefficient of friction and wear than TiN [41].

Ball-on-disc geometry was used for testing ternary steered arc ion plated Ti*X*N coatings on HSS with humidity of 50–70% against α-Al_2O_3 balls, where *X* is the variable element, and it was found that wear debris phase obtained when $X \equiv$ Al corresponds mainly to $AlTiO_5$, detected by transmission electron microscopy (TEM) with some TiO and AlN [42].

Quaternary systems were also studied in order to improve wear of Ti–Al-based coatings. For example, in ball-on disc tests of Ti–Al–V–N against alumina the introduction of vanadium (ranging from 0 to 25 at.%) reduces the wear rate at low temperatures, which is mostly affected by the oxide formation, due to structural changes from *hcp* to the harder *fcc* [43]. Another quaternary coating considered is Cr–Ti–Al–N. Tam et al. investigated the coating tribological behavior by a pin-on disc tribometer, against a 5 mm diameter WC-6%Co ball at a linear speed of 0.20 ms^{-1} for 200,000 cycles (~5 km) at room temperature and about 45% of humidity. They detected the presence of CrN and TiN phases and (111) reflection corresponding to a *fcc* structure by XRD, and evidenced the chemical bonding of each metal with nitrogen by XPS. The coating as a large prevalence of Cr with the atomic percent between 50 and 35% depending on the sample, and a 50% of N. The authors claimed for a wear rate much lower than that of CrN [44].

Meneau et al. growth an AlN coating on aluminum substrates by laser-induced plasma in both nitrogen and air [45]. Friction tests were performed by a pin-on-disc machine, with an aluminum pin of 5 mm diameter sliding against the coated disc, at a rotation speed of 10 rpm, load of 5 N, in ambient air. A short transient behavior led to a high coefficient of friction in steady state, around 0.7 (Fig. 2.65a) when coverage of nitride is the very low. A low and steady coefficient of friction was found when AlN coverage is increased (Fig. 2.65b). The specimen laser-treated in pure nitrogen but with lower nitride coverage exhibits a transient low coefficient of frictions growing up to 0.7 at the test ending (Fig. 2.65c). This behavior can be due to the formation of aluminum oxide. Nevertheless, it is difficult to say that is due either to aluminum nitride reaction or direct metallic aluminum reaction with ambient water or oxygen. A low coefficient of friction was found for the sample treated in air (Fig. 2.65d) probably due to the phase identified by X-ray diffraction (XRD) as AlO_xN_y.

Aluminum nitride and its compounds often in combination with Ti are studied for coatings of aluminum and titanium parts. Although, there is evidence of surface oxidation and hydroxylation, their tribochemistry is not yet well known. Presumably, the mechanism is similar to that of silicon nitride. However, silicon oxides and alkoxides can polymerize forming lubricating soft oxide layers. Aluminum oxides are much harder than silicon oxides, but titanic acid can play a similar role to that of silicic acid. Nitrides always expose an oxidized surface, indicating that silicon nitride tribochemistry is a paradigm for this class of engineering ceramics, and many of the achievements in lubrication can be probably extended to aluminum and titanium nitrides.

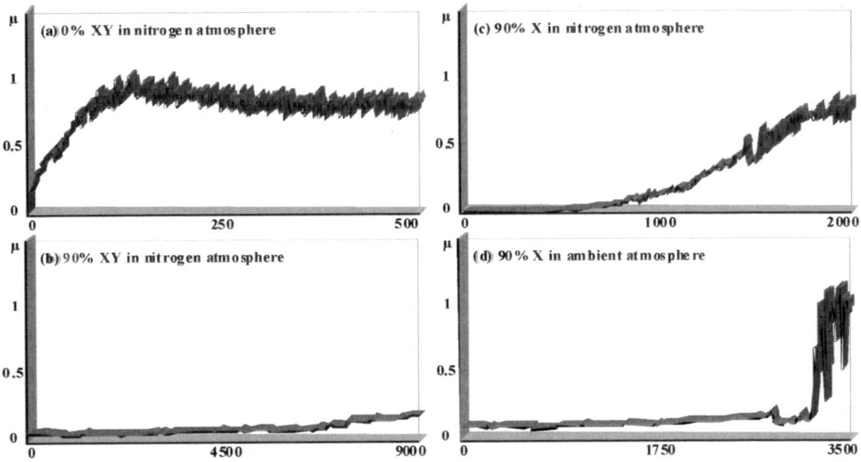

Figure 2.65 Coefficient of friction evolution vs. revolutions in different samples. XY indicates the overlapping of the laser treatment by scanning in XY directions, % indicates the percent of overlapping in the indicated direction. Reprinted from *Surf. Coatings Tech.*, 100–101, C. Meneau, P. Andreazza, C. Andreazza-Vignolle, P. Goudeau, J.-P. Villain, C. Boulmer-Leborgne, Laser surface modification: Structural and tribological studies of AlN coatings, 12–16, Copyright (1998), with permission from Elsevier.

2.4.6.3 Synthesis of AlN powder

Work [46] is on aluminum nitride (AlN) formation and [47] on synthesis of (AlN). The latter is performed thermochemically. The amorphous alumina in the precursor was first transformed into γ-Al_2O_3, and then directly nitrided by air to yield AlN during the calcination process. The reaction temperature needed for a complete conversion for the precursor was about 1400°C, which is much lower than that when using alumina and carbon black as starting materials. The synthesized AlN powder was composed of very fine particles and had good dispersability. The commercial AlN powders are usually synthesized by either carbothermal [47], direct nitridation of aluminum or reduction and nitridation of Al-polynuclear complex [48]. The most recent paper describes preparation of nanometer AlN powders by combining spray pyrolysis with carbothermal reduction and nitridation [49].

2.4.6.4 General information on tribology, tribology of composites and additional references on tribochemistry

There are many books providing and summarizing fundamental tribology concepts of engineering materials. Examples represent references [50–55]. Other books of note consider flat surfaces under contact/rubbing [56], contact mechanics [57] and tribophysics [58]. Focus on silicon nitride and other ceramic nitrides' tribochemistry relates to this book chapter; therefore, all the information on tribochemistry is described in previous sections.

The most recent tribology of ceramic perspectives combined with tribology of composites is detailed in the book by Basu and Kalin [59]. It also largely discusses the case studies of ceramics and composites plus mechanical properties of ceramics. Other factors include load, relative speed, lubrication, temperature, ambient, inert atmosphere, relative humidity, the chemical nature and compatibility of mating materials with environment and lubricants. Book [59] presents the key material properties, which are available to a material scientist to design a material, and have a critical influence on the material's tribological behavior. It also provides examples of the microstructural control of advanced ceramic and composite materials for optimal tribological performance, and therefore provides adequate background for the tribochemistry of silicon nitride and other ceramics. Papers [60–64] enhance references concerning tribochemistry of ceramics.

2.4.7 Conclusions

Generally, it is well accepted that ceramic materials are of particular importance for practical applications and their tribochemistry is very complex. Despite of very significant advance in research on physics and chemistry of engineering ceramics, even well progressed tribochemistry of silicon nitride still is not fully elucidated. The present chapter aimed at providing better understanding of some nitrides' tribology, especially of Si_3N_4, AlN, and TiN, by accounting for the mechanisms of their real tribochemistry.

Our working model, based on the triboemission phenomenon is here considered and discussed in detail. We demonstrate that during friction of a silicon nitride material, positively charged silicon sites and nitrogen free radicals are produced. Detailed tribochemistry of ammonia generation from the milled silicon nitride is developed.

Accordingly, tribochemistry of silicon nitride has been taken into account as the paradigm for ceramic nitrides. Thereafter, the NIRAM-HSAB derived tribochemistry of silicon carbide and silica (see Figs. 9 and 11 in ref. [7]) is compared with the tribochemistry approach of these ceramics based on the model suggested in work [7]. The latter type of tribochemistry is depicted in Fig. 10 of ref. [7].

The mechanical stresses are of particular importance for tribochemistry of ceramics, e.g., work [61] demonstrates that oxidation of silicon nitride is primarily related to the sliding distance. This fact corroborates the NIRAM-HSAB approach to ceramic tribochemistry. Therefore, we conclude that the silicon carbide tribochemistry is also in line to silicon nitride tribochemistry paradigm.

References

1. R. C. Dante, C. K. Kajdas, A review and a fundamental theory of silicon nitride tribochemistry, *Wear*, 288 (2012) 27–38.
2. K. H. Jack, Review: Sialons and related nitrogen ceramics, *J. Mater. Sci.*, 11 (1976) 1135–1158.
3. R. S. Gates, S. M. Hsu, National Institute of Standards and Technology (NIST) Special Publication 876. NIST, Gaithersburg, MD 20899-0001, USA, February (1995).
4. T. E. Fischer, H. Tomizawa, Interaction of tribochemistry and microfracture in the friction and wear of silicon nitride, *Wear*, 105 (1985) 29–45.
5. H. Tomizawa, T. E. Fischer, Friction and wear of silicon nitride and silicon carbide in water, *ASLE Trans.*, 30 (1986) 41–46.
6. R. S. Gates, S. M. Hsu, Effect of selected chemical compounds on the lubrication of silicon nitride, *Tribol. Trans.*, 34 (1991) 417–425.

7. K. Mizuhara, S. M. Hsu, Tribochemical reaction of oxygen and water on silicon surfaces, in: *Wear Particles* (D. Dowson, ed.), Elsevier Publishers B. V. (1992).
8. M. Volante, B. Fubini, E. Giamello, V. Bolis, Reactivity induced by grinding in silicon nitride, *Mater. Sci. Lett.*, 8 (1989) 1076–1078.
9. C. Kajdas, Hydrolysis, in: *Surface Modifications and Mechanisms* (G. E. Totten, H. Liang, eds.), chapter 8, Marcel Dekker, Inc., New York, Basel (2004).
10. S. Loughin, R. H. French, W. Y. Ching, Y. N. Xu, G. A. Slack, Electronic structure of aluminium nitride: Theory and experiment, *Appl. Phys. Lett.*, 63 (1993) 1182–1184.
11. M. Berger, S. Hogmark, Tribological properties of selected PVD coatings when slid against ductile materials, *Wear*, 252 (2002) 557–565.
12. J. Bienk, H. Reitz, N. J. Mikkelsen, Wear and friction properties of hard PVD coatings, *Surf. Coatings Technol.*, 76–77 (1995) 475–480.
13. J. M. Lackner, W. Waldhauser, R. Ebner, R. J. Bakker, T. Schöberl, B. Major, Room temperature pulsed laser deposited (Ti, Al)C_xN_{1-x} coatings—chemical, structural, mechanical and tribological properties, *Thin Solid Films*, 468 (2004) 125– 133.
14. Yu. N. Drozdov, S. D. Ivanov, T. M. Savinova, Wear Prediction of Ceramics, AITC-AIT 2006 Proceedings International Conference on Tribology, 20–22 September 2006, Parma, Italy.
15. Z. Jiang, F. Kragh, D. J. Frost, K. Ståhl, H. Lindelov, Hardness and thermal stability of cubic silicon nitride, *J. Phys. Condens. Matter*, 13 (2001) L515–L520.
16. J. A. Wendel, A. Goddard III, The Hessian biased force field for silicon nitride ceramics: Predictions of thermodynamic and mechanical properties for α– and β–Si_3N_4, *J. Chem. Phys.*, 97 (1992) 5048–5062.
17. T. Nishimura, X. Xu, K. Kimoto, N. Hirosaki, H. Tanaka, Fabrication of silicon nitride nanoceramics–Powder preparation and sintering: A review, *Sci. Technol. Adv. Mater.*, 8 (2007) 635–643.
18. X. Dong, S. Jahanmir, Wear transition diagram for silicon nitride, *Wear*, 165 (1993) 169–180.
19. S. Novak, G. Dražic, Z. Samardžija, M. Kalin, J. Vižintin, Wear of silicon nitride ceramics under fretting conditions, *Mater. Sci. Eng.*, A215 (1996) 125–133.

20. M. Kalin, J. Vižintin, S. Novak, G. Dražic, Wear mechanisms in oil-lubricated and dry fretting of silicon nitride against bearing steel contacts, *Wear*, 210 (1997) 27–38.

21. X. Z. Hud, J. M. Dell, L. Faraone, Effect of deposition conditions on mechanical properties of low-temperature PECVD silicon nitride films, *Mater. Sci. Eng. A*, 435–436 (2006) 453–459.

22. M. Kalin, J. Vižintin, J. Vleugels, O. Van Der Biest, Chemical reactivity of silicon nitride with steel and oxidised steel between 500 and 1200°C, *Mater. Sci. Eng. A*, 281 (2000) 28–36.

23. S. Sasaki, The effects of the surrounding atmosphere on the friction and wear of alumina, zirconia, silicon carbide and silicon nitride, *Wear*, 134 (1989) 185–200.

24. T. Saito, Y. Imada, F. Honda, An analytical observation of the tribochemical reaction of silicon nitride sliding with low friction in aqueous solutions, *Wear*, 205 (1997) 153–159.

25. V. A. Muratov, T. Luangvaranunt, T. E. Fischer, The tribochemistry of silicon nitride: Effects of friction, temperature and sliding velocity, *Tribol. Int.*, 31 (1998) 601–611.

26. S. Kitaoka, T. Tsuji, Y. Yamaguchi, K. Kashiwagi, Tribochemical wear theory of non-oxide ceramics in high-temperature and high-pressure water, *Wear*, 205 (1997) 40–46.

27. B. Fubini, E. Giamello, M. Volante, V. Bolis, Chemical Functionalities at the silica surface determining its reactivity when inhaled. Formation and reactivity of surface radicals, *Tech. Ind. Health*, 6 (1990) 571–597.

28. B. Fubini, V. Bolis, A. Cavenago, M. Volante, Physicochemical properties of crystalline silica dusts and their possible implication in various biological responses, *Scand. J. Work. Environ. Health*, 21 (1995) 9–14.

29. B. Fubini, V. Bolis, E. Giamello, M. Volante, Chemical functionalities at the broken fibre surface relatable to free radicals production, in: *Mechanisms in Fibre Carcinogenesis* (R. C. Brown, et al., eds.), Plenum Press, New York (1991).

30. B. Fubini, M. Volante, V. Bolis, E. Giamello, Reactivity towards water of silicon nitride: Energy of interaction and hydration and dehydration mechanism, *J. Mater. Sci.*, 24 (1989) 549–556.

31. C. K. Kajdas, Physical and chemical phenomena related to tribochemistry, in: *Advances in Contact Mechanics: Implications for Material Science, Engineering and Biology* (R. Buzio, U. Valbusa, eds.), chapter 12, Transworld Research Network, Kerala, India (2006).

32. C. K. Kajdas, Tribochemistry of selected ceramic materials, *Solid Sate Phenomena*, 113 (2006) 339–347.
33. R. C. Dante, C. Kajdas, A. Kulczycki, Theoretical advances in the kinetics of tribochemical reactions, Reaction Kinetics, *Mech. Catal.*, 99 (2010) 37–46.
34. Y. Hibi, Y. Enomoto, Tribochemical wear of silicon nitride in water, *n*-alcohols and their mixtures, *Wear*, 133 (1989) 133–145.
35. I. Minami, Ionic liquids in tribology, *Molecules*, 14 (2009) 2286–2305.
36. X. Ding, C. T. Bui, X. T. Zeng, Abrasive wear resistance of $Ti_{1-x}Al_xN$ hard coatings deposited by a vacuum arc system with lateral rotating cathodes, *Surf. Coatings Tech.*, 203 (2008) 680–684.
37. M. Z. Huq, J. P. Celis, Fretting wear of multilayered Ti, Al/NrTiN coatings in air of different relative humidity, *Wear*, 225–229 (1999) 53–64.
38. Z. P. Huang, Y. Sun, T. Bell, Friction behaviour of TiN, CrN and (TiAl) N coatings, *Wear*, 173 (1994) 13–20.
39. P. W. Shum, W. C. Tam, K. Y. Li, Z. F. Zhou, Y. G. Shen, Mechanical and tribological properties of titanium–aluminium–nitride films deposited by reactive close-field unbalanced magnetron sputtering, *Wear*, 257 (2004) 1030–1040.
40. G. Meier zu Köcker, T. Gross, E. Santner, Influence of the testing parameters on the tribological behaviour of self-mated PVD, *Wear*, 179 (1994) 5–10.
41. I. Efeoglu, R. D. Arnell, S. F. Tinston, D. G. Teer, The mechanical and tribological properties of titanium aluminium nitride coatings formed in a four magnetron closed-field sputtering system, *Surf. Coatings Tech.*, 57 (1993) 117–121.
42. E. Vancoille, J. P. Celis, J. R. Roos, Dry sliding wear of TiN based ternary PVD coatings, *Wear*, 165 (1993) 41–49.
43. M. Pfeile, K. Kutschej, M. Penoy, C. Michotte, C. Mitterer, M. Kathrein, The effect of increasing V content on structure, mechanical and tribological properties of arc evaporated Ti–Al–V–N coatings, *Int. J. Refractory Metals Hard Mater.*, 27 (2009) 502–506.
44. P. L. Tam, Z. F. Zhou, P. W. Shum, K. Y. Li, Structural, mechanical, and tribological studies of Cr–Ti–Al–N coating with different chemical compositions, *Thin Solid Films*, 516 (2008) 5725–5731.
45. C. Meneau, P. Andreazza, C. Andreazza-Vignolle, P. Goudeau, J.-P. Villain, C. Boulmer-Leborgne, Laser surface modification: Structural

and tribological studies of AlN coatings, *Surf. Coatings Tech.*, 100–101 (1998) 12–16.

46. A. Gromov, V. Vereshchagin, Study of aluminum nitride formation by superfine aluminum powder combustion in air, *J. Euro. Ceram. Soc.*, 24 (2004) 2879–2884.
47. M. Qin, X. Du, Z. Li, I. S. Humail, X. Qu, Synthesis of aluminum nitride powder by carbothermal reduction of a combustion synthesis precursor, *Mater. Res. Bull.*, 43 (2008) 2954–2960.
48. N. Hashimoto, Y. Sawada, T. Bado, H. Yoden, S. Deki, Preparation of aluminum nitride from aluminum polynuclear complexes, *J. Am. Ceram. Soc.*, 74 (1991) 1282–1286.
49. S. H. Lee, J. H. Yi, J. H. Kim, Y. N. Ko, Y. J. Hong, Y. C. Kang, Preparation of nanometer AlN powders by combining spray pyrolysis with carbothermal reduction and nitridation, *Ceram. Intern.*, 37 (2011) 1967–1971.
50. H. Czichos, *Tribology*, Elsevier, Amsterdam (1978).
51. K. H. Zum Gahr, *Microstructure and Wear of Materials*, Elsevier, Oxford (1987).
52. A. D. Sarkar, *Friction and Wear.* Academic Press, London (1980).
53. G. W. Stachowaik, A. W. Batchelor, *Engineering Tribology, Tribology Series*, vol. 24, Elsevier, Amsterdam (1993).
54. E. Rabinowicz, *Friction and Wear of Materials*, 2nd ed. Wiley, New York (1995).
55. I. M. Hutchings, *Tribology: Friction and Wear of Engineering Materials*, Butterworth–Heinemann Publications, Guernsey, UK (1992).
56. B. Bhushan, *Principles and Applications of Tribology*, A Wiley–Interscience Publication, John Wiley & Sons, New York (1999).
57. F. Archard, Contact and rubbing of flat surfaces, *J. Appl. Phys.*, 24 (1953) 981–988.
58. N. P. Suh, *Tribophysics*, Prentice–Hall (1986).
59. B. Basu, M. Kalin, *Tribology by Ceramics and Composites: Materials Science Perspective*, Wiley-VCH (2011).
60. Y. Tsunai, Y. Enomoto, Tribochemical wear of silicon nitride in water, n-alcohol and their mixtures, in: *Wear of Materials* (Ludema, K. C., ed.), ASME, New York (1989).
61. D. C. Cranmer, Wear surface analysis of silicon nitride, *STLE Lubrication Eng.*, 44 (1988) 975–980.

62. T. E. Fischer, Z. Zhu, H. Kim, D. S. Shin, Genesis and role of wear debris in sliding wear of ceramics, *Wear*, 245 (2000) 53–60.

63. W. Maw, F. Stevens, S. C. Langford, J. T. Dickinson, Single asperity tribochemical wear of silicon nitride studied by atomic force microscopy, *J. Appl. Phys.*, 92 (2002) 5103–5109.

64. Q. Xue, J. Weii, Tribochemistry in ceramic lubrication, *Lubrication Sci.*, 8 (1996) 369–377.

Chapter 3

Tribocatalysis

Ken'ichi Hiratsuka,[a] Czesław Kajdas,[b] Andrzej Kulczycki,[c,d] and Roberto C. Dante[e,f]

[a]*Department of Innovative Mechanical and Electronic Engineering, Chiba Institute of Technology, Japan*
[b]*Institute of Chemistry in Płock, Warsaw University of Technology, Poland*
[c]*Air Force Institute of Technology, Warsaw, Poland*
[d]*Cardinal Stefan Wyszynski University, Warsaw, Poland*
[e]*Department of Rheology and Complex Materials, Instituto de Investigaciones en Materiales, Universidad Nacional Autónoma de México, Mexico*
[f]*Laboratory of Advanced Materials ETSIIA, University of Valladolid, Spain*

3.1 Tribocatalysis

Ken'ichi Hiratsuka[a] and Czesław Kajdas[b]

[a]*Department of Innovative Mechanical and Electronic Engineering, Chiba Institute of Technology, Japan*
[b]*Automotive Industry Institute in Warsaw, Poland*

hiratsuka@sea.it-chiba.ac.jp

3.1.1 Introduction

3.1.1.1 Mechanocatalysis

Recent innovations of nanotechnology have hybridized different fields of science and technology to enrich both fields and even create new ones. Mechanochemistry is an example, which has

Tribocatalysis, Tribochemistry, and Tribocorrosion
Edited by Czesław Kajdas and Ken'ichi Hiratsuka

ISBN 978-981-4316-95-8 (Hardcover), 978-1-315-36498-8 (eBook)
www.panstanford.com

rapidly progressed nowadays because of the various developments in nanotechnology. Atomic force microscopy (AFM) and scanning tunneling microscopy (STM) are representative devices to apply force and manipulate individual atoms and molecules. With these devices, chemical reactions can be modulated mechanically. It is symbolized by the term "Mechanically Induced Chemistry" [1] or "Mechanically-Induced Chemical Changes [2]." Furthermore, especially when catalysts are used in mechanochemical reactions, they are called mechanocatalytic reactions.

It is well known that in both homogeneous and heterogeneous catalytic reactions, mechanical force can activate a substance to give rise to the catalytic activity.

In homogeneous catalytic reaction, Piermattei et al. showed that the dissociation of an organo-metallic compound exposes the active centers at the cleaved sites to induce chemical reactions [3]. This organo-metallic compound is called latent catalyst because catalytic activity has been concealed in the molecule. The activation is triggered by mechanical force. Therefore, this system is expressed: "catalysts feel the force [4]." Gilman reviewed how strain in molecules triggers chemical reactions [5]. He showed how the gap between lowest unoccupied molecular orbital (LUMO) and highest occupied molecular orbital (HOMO) is decreased when strain is applied to molecules to shear the covalent bonds. This results in the decrease in internal stability of a molecule, leading to a higher tendency for chemical reactions than the state without any strain. LUMO-HOMO gap corresponds with the band gap in crystalline solids. The shear stress to the solid changes the interatomic bond angles, which consequently reduces the band gap. Reducing the gap enables bonding electrons to move easily, which can initiate the reaction without any help of heat.

In heterogeneous catalytic reaction, when catalyst surface is covered by contaminants and native oxides, its catalytic activity is buried. Mechanical action removes these layers and exposes the catalyst with surface bonds unsaturated. This is the first step of activating latent heterogeneous catalyst driven by mechanical force. Mechanically produced surfaces have a number of lattice defects with kinks and steps, as compared to evaporated films. This disordered lattice structure has inherently higher energy than the ordered one, which leads to the enhanced surface activity in chemical reactions. This process is called mechanochemical activation (MCA) of catalyst [6]. Apart from the activation of the nascent surface, there is another excitation process that takes place specifically in surrounding molecules when energy is brought by the attachment of electrons [7]. The effect of exoelectrons from the friction surface on the activation of chemical reactions will be discussed in detail in later sections.

3.1.1.2 Tribocatalysis

Tribocatalysis was first defined by Heinicke in his book of tribochemistry as "the subsection of tribochemistry dealing with the changes of catalytic properties of solids due to the effect of mechanical energy [8]." Although he introduced a number of examples of tribocatalytic reactions, the term has not been widely used as tribochemical reactions. This is because catalysts are hardly applied to tribological components. On the contrary, the term mechanocatalysis has been the target of research more often than tribocatalysis, even when solids are under sliding contact [9]. Mechanocatalysis has now been widely used to refer to the mechanical stimulations generated by all kinds of mechanical actions including compression, extension, impulse, shear, etc. [10]. Despite a limited number of research works, tribocatalysis has been known to be the important reaction schemes [11]. The reason is that precise experiments of tribochemical reactions, where tribological conditions were kept constant, have revealed how surface active sites operate to enhance reactions [12]. Based on the aforementioned background, this chapter/section is devoted to discuss tribocatalysis being focused on its mechanisms and activation energy.

Although tribocatalysis was clearly defined in Heinicke's book, that definition is sometimes taken in various contexts. Therefore, to begin with, it is necessary to redefine tribocatalysis and related words by analogy to other fields such as photocatalysis. The glossary of terms relevant to photocatalysis has already been presented in [13] by the International Union of Pure and Applied Chemistry (IUPAC). To take the definitions in photochemistry as a guide would provide a reliable approach to establish the right terms in tribocatalysis. Definitions of photocatalyst and photocatalysis are presented below:

Photocatalyst: Substance able to produce, by absorption of ultraviolet, visible, or infrared radiation, chemical transformations of the reaction partners, repeatedly coming with them into intermediate chemical interactions and regenerating its chemical composition after each cycle of such interactions.

Photocatalysis: Change in the rate of a chemical reaction or its initiation under the action of ultraviolet, visible, or infrared radiation in the presence of a substance—the photocatalyst—that absorbs light and is involved in the chemical transformation of the reaction partners.

As for catalysis and catalyst, IUPAC defined catalyst first, and catalysis was merely stated as "Catalysis is the action of catalyst" [13]. The terms of photocatalysis and photocatalyst were defined taking after this approach. In this regard, tribocatalyst and tribocatalysis shall be defined as follows:

Tribocatalyst: Substance able to produce, by receiving normal and tangential forces, chemical transformations of the reaction partners, repeatedly coming with them into intermediate chemical interactions and regenerating its chemical composition after each cycle of such interactions.

Tribocatalysis: Change in the rate of a chemical reaction or its initiation under the action of friction in the presence of a substance—the tribocatalyst—that receives frictional force and is involved in the chemical transformation of the reaction partners.

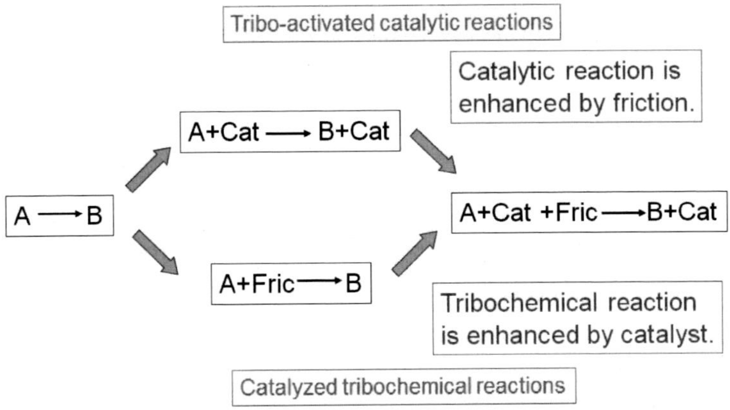

Figure 3.1 Two types of tribocatalytic enhancement of chemical reactions [11].

There are two ways to realize tribocatalytic reactions as shown in Fig. 3.1 [11]. One is to enhance the catalytic activity of native catalyst by frictional stimulation. The other is to generate catalytic activity from the latent catalyst by frictional force. The two processes correspond with triboactivated catalytic reaction and catalyzed tribochemical reactions, respectively.

3.1.2 Mechanisms of Tribocatalysis

3.1.2.1 Overall approach

A typical scheme of the three stages of heterogeneous catalytic reactions is shown in the upper part of Fig. 3.2, under which their corresponding tribocatalytic reactions are illustrated. Actually, the upper part is thermally activated catalytic reaction, while the lower is tribologically activated one. Generally, heterogeneous catalytic reaction comprises three stages:

(1) Adsorption on catalyst
(2) Surface reaction
(3) Desorption from catalyst

Although Langmuir–Hinshelwood and Eley–Rideal mechanisms are different in terms of adsorption and surface reaction processes, both involve the adsorption of at least one of the molecules on the catalyst surface. The central subject in tribocatalysis is how frictional energy is injected to the adsorbents and adsorbates for adsorption and surface reaction, and to the products for desorption. These tribological effects in the three stages are discussed as follows:

The first tribological effect is acting at the adsorption process on catalyst. Shearing force cleaves the interatomic bonds beneath the adsorbent surface, exposing the nascent surface. It has three kinds of active sites such as two dangling bonds, an acid and a base site, and an acid and a dangling bond with electron emission, as drawn in the lower left part of the figure under (1) triboadsorption on catalyst. Such sites have a disordered lattice structure with kinks and steps, which also promotes activity.

The second kind, (2) tribosurface reaction, utilizes friction force to stretch and break the bonds in adsorbate molecules and to promote chances of creating new bonds by frictional enhancement of surface diffusion. The original idea of "friction catalysis" for the synthesis of water [14] and carbon dioxide [15] was based mainly on the surface diffusion enhanced by friction.

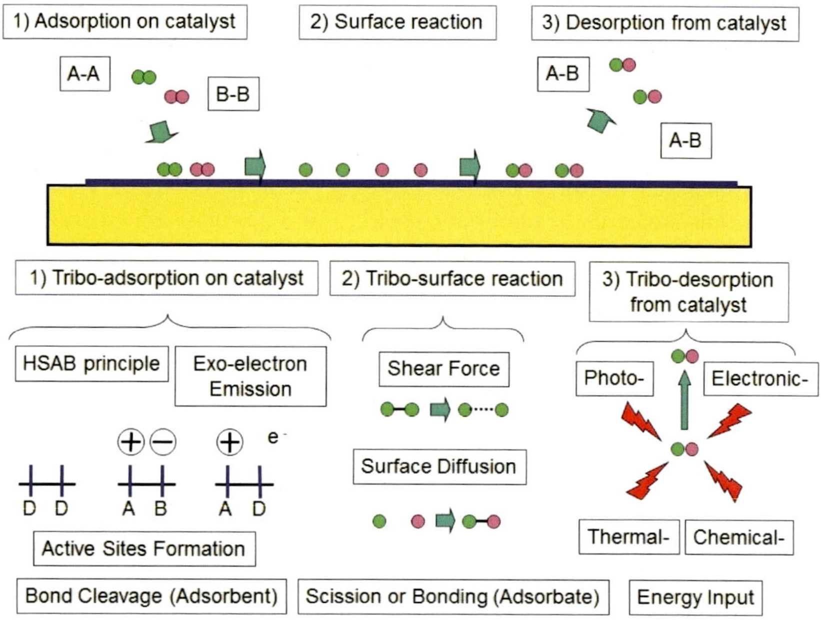

Figure 3.2 Three stages of heterogeneous catalytic reactions and their corresponding tribocatalytic reactions.

The last process is to give the new molecules energy to be desorbed from the surface. The energy sources are (1) heat by friction and chemical reactions, (2) electric potential or discharge and (3) photons from discharge. Heat is supplied from the dissipated energy and the instant increase/decrease in temperature is called flash temperature.

3.1.2.2 Electronic approach

Heterogeneous catalysts are environmentally friendly and are easily operated in continuous processes. They are widely applied in industry. On the other hand, mechanisms by which these catalysts lower activation energy are still under consideration. It is generally accepted that heterogeneous catalytic reactions take place at centers where electron exchange in the catalyst–substrate system is facilitated, and that electrons are produced from sites where electrons are weakly bound.

The transition state along a reaction path is the point of maximum free energy, where bond-making and bond-breaking are balanced. Heterogeneous catalysis accelerates the movement of atoms on the passage through the transition state; it provides the link between reactants and products on a reaction pathway, which involves simultaneous motion of several to very many atoms [16]. It is also known that heterogeneous catalytic reactions take place at sites (centers) where the electron exchange in the catalyst–substrate system is facilitated, and EEE is produced from sites where electrons are weakly bound. Sato and Seo [17] pioneered experiments to show that thermally stimulated exoelectron-emission (EEE) occurred in the same range of temperatures as that typical for heterogeneous catalysis. Experiments performed by Sato and Seo [18] show that thermally stimulated EEE occurs in the same range of temperatures as that of typical of heterogeneous catalysis. Figure 3.3 shows this relationship between the formation rate of ethylene oxide on silver and the thermally stimulated EEE rate. "Chemically stimulated" EEE occurs simultaneously during the partial oxidation process of ethylene and that the emission rate is proportional to the rate of ethylene oxide formation.

Emission of negative charges (electrons and ions) also was observed by Krylova et al. [19] during catalytic decomposition of methanol and formic acid, and for hydrogen oxidation on adsorbed layers of Al_2O_3, NaCl, and NiO catalysts. Research by Hoenig and Tamjidi [18] found that the EEE level correlates to the rate of reaction for catalytic oxidation of carbon oxide, hydrogen, and ammonia over a heated platinum wire. Back few years, good progress was made to clarify how reactant molecules that adsorb in the catalyst could reach the activation energy [20].

Krylova et al. [19] claimed that the measured EEE thermal emission curves would characterize the energy spectrum of surface trapping levels for electrons participating in the catalysis process. Hoenig and Utter [22] suggested that the electrons from the catalyzed reaction may be used to monitor the rate of oxidation of carbon monoxide and methane over palladium catalysts. More recently, Chrzanowski et al. [23] presented results on EEE from the surfaces of catalysts on tin dioxide substrate (which are used in industry for resistive gas-component sensors of high sensitivity and low selectivity); they showed that the rate of EEE-decay from catalysts surfaces correlated with oxygen adsorption, which is known to be required for heterogeneous catalysis.

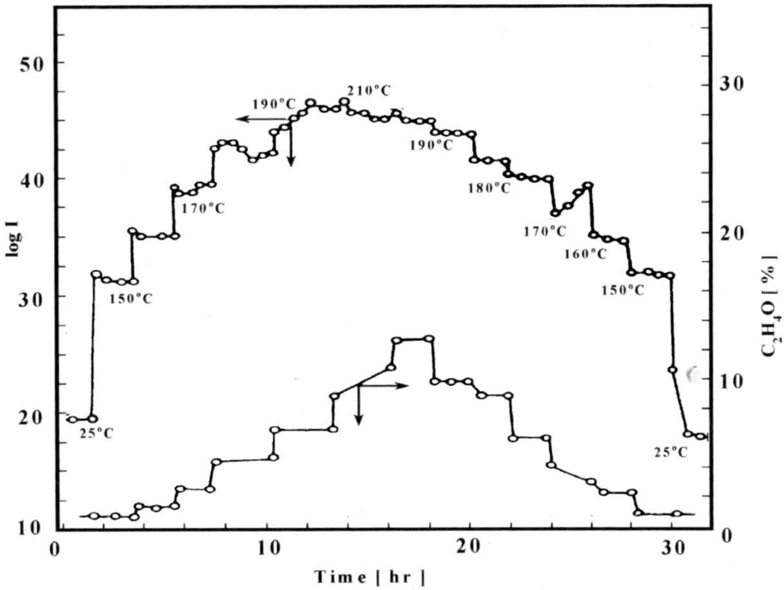

Figure 3.3 Effect of temperature rise on the exoelectron emission rate (log of EEE current) and on the formation rate of ethylene oxide (C_2H_4O), for silver catalyst. Reprinted from *J. Catal.*, 24, N. Sato, M. Seo, Chemically stimulated exoelectron emission from silver catalyst during partial oxidation of ethylene, 224–232, Copyright (1972), with permission from Elsevier.

Tomishko [24] found the similar effect for the sulfur dioxide oxidation in the cases of vanadium, iron oxide and chromium oxide catalysts. The charge sign of the catalyst surface remained positive over the entire temperature range of catalyst operation, and such polarization was connected with the emission of negative charge to the gas reactant phase. Grinevich et al. [25] found that in a combined-plasma catalytic process sulfur dioxide could oxidize on vanadium catalyst at just 60–80°C but not at the usual 400–600°C.

Understanding the role of charged particle emission can also help explaining the recently discovered enhanced catalytic performance of some noble metals clustered in nanocatalytic clusters. Such nanoclusters containing approximately 10 atoms, of about 1 nm in size, exhibit enhanced catalytic activity when supported on metal oxides or as charged gas-phase clusters [26, 27]. Models of gold nanoparticles deposited on thin magnesia (MgO) layers for various reactions are presented in paper [28]. Work [29] discusses high catalytic activity of nanosize-gold for ethylene formation. More four papers [30–33] relate to carbon mono-oxide (CO) oxidation processes.

Interestingly, Wu et al. [34] simulated CO oxidation applying density functional theory for different CO adsorption sites on small cationic, neutral, and anionic gold clusters as adsorption points. The calculated findings demonstrate that adsorption energies of CO on cationic clusters were generally greater than those on the neutral and anionic complexes, and that energy would decrease with size. Another simulation work [35] concluded that some of the nanocluster Au atoms, independent of the substrate, became positively charged (i.e., cationic) in the oxidation process. This positive surface charge would correspond to an exchange of charges (e.g., electrons) with the reactants. Support for the above nanoclusters has been demonstrated by Fierro-Gonzalez and Gates work [36].

Enhanced catalytic activity was observed by Socaciu et al. [37] for the oxidation of carbon monoxide on silver nanoclusters, but they also reported a strongly discontinuous size dependence on the reaction behavior. Recent experimental evidence indicates that for heterogeneous catalysis active sites on the oxide support-surface are required, and that these active sites control the charge state and exchange of the catalytic clusters [38, 39] for promoting the activation of adsorbed reactant molecules.

Basing on the above-discussed research works/papers, in summary we are stating that there is an extensive body of work demonstrating and/or suggesting that heterogeneous catalysis involves various electro-physical phenomena. Such electronic interactions play an important role in controlling catalysis dynamics, and that the involved electrons are of low-energy and they are produced from the excited catalytic surfaces and/or from charged active sites on the oxide support-surfaces.

In 2010, G. J. Molina et al. [40] presented research on the triboemission of electrons, charged particles and photons from sliding contacts under high vacuum. The research was focused on the design of an instrument to detect charged emissions in controlled atmospheres and gas flows. At that time, they scheduled to use this instrument for further investigation of possible mechanisms of low-energy plasma formation, and to study feasibility of employing such triboemissions for monitoring surface phenomena, or as precursor signals for material failure. Presently, the apparatus is being further developed in G. J. Molina's laboratory. Details of the apparatus can be found in Section 2.2.

Bonn et al. explored the precise mechanisms of desorption kinetics of carbon monoxide and dioxide on ruthenium (0001) surface [41]. They clearly evidenced that electrons in substrate activate the oxidation of carbon dioxide by femtosecond laser irradiation. It is contrasted to the conventional thermal excitation where carbon monoxide alone is desorbed. They indicated that femtosecond laser induces the excitation of electrons in the substrate, which are transferred to the adsorbate with conveying energy. Figure 1 in ref. [41] shows that the very short (ps range) yet high energy pulse, estimated to be around 6000 K of electron temperature, drives the oxidation of carbon monoxide to produce carbon dioxide.

It reminds us now of the flash temperature in friction processes [42]. It was demonstrated that surface temperature at small contact area less than 0.1 mm of steel/steel interface exceeded 1100°C during very short time at the sliding velocity of 34 m/s under the normal pressure of 90 MPa [43]. At this area, the electron temperature remains unknown, however, it can be higher than that. In this case, hot electrons can be transferred from the substrate to the adsorbates. Not only high temperature but also this electronic excitation would accelerate chemical reactions. Thus, we should have in mind that large energy input with electrons, in spite of femtosecond duration, provides different reaction paths to lead to the enhancement of expected or unexpected reactions.

Figure 4 of ref. [41] presents the CO_2 formation and the CO desorption after the laser pulse excitation. It demonstrates well that the oxidation reaction requires the electron-mediated excitation of the Ru-O bonding that is most efficient during the short time in which electrons are at their peak temperature. The thermal and laser-induced processes are depicted in the figure. It is of note that the 1.8 eV barrier for the oxygen activation is overcome by coupling to the hot electrons, so that CO_2 can be generated.

3.1.3 Activation Energy

3.1.3.1 Definition

Catalysis is the phenomenon of a catalyst in action modifying the rate of a chemical reaction induced by material unchanged chemically at the end of the reaction, and lowers the activation energy (E_a). In chemical kinetics, E_a is the height of the potential barrier separating the products and reactants. It is usually assumed that a catalytic reaction has a lower E_a than that of the thermally activated one.

3.1.3.2 Model approach

A first hypothesis of the considered model was described by Kulczycki [44] and presently is known as α_i model. Further consideration of the model was discussed in work published in 2008 [45]. Then the model had been further developed and theoretically supported [46]. It has been considered in terms of phenomenological data from the surface chemistry under friction, work function modification by mechanical input, possible electron and photon emissions, and the NIRAM approach [47, 48]. The α_i model correlates mechanical work performed on a solid with its catalytic activity. The tribological system contains fluid reagents and a solid surface (a specific material) that plays the role of catalyst (or "tribocatalyst") because its energy state is enhanced by the mechanical input [49].

3.1.3.3 Experimental evidence

The activation energies of some heterogeneous catalytic reactions, where catalysts are under friction and wear conditions, have been calculated. Figure 3.4 is one example in the oxidation of carbon monoxide from the rubbing of palladium catalyst against aluminum oxide. It was revealed that with the catalyst under friction, the activation energy of such reaction is 2.5 kJ/mol [50]. It is much lower as compared with those of the well-defined palladium surfaces without friction ranging from 105 to 134 kJ/mol.

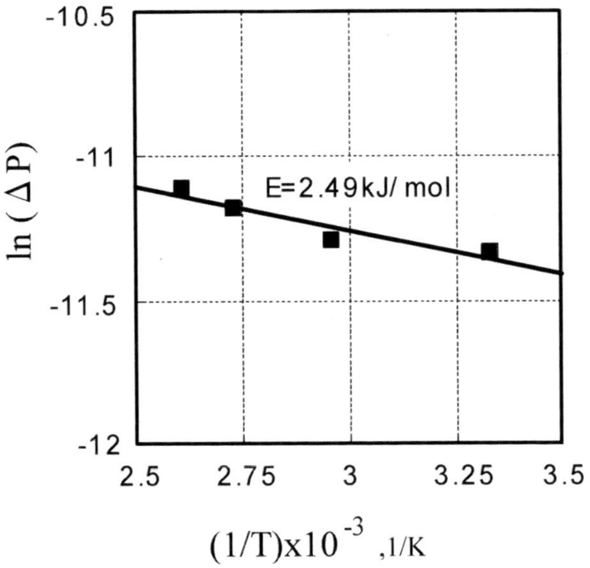

Figure 3.4 Arrhenius plot of the desorption of carbon dioxide. Reprinted from Tribo-catalysis in the synthesis reaction of carbon dioxide, K. Hiratsuka, C. Kajdas, M. Yoshida, *Tribol. Trans.*, 47, 86–93, 2004, by permission of Taylor & Francis Ltd.

Another example of Fig. 3.5 is taken from the adsorption of benzene. In this case, the adsorption activity was defined by the authors and set as a measure of reaction rate [51]. The Arrhenius plot of the adsorption activity shows that the activation energy is 4 kJ/mol, which is also much lower as compared to typical catalytic reactions.

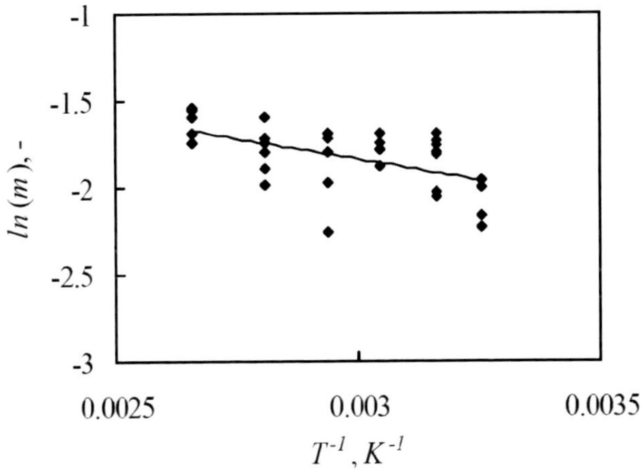

Figure 3.5 Arrhenius plot of the adsorption activity of benzene on nascent steel surfaces. Reproduced from N. Shimotomai, H. Nanao, S. Mori, Tribochemical reaction of benzene on nascent steel surface and effect of temperature, *Tribol. Online*, 7(2) (2012) 54–59, with permission of the Japanese Society of Tribologists.

The temperatures of the two examples mentioned above both ranged from 25 to 80°C, which are less than the temperatures where catalysts are normally used. Then, the activation energy of catalysts under friction within standard temperature ranges must be checked.

In the oxidation of ethylene, palladium catalyst was under friction with aluminum oxide [52]. Figure 3.6 shows the Arrhenius plot with actual variations of the ion current of m/e = 44, representing carbon dioxide, in the temperature ranges: 60–100°C, 150–230°C, and 300–390°C [53]. Each friction period is marked with a red bar above the current signal. It is apparent that the current during friction period is increased as the temperature is increased. It should also be noted that the base line before and after friction slightly increases with temperature. The base line corresponds to the reaction without friction, whereas the friction effect is the increment between the values with and without friction. In the temperature ranges from 60 to 100°C, and from 150 to 230°C, without friction, the reaction rates are almost constant, i.e. too low, so there is no notable reaction to discuss. On the other hand, with friction, there is a clear and linear increase in the logarithm of reaction rate. The activation energies are calculated as 9 and 10 kJ/mol for two temperature ranges. In 300 to 390°C, even without friction, the reaction rate is increased with an activation energy of 39 kJ/mol. With friction, that the activation energy is less than that without friction: 5 kJ/mol. However, it is to be noted that in this high temperature range, a part of CO_2 is decomposed to CO and O_2, which should affect the rate of CO_2.

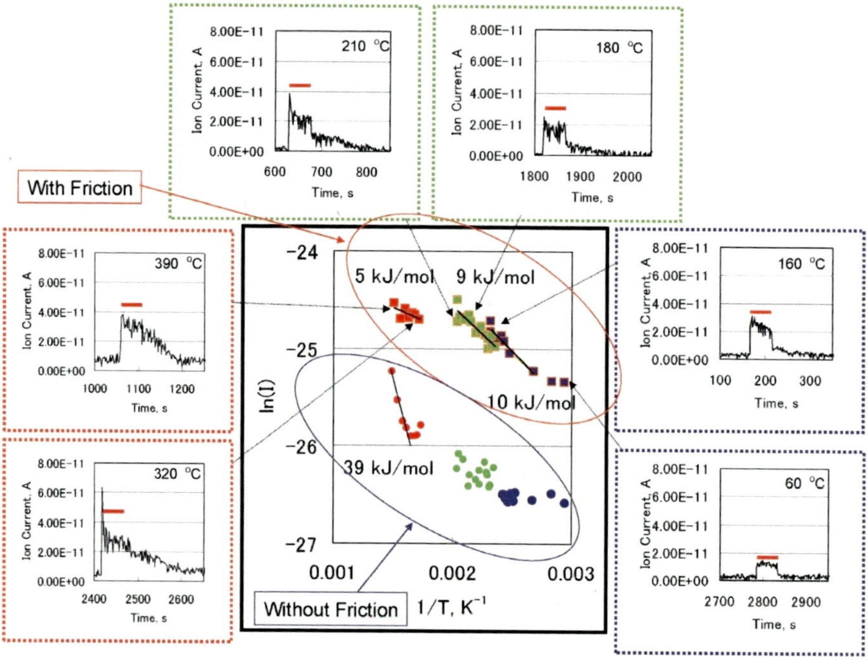

Figure 3.6 Actual ion currents of m/e = 44 before, during and after friction, and Arrhenius plots of the increments of ion current of m/e = 44 with and without friction. Reproduced from S. Fujiya, T. Tsutsumi, K. Hiratsuka, Effect of Temperature on the Ethylene Oxidation during Rubbing of Palladium, *Proc. Intl. Trib. Conf.* Tokyo (2015) 19pE-01, 410.

In all the cases mentioned above, the activation energy of catalytic reaction under friction condition is lower than 10 kJ/mol, which is far less than that in normal catalytic reactions.

3.1.4 Wear Particles and Catalysts

There are some common points between wear particles and catalysts listed below:

(1) Nanoparticles
(2) Metal–oxide interface
(3) Electron-activated

The size of wear elements, the original unit that composes wear particle, is around 10 to 40 nm [54]. It is equivalent to the size of a typical catalyst. This means that tribological interaction between surfaces naturally produces catalyst powders. Additionally, the lattice structure is so disordered that the surface is highly reactive by itself. The surfaces of wear particles are oxidized when they are generated in air, resulting in a mixture of metal and oxide. The metal–oxide interface acts as nanodiode to contribute to the production of hot electrons [55]. From the discussion above, it can be deduced that wear processes can naturally enhance chemical reactions through the production and enhancement of activity of catalysts. Wear and catalysis are, thus, related to each other.

3.1.5 Concluding Remarks

Importance of the catalysis surface science from single crystals to metal, recently (2008) was presented and described by Somorjai and Park [56]. It is supposed that a pulse of hot electrons of kinetic energy (1–3 eV), in metals might be produced in atomic and/or molecular processes. According to Park and Somorjai work published in 2006, detection of the electron flow became one of the frontier areas of research in the surface physics and chemistry community [55]. Such electrons, also called "ballistic electrons" are detected upon excitation of metal electrons by photons. Specific probe experiments carried out on femtosecond timescales could detect these electrons, which have an elastic mean free path in the metals with a range of about 10 nm [56].

Hot electrons, when emitted, can be further accelerated by the surface charge due to tribocharging. Then they attack and attach to the molecules to supply energy for reaction. Metal–oxide interface is a key to enhance catalytic activity, which has been called "strong metal-support interaction" [55].

From a perspective of industrial and molecular scientific viewpoint presented by Somorjai and Park, the field of catalysis is somehow similar to an onion with shells and a core, as depicted in Fig. 3.7. The outer shell represents all parts concerning any technology, along with process control, fabrication of catalysts and, the techniques/instrumentation to characterize catalysts. There are three types of catalyst systems (i) enzyme, (ii) heterogeneous and (iii) homogeneous. A common property of all these catalyst systems is that they are all nanoparticles.

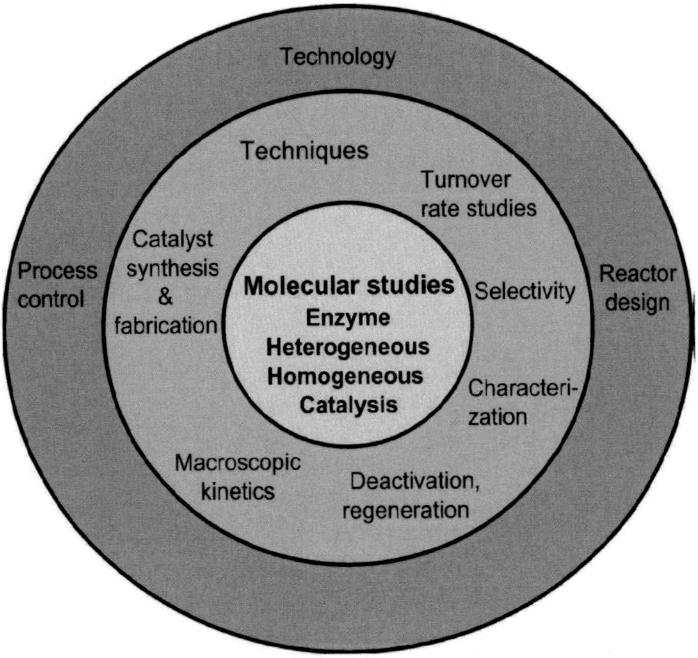

Figure 3.7 Layers of catalysis from technologies to molecular sciences. Reprinted from, G. A. Somorjai, J. Y. Park, Evolution of the surface science of catalysis from single crystals to metal, *J. Chem. Phys.*, 128 (2008) 182504. with the permission of AIP Publishing.

Very sophisticated work by Doshi et al. published in 2000, found efficient excitation of electrons in gold on the sub-picosecond timescale by NO molecules prepared in high vibrational states (n = 15) impinging on the metal surface [57]. Work [58] demonstrates laser-excited electron distribution in silicon.

Work [59] is an example of convincing approach to ecological and energy saving aspects of the catalytic converter in internal combustion engines.

Last but not least, it should be here stressed that femtosecond laser induces the excitation of electrons in the substrate, which are transferred to the adsorbate with conveying energy, as presented in Fig. 4 of ref. [41], showing that the 1.8 eV barrier for the oxygen activation is overcome by coupling to the hot electrons.

Summarizing all the above data and information, it seems that the direction for future perspective trends in tribocatalysis, follow the most effective generation of picosecond high energy pulse. It should also be noted at this point that the most recent publication [60], additionally presents effects of electro-catalysis, enhancing processes of tribocatalysis.

References

1. I. Frank, Mechanically induced chemistry: New perspectives on the nanoscale, *Angew. Chem. Intern. Ed.*, 45 (2006) 852–854.
2. M. M. Caruso, D. A. Davis, Q. Shen, S. A. Odom, N. R. Sottos, S. R. White, and J. S. Moore, Mechanically-induced chemical changes in polymeric materials, *Chem. Rev.*, 109 (2009) 5755–5798.
3. D. Aktah, I. Frank, Breaking bonds by mechanical stress: When do electrons decide for the other side? *J. Am. Chem. Soc.*, 124 (2002) 3402–3406.
4. J. S. Rathore, A. Nelson, Catalysts feel the force, *Nat. Chem.*, 1 (2009) 102–103.
5. J. J. Gilman, Mechanochemistry, *Science*, 274(4) (1996) 65.
6. R. A. Buyanov, V. V. Molchanov, and V. V. Boldyrev, Mechanochemical activation for resolving the problems of catalysis, *KONA Powder Particle J.*, 27 (2009) 38–54.
7. E. Illenberger and J. Momigny, *Gaseous Molecular Ions: An Introduction to Elementary Processes Induced by Ionization*, Springer Verlag (1992).

8. G. Heinicke, *Tribochemistry*, Berlin, Academie Verlag (1984).
9. K. Domen, S. Ikeda, T. Takata, A. Tanaka, M. Hara, J. N. Kondo, Mechano-catalytic overall water-splitting into hydrogen and oxygen on some metal oxides, *Appl. Energy*, 67 (2000) 159–179.
10. R. Groote, R. T. M. Jakobs, R. P. Sijbesma, Mechanocatalysis: Forcing latent catalysts into action, *Polym. Chem.*, 4 (2013) 4846–4859.
11. C. Kajdas, K. Hiratsuka, Tribochemistry, tribocatalysis, and the negative-ion-radical action mechanism, *Proc. Inst. Mech. Eng. Part J: J. Eng. Tribol.*, 223 (2009) 827–848.
12. R. Lu, I. Minami, H. Nanao, S. Mori, Investigation of decomposition of hydrocarbon oil on the nascent surface of steel, *Tribol. Lett.*, 27(1) (2007) 25–30.
13. S. E. Braslavsky, A. M. Braun, A. E. Cassano, A. V. Emeline, M. I. Litter, L. Palmisano, V. N. Parmon, N. Serpone, Glossary of terms used in photocatalysis and radiation catalysis (IUPAC Recommendations 2011), *Pure Appl. Chem.*, 83(4) (2011) 931–1014.
14. K. Hiratsuka, M. Kuzuya, T. Sasada, Friction catalysis in the synthesis of H_2O during adhesive wear, *Proceedings of the 33rd Japan Congress on Materials Research*, March (1990) 191–195.
15. K. Hiratsuka, M. Kuzuya, T. Sasada, Friction catalysis in the synthesis of CO_2 during adhesive wear, *Proceedings of the 34th Japan Congress on Materials Research*, March (1991) 119–123.
16. K. Klier, The transition state in heterogeneous catalysis, *Topics Catal.*, 18(3–4) (2002), 141 –156.
17. N. Sato, M. Seo, Chemically stimulated exo-emission from silver catalyst, *Nature*, 216 (1967) 361–362. DOI:1.1038/216361a0.
18. N. Sato, M. Seo, Chemically stimulated exoelectron emission from silver catalyst during partial oxidation of ethylene. *J. Catal.*, 24 (1972), 224–232.
19. V. Krylova, N. I. Konyushkina, I. A. Rodina, V. I. Svitov, Exoelectron emission accompanying catalytic reactions in adsorbed layers, *J. Catal.*, 60(1) (1979) 8–14.
20. S. A. Hoenig, F. Tamjidi, EEE during heterogeneous catalysis (the effect of external electric potential), *J. Catal.*, 28 (1973) 200–208.
21. A. Kulczycki, C. Kajdas, A new attempt to better understand Arrhenius equation and its activation energy, in: *Tribology in Engineering* (Pihtili, H., ed.), Chapter 4, InTech Open (May 2012).

22. S. A. Hoenig, M. G. Utter., Observations of exoelectron emission associated with heterogeneous catalysis, *J. Catal.*, 47(1977) 210–213.

23. J. Chrzanowski, H. Teterycz, R. Klimkiewicz, Exoelectron emission from tin oxide materials, *J. Electrochem. Soc.*, 150(6) (2003), H140–H146.

24. M. M. Tomishko, Plasma in catalytic processes, *High Energy Chem.*, 37(5) (2003) 333–340.

25. V. I. Grinevich, N. V. Kolobova, V. V. Kostrov, Influence of catalysts on sulfur dioxide and carbon monoxide oxidation in a barrier-plasma discharge, *High Energy Chem.*, 31(6) (1997) 400–441.

26. U. Landman, B. Yoon, C. Zhang, U. Heiz, M. Arenz, Factors in gold nanocatalysis: Oxidation of CO in the non-scalable size regime, *Topics Catal.*, 44(1–2) (2007) 145–158.

27. B. Yoon, H. Hakkinen, U. Landman, A. S. Worz, J. M. Antonietti, S. Abbet, K. Judai, U. Heiz, Charging effects on bonding and catalyzed oxidation of CO on Au_8 Clusters on MgO, *Science*, 307(5708) (2005) 403–407.

28. T. M. Bernhardt, R. N. Barnett, U. Landman, S. M. Lang, Methane activation and catalytic ethylene formation on free Au_2, *Angew. Chem. Int. Ed.*, 49/5 (2010) 980–983.

29. H. Hakkinen, U. Landman, Gas-phase catalytic oxidation of CO by Au_2, *J. Am. Chem. Soc.*, 123 (2001) 9704–9705.

30. C. Zhang, B. Yoon, U. Landman, Predicted Oxidation of CO Catalyzed by Au Nanoclusters on a Thin Defect-Free MgO Film Supported on a Mo(100) Surface, *J. Am. Chem. Soc.*, 129 (2007) 2228–2229.

31. M. Amft, N. V. Skorodumova, Catalytic activity of small Mg0-supported Au clusters towards CO oxidation: A density functional study, *Phys. Rev. B*, 81 (2010) 195443.

32. S. M. Lang, T. M. Bernhardt, R. N. Barnett, B. Yoon, U. Landman, Hydrogen-promoted oxygen activation by free gold cluster cations, *J. Am. Chem. Soc.*, 131 (2009) 8939–8951.

33. L. M. Molina, B. Hammer, Theoretical study of CO oxidation on Au nanoparticles supported by MgO(100), *Phys. Rev. B*, 69/15 (2004) 155424.

34. X. Wu, L. Senapati, S. K. Nayak, A. Selloni, M. Hajaligol, A density functional study of carbon monoxide adsorption on small cationic, neutral, and anionic gold clusters, *J. Chem. Phys.*, 117 (2002) 4010–4015.

35. L. M. Molina, B. Hammer, Some recent theoretical advances in the understanding of the catalytic activity of Au, *Appl. Catal. A Gen.*, 291(21) (2005) 21–31.

36. J. C. Fierro-Gonzalez, B. C. Gates, Evidence of active species in CO oxidation catalyzed by highly dispersed supported gold, *Catalysis Today*, 122(3–4) (2007) 201–210.

37. L. D. Socaciu, J. Hagen, J. Le Roux, D. Popolan, T. M. Bernhardt, L. Woste, S. Vajda, Strongly cluster size dependent reaction behavior of CO with O_2 on free silver cluster anions, *J. Chem. Phys.*, 120(5) (2004) 2078.

38. U. Landman, B. Yoon, C. Zhang, U. Heiz, M. Arenz, Factors in gold nanocatalysis: Oxidation of CO in the non-scalable size regime, *Topics Catal.*, 44(1–2) (2007) 145–158.

39. B. Yoon, H. Hakkinen, U. Landman, A. S. Worz, J. M. Antonietti, S. Abbet, K. Judai, U. Heiz, Charging effects on bonding and catalyzed oxidation of CO on Au_8 clusters on MgO, *Science*, 307(21) (2005) 403–407.

40. G. J. Molina, C. Kajdas, A. Kulczycki, Development of an instrument for studies of charged tribo-emission transport in gas atmospheres, *2011 STLE Annual Meeting & Exhibition*, Atlanta, Georgia, USA (2011).

41. M. Bonn, S. Funk, Ch. Hess, D. N. Denzler, C. Stampfl, M. Scheffßer, M. Wolf, G. Ertl, Phonon-versus electron-mediated desorption and oxidation of CO on Ru(0001), *Science*, 285 (1999) 1042–1045.

42. F. P. Bowden, D. Tabor, *The Friction and Lubrication of Solids*, Oxford Clarendon Press (1950).

43. G. Sutter, N. Ranc, Flash temperature measurement during dry friction process at high sliding speed, *Wear*, 268 (2010) 1237–1242.

44. A. Kulczycki, The correlation between results of different model friction test in terms of an energy analysis of friction and lubrication, *Wear*, 103 (1985) 67–75.

45. C. K. Kajdas, A. Kulczycki, A new idea of the influence of solid materials on kinetics of chemical reactions, *Mater. Sci. Poland*, 26(3) (2008) 787–796.

46. C. K. Kajdas, A. Kulczycki, K. J. Kurzydłowski, G. J. Molina, Activation energy of tribochemical and heterogeneous catalytic reactions, *Mater. Sci. Poland*, 28(2) (2010) 523–533.

47. C. Kajdas, Importance of ionic reactive intermediates for lubricant component reactions with friction surfaces, *Lubrication Sci.*, 6 (1994) 203–228.

48. C. Kajdas, *Proceedings of the International Tribology Conference*, Yokohama 1995, Satellite Forum on Tribochemistry, JAST, Tokyo (1995) 31–35.

49. C. K. Kajdas, A. Kulczycki, A new idea of the influence of solid materials on kinetics of chemical reactions, *Mater. Sci. Poland*, 26 (2008) 787–796.

50. K. Hiratsuka, C. Kajdas, M. Yoshida, Tribo-catalysis in the synthesis reaction of carbon dioxide, *Tribol. Trans.*, 47 (2004) 86–93.

51. N. Shimotomai, H. Nanao, S. Mori, Tribochemical reaction of benzene on nascent steel surface and effect of temperature, *Tribol. Online*, 7(2) (2012) 54–59.

52. K. Hiratsuka, T. Abe, C. Kajdas, Tribocatalytic oxidation of ethylene in the rubbing of palladium against aluminum oxide, *Trib. Intern.*, 43 (2010) 1659–1664.

53. S. Fujiya, T. Tsutsumi, K. Hiratsuka, Effect of Temperature on the Ethylene Oxidation during Rubbing of Palladium, *Proc. Intl. Trib. Conf.* Tokyo (2015) 19pE-01, 410.

54. A. Hase, H. Mishina, Wear elements generated in the elementary process of wear, *Trib. Intern.*, 42 (2009) 1684–1690.

55. J. Y. Park, G. A. Somorjai, The catalytic nanodiode: Detecting continuous electron flow at oxide–metal interfaces generated by a gas-phase exothermic reaction, *Eur. J. Chem. Phys. Phys. Chem.*, 7 (2006) 1409–1413.

56. G. A. Somorjai, J. Y. Park, Evolution of the surface science of catalysis from single crystals to metal, *J. Chem. Phys.*, 128 (2008) 182504.

57. D. A. Doshi, N. K. Huesing, M. Lu, H. Fan, Y. Lu, K. Simmons-Potter, B. G. Potter Jr., A. J. Hurd, C. J. Brinker, Optically defined multifunctional patterning of photosensitive thin-film silica mesophases, *Science*, 290(5489) (2000) 107–111; DOI: 10.1126/science.290.5489.107.

58. J. R. Goldman, J. A. Prybyla, Ultrafast dynamics of laser-excited electron distributions in silicon, *Phys. Rev. Lett.*, 72 (1994) 1364–1367.

59. Z. Zmudka, S. Postrzednik, Ecological and energy aspects of catalytic converter operation in internal combustion engines; http://www.combustion.org.uk/ECM_2009/P810091.pdf.

60. B. Yildiz, "Stretching" the energy landscape of oxides. Effects on electro-catalysis and diffusion, *MRS Bull.*, 39 (2014) 147–156; DOI: 10.1557mrs2014.8; www.mrsorg/bulletin.

3.2 Activation Energy (E_a)

Andrzej Kulczycki[a,b] and Czesław Kajdas[c]

[a]*Air Force Institute of Technology, Warsaw, Poland*
[b]*Cardinal Stefan Wyszynski University, Warsaw, Poland*
[c]*Institute of Chemistry in Płock,*
Warsaw University of Technology, Poland

andrzej.kulczycki@itwl.pl

3.2.1 Brief Description of Terms Related to Activation Energy

A catalyst is a substance that increases the rate of chemical reaction, and for equilibrium reactions, it increases the rate at which a chemical system approaches equilibrium, without being consumed in the process [1, 2]. It is applied in small amounts relative to the reactants, which enable a chemical reaction not only to proceed faster but also at a lower temperature than otherwise possible.

The solid heterogeneous catalyst mechanism that would lower activation energy is still under discussion. It is known that the effect of catalysts is intrinsically connected to the material surface states. However, the connection of catalysts material states to their action is not yet fully clear and the specific agents/stimulators of this action are unknown.

As Sato and Seo showed in Fig. 3.8 [3], a "chemically stimulated" EEE occurs simultaneously during the partial oxidation process of ethylene and that the emission rate is proportional to the rate of ethylene oxide formation. However, it is not clear whether the electron emission is produced by the reaction, or whether the former would drive or initiate the latter.

In summary, there is an extensive body of work suggesting that heterogeneous catalysis involves electrophysical phenomena, electronic interactions play an important role in controlling catalysis dynamics, and such electrons are of low-energy and are produced from the excited active catalytic surfaces.

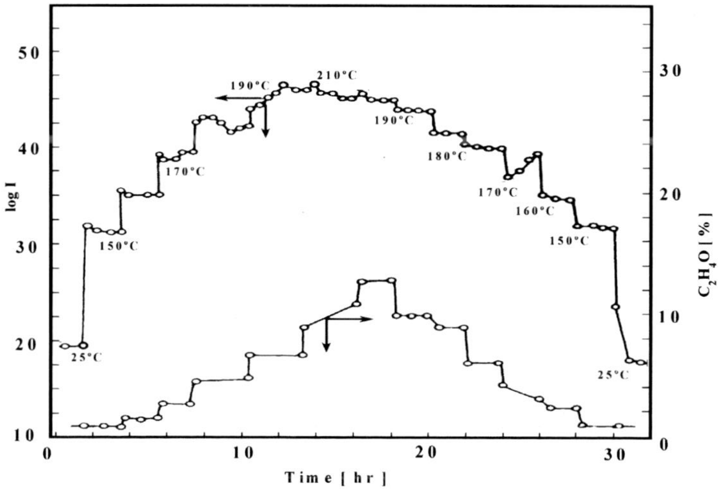

Figure 3.8 Effect of temperature rise on the exoelectron emission rate (log of EEE current) and on the formation rate of ethylene oxide (C_2H_4O), for silver catalyst. Reprinted from *J. Catal.*, 24, N. Sato, M. Seo, Chemically stimulated exoelectron emission from silver catalyst during partial oxidation of ethylene, 224–232, Copyright (1972), with permission from Elsevier.

3.2.2 Main Assumptions of α_i Model and Its Importance for Tribology

3.2.2.1 The α_i model hypothesis

This hypothesis was described over 33 years ago [5] and presently is known as the α_i model. The model being recently further developed, considers phenomenological data from the surface chemistry under friction, work function modification by mechanical input, possible electron and photon emissions, and Kajdas' negative ion–radical action mechanism [6, 7].

The α_i model is explained below. The term known as α_i is very specific, and it assumes forms that suggest the accompanied triboreactions, which would follow thermodynamics paths typical of irreversible processes [8]. The relationship between E_a and enthalpy in solids is also of importance in this approach, because the energy required for the reactants in fluid phase to go into transition state decreases, thereby decreasing the energy required to initiate the tribocatalyzed reaction.

The detailed mechanism of protective layer formation by lubricants in space near solid materials during a tribological process was analyzed by using the α_i method. The principle of the α_i method (model) was based on current knowledge about the mechanism of Antiwear (AW) and EP protective layers formation by lubricants containing additives. These were presented in detail by Kulczycki [4, 5]. The model can be applied to tribological systems where the mechanical work done can be treated as input and dissipated energy—predominantly heat—as output. The functional input–output relation may also be related to tribochemical reactions which are known to proceed much faster than thermochemical ones. On the other hand, one may compare tribochemical reactions with heterogeneous catalytic reactions [5]. The kinetics of tribochemical reactions depends on conditions of the tribological process; therefore, it is necessary to find more than one function of the mechanism of a tribochemical reaction [5, 9].

Based on the Gibbs phase rule, $k_r(f - 1) + r + f$ = number of equation describing chemical process, where k_r is number of reagents, f the number of phases, and r the number of equation describing chemical equilibrium.

In the model shown in Fig. 3.9, the following approach was applied:

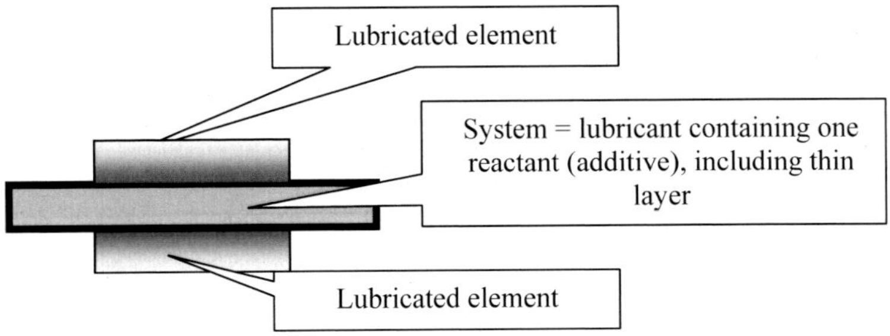

Figure 3.9 The tribological system described by α_i model.

(a) It was assumed that triboreactions occur in one phase—lubricant and lubricated elements are treated as environment; taking into account the following:

$f = 1; k_r(f - 1) = 0; r = 1$

(b) The number of equations describing tribochemical process is 2 $\{k_r(0) + 1 + 1 = 2\}$, thus tribochemical process can be described by two equations.

The Gibbs phase rule is related to equilibrium state of system. The determined system, related to the protective layer formed by lubricant during tribological process, remains in equilibrium until seizure or weld. Consequently, the critical state of the system is therefore equivalent to equilibrium disorder, which takes place during seizure or weld.

One of two functions of the same variable y, describing the above system can be related to the mechanics of a tribological process, the other one to its chemical reactions. It was assumed that the relationship between these functions makes it possible to analyze the mechanism of a tribological process on a "basic level."

The relation between these two functions was determined using principles of differential calculus. Based on a Lagrange theorem,

$$[f(b) - f(a)]/(b - a) = f'(c)$$

$$[\phi(b) - \phi(a)]/(b - a) = \phi'(c),$$

where a and b are values of y parameter, and consequently, according to Cauchy theorem,

$$[f(b) - f(a)]/[\phi(b) - \phi(a)] = f'(c)/\phi'(c)$$

The values of parameter y—a and b—can be determined experimentally, while the value of c is dependent on the behavior of the functions $f(y)$ and $f(y)$ and, on the other hand, on values of parameters a and b. To replace $f'(c)$ and $\phi'(c)$ the following relationships were introduced, assuming that value a is constant:

$$f'(c) = \beta_1 d[f(b) - f(a)]/d(b)$$

$$\phi'(c) = \beta_2 d[\phi(b) - \phi(a)]/d(b)$$

Proportionality coefficients β_1 and β_2 depend on the value of (a) and (b) and behavior of each of these functions. Applying the Cauchy theorem to the above equations resulted in the following:

$$[f(b) - f(a)]/[\phi(b) - \phi(a)] = \beta_1/\beta_2\{d[f(b) - f(a)]/d[\phi(b) - \phi(a)]\}$$

Equation (3.1) represents α_i as follows:

$$\alpha_i = \{[f(b) - f(a)]/[\phi(b) - \phi(a)]\}/[f'(b)/\phi'(b)], \quad (3.1)$$

where a and b are values of y parameter. The value of $\beta_1/\beta_2 = \alpha_i$ is therefore dependent on behavior of functions $f(y)$ and $f(y)$. When functions $f(y)$ and $\phi(y)$ are both exponential (y^n type), and $f(a) = 0$ and $\phi(a) = 0$, the value of $\beta_1/\beta_2 = \alpha_i$ does not depend on the values of b. When functions $f(y)$ and $\phi(y)$ are exponential, but of different type, for example $f(y) = y^n$ and $(y) = e^y$, the values of $\beta_1/\beta_2 = \alpha_i$ depend on the values of b. In this case α_i is linear function of $y = b$ and the parameters of this dependence are different for different functions as shown on Fig. 3.10. The relation between values b and a determines the character of proportionality. In case $b > a$, α_i is proportional to y, while in case $b < a$, α_i is inversely proportional to y.

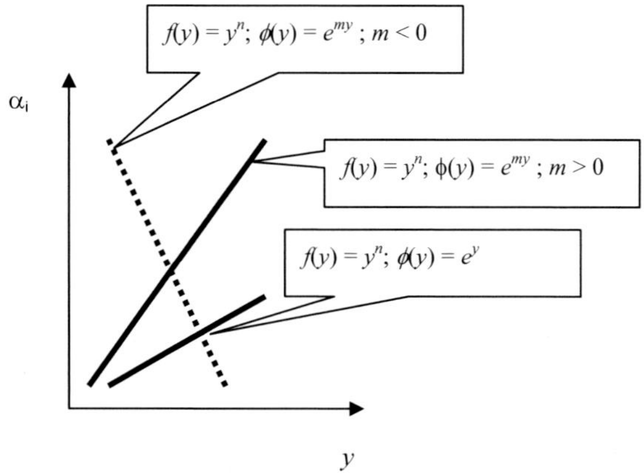

Figure 3.10 Relationship between α_i and y for different types of function describing tribological process.

For different reagents value b is the only one variable (assumption of Cauchy theorem) and a is constant. According to these assumptions the number of independent variables is $[2 + k_r f] = [2 + 1] = 3$. ($k_r = 1$; the reaction of one reagent–additive takes place in lubricant phase and is described by the model). The value of the independent variable $y = b$ or $y = a$, is fixed for a number of tested additives and the value of another parameter, for example lubricants viscosity, is fixed for tested lubricants as well. Variable b was related to tribological process conditions—for example, applied load P.

As in tribological tests usually the only one parameter is used for evaluation of lubricating additives effectiveness and this parameter characterizes the critical test conditions, resulting in seizure or weld; this parameter can be treated as variable y in above equations. Variable α_i is related to lubricant/additive as reagent in triboreactions that occur during tribological process. It makes possible quantitative description of lubricant/additive reactivity under conditions of a tribological process. The model assumes that $f(y)$ relates to the energy introduced to the tribological system, and $\phi(y)$ relates to the additive/reactant reaction initiated by impact of the introduced energy. Consequently α_i is the relative [related to $f(a)$ and $\phi(a)$] measure of the additive/reactant ability to the chemical reaction process.

The relationship between these two functions makes it possible to relate test result to the mechanism of this tribological process. The same mechanism can take place during different tests. α_i is non-dimensional parameter of value corresponding with lubricants properties related to a given mechanism of tribological process/tribochemical reactions. Thus, it can be concluded that the ability of the lubricants and/or reagents to create boundary layer and/or reaction products may be expressed by this coefficient $\alpha_{i.}$ Functions $f(y)$ and $\phi(y)$ should be defined and described using parameters measured during test. In tribological tests y is mainly applied load P. It was found for a number of lubricants the linear dependence of α_i and y. In this case, based on conclusions from Fig. 3.10, Eq. (3.1) can be described as

$$\alpha_i = [f(b) - f(a)]C, \tag{3.2}$$

where $$C = [1/\phi(b) - \phi(a)]\,\phi'(b)/f'(b). \tag{3.3}$$

C is a function of applied load P and is a constant for a number of tested lubricants undergoing the same mechanism of tribological process with different lubricating additives and/or reagents. Concluding, C is a function of y and constant value of C for various values of $y = P$ (critical value of applied load) require C to be a harmonic function of $y = P$, described by the following equation:

$$C = B \exp(-k_1 y) \cos (k_2 y + k_3), \tag{3.4}$$

where $y = P$ or T and B, k_1, k_2 and k_3 are constant values.

Consequently for given constant value of C only some of values P meet Eq. (3.2) (red points on Fig. 3.11).

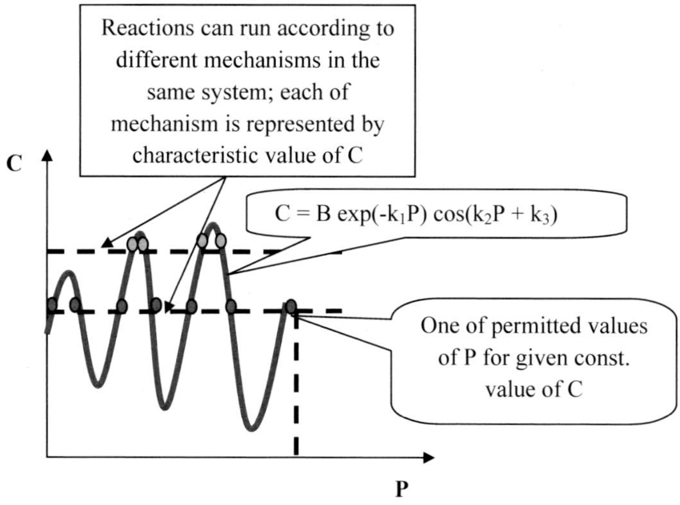

Figure 3.11 Dependence between C and applied load P.

What is physical and chemical meaning of those dependences? Answer to this question seems to be also a key to tribocatalysis and heterogeneous catalysis. Exponential part of this dependence can be connected with kinetics of chemical reactions of additive/reactant by Arrhenius equation:

$$k = A \exp (B_k - E_a/RT), \tag{3.5}$$

where temperature T may be connected with load P applied in tribological process

$$T = T_b + \mu P^{0,5} D; \tag{3.6}$$

here $D = 10^{-5} V_s P_p^{0,5} (\kappa_1 + \kappa_2)^{-1}$, where V_s is the sliding speed, P_p the unit pressure of the metal flow, and κ_1 and κ_2 the coefficients of thermal conductivity of cooperated elements of tribological system [42].

E_a cannot be a function of applied load P, which is the only one variable in Eqs. 3.2 and 3.4. Analysis of described above model leads to the conclusion that E_a value is constant for different additives which give different values of critical load P in given tribological process (given mechanism of protective layer formation).

Protective layer formation process, already known and described, assumes adsorption and chemical reactions of additive inside the lubricants layer and between additive or product of its reaction inside lubricant layer and on lubricated surface. Conventional models assume that additives of different chemical structure undergo different reactions, each of different E_a.

Conclusions resulting from α_i model are that a part of additives undergo similar reactions during tribological process and the resultant value of E_a for these reactions in this process is the same one.

$$k = A \exp\left[B_k - E_a/R(T_b + \mu P^{0,5} D)\right] \quad (3.7)$$

The trigonometric part of Eq. (3.4) can be connected with emission of electrons, photons from metal and solid body surface. The density of energy stream, dissipated to reaction space, is different in various angles and can be expressed as follows:

$$e_\gamma = e_0 \cos\gamma, \quad (3.8)$$

where γ is the function of P; e_0 is the density of energy stream in perpendicular direction to solid body surface. The anisotropic electron emission was observed for cathodes [10–13] as well as anisotropy dependence on cathodes temperature. Temperature in friction space is a function of applied load (see Eq. (3.6)) and consequently the following hypothesis was formulated: anisotropy of electron/photon emission during tribological process depends on applied load P as only one variable in α_i model (other parameters like friction coefficient should be constant). The influence of applied load and consequently generated temperature on anisotropy was partly confirmed by Hrach research results

[12–13] (see Fig. 3.14). Anisotropy depends on the solid type and physical and chemical structure of its surface.

This conclusion results from mathematical assumptions of α_i model, which requires only one variable—in the model this variable is applied load *P*. It was experimentally found that in four-ball tests under conditions of critical state of protective layer (e.g. seizure) values of friction coefficient is the same for different lubricants. Consequently variables in Eq. (3.4) (*T* and angle γ) should be functions of applied critical load *P*. The relationship between temperature in friction space *T* can be connected with *P* by Eq. (3.6). There is no general relationship between *P* and angle γ. Admitting this physical meaning of exponential and trigonometric parts of Eq. (3.4). *C* can be expressed by the following dependence:

$$C = B\exp(-k_1E_a/P^{0,5})\cos(k_2P + k_3) \qquad (3.9)$$

Concluding:

- *C* describes additive/reactant chemical reactions induced by introduced energy and is the function of reaction rate, which determines protective layer durability.
- *C* is the function of *P* only, so activation energy E_a has to be constant for different additives/reactants and different *P*.
- The slowest reaction in the chain of reactions leading to protective layer creation proceeds by the same mechanism in case the lubricating additives are of similar chemical structure. Consequently for all these additives E_a value is the same.
- The mathematical analysis of α_i model (see Fig. 3.10) leads to the conclusion that critical state (for example seizure) of tribological systems containing lubricants with additives of similar chemical structure, which proceeds reaction by given mechanism, can be reached only under permitted values of the critical load. In fact, it relates to values allowed by Eq. (3.3) at the given value of *C*.

On the basis of this mathematical model, physical model of tribocatalysis/catalysis was created. The model assumes that mechanical work done on the system (containing liquid reagents

= lubricants), is transformed to internal energy increase and dissipated energy. Internal energy is distributed in the system: one part is distributed to the liquid phase and is responsible for ambient temperature increase of the lubricant (T), the second part concerns energy of electrons/photons emitted to liquid phase from the solid surface. This part of energy, done as mechanical work, had been cumulated in the solid body. The cumulated energy in the solid body is emitted as electrons or photons from its surface as impulses of high intensity. Energy cumulated in the liquid phase is not sufficient to reach value of E_a. Energy emitted from the surface as impulses can reach value of E_a and reaction/triboreaction begins to proceed.

The value of the angle γ depends on the system energy flux. The critical state of a tribological system appears at conditions resulting in destruction of the protective film. It has been observed that for different reactants, the critical rate of reaction leading to protective layer destruction was achieved for different values of energy flux into the system (different values of applied load P). The same value of C obtained for different reactants and different densities of energy streams introduced into the system (characteristic for each reactant) leads to the conclusion that the same critical rate of destruction reaction was achieved and thus for each reactant a different angle γ is connected with the critical rate of reaction (for different reactant protective layer destruction appears under different applied load P and angle γ is characteristic for applied load P).

Consequently, for each reactant there is a specific value of the energy flux density (e_γ) emitted by the solid/catalyst and the value of activation energy E_a is constant. Accordingly, it is possible to say that

(i) E_a is constant for a given type of reaction and the critical rate of reaction depends not only on the energy quantity added to reactants but on the density of the introduced energy stream (time of the tribological process is constant for each load).

(ii) The catalyst emits pulses of energy flux of high density. At the angle γ the value of energy emitted in short time is equal to activation energy E_a.

(iii) The catalyst collects energy done as mechanical work and emits it as pulses of the high energy density flux, thereby initiating the reaction or increasing its rate.

The value of parameter C is constant for a given mechanism of reaction. For different mechanisms, different values of C are obtained. Consequently, for each mechanism of reaction (triboreaction) the critical rate of reaction can be achieved only under certain values of energy stream introduced into the system (T or P). For each of these values, relation between the rate constant calculated for temperature T in the friction area (exponential part of Eq. (3.4)) and stream of energy emitted in angle γ by the solid (trigonometric part of Eq. (3.4)) equal to or higher than E_a is constant for a given mechanism of reaction. For another mechanism, this relation changes. It results from analysis of the mathematical model in that it is feasible to say that the tribocatalytic effect is observed due to accumulation of energy introduced into the solid body as mechanical work—a function of applied load P.

3.2.3 Thermodynamic Interpretation of α_i Model

One of the possibilities is to describe tribological process using thermodynamic relationships. Then $f(y)$ can be the work done on tribological system during test, L, and $\phi(y)$ can be internal energy change, Δu. Then Eq. (3.10) is as follows:

$$\alpha_i = [(L - L_o)/(\Delta u - \Delta u_o)]\, x(d\Delta u/dL); \qquad (3.10)$$

here $L = \mu Pvt$, where μ the friction coefficient, v the speed; t the time, P applied load (test result), and

$$L_0 = f(a),\ \Delta u_0 = \phi(a).$$

On the other hand, it is difficult to define relationship $\Delta u = \phi(P)$, because Δu is not linear dependence of the applied load P. On the basis of the first law of thermodynamics, L can be expressed as a function of Δu:

$$L = Q + \Delta u \qquad (3.11)$$

Q is energy dissipated by system during tribological process; mainly it is a dissipated heat, which in relation to tribological process can be described by the following dependence:

$$Q = c_h(T - T_{ot}) \tag{3.12}$$

$$T = T_b + A\,P^{0,5}, \tag{3.13}$$

where c_h is the average specific heat capacity, T_b the temperature of lubricant out of friction area, T_{ot} the temperature of environment, A the constant coefficient.

Assuming that both average specific heat capacity and T_b are constant for different oils, Q can be expressed as $Q = c_h\,[T_b + AP^{0,5} - T_{ot}]$ and approximately $Q = A_1P^{0,5}$; A_1 is the constant coefficient. Accordingly, the Δu can be presented as

$$\Delta u = L - Q = \mu vtP - A_1P^{0,5}$$

In this case C can be

$$C = (d\Delta u/dL)/(\Delta u - \Delta u_o)$$

$$C = [(\mu vt\,dP - A_1\,dP^{0,5})/\mu vt\,dP]/(\mu vtP - A_1P^{\,0,5} - \Delta u_o). \tag{3.14}$$

C is the function of P and, referring this function to Cauchy's theorem, P can be the only one variable in functions $f(y)$ and $\phi(y)$. The dependence (3.14) can be more complex in case μ and A_1 are not fixed and they are a function of P.

R. C. Dante found that the α_i model can be related to thermodynamics of irreversible processes (TIP) [19, 20]. Most recently, the results were reported by Dante, Kajdas, and Kulczycki [8]. In the α_i model, there are two independent variables: applied load L and chemical structure of lubricant (CSL), i.e., the reagent as described and discussed by Kajdas and Kulczycki in paper [4]. The latter variable was often given qualitatively, but from the tribochemical process point of view it was necessary to express it as a quantitative variable by mathematical dependence. In this model, it was assumed that CSL is expressed by α_i, which is a measure of lubricants ability to boundary layer creation. It was assumed that value of α_i is a linear function of critical load L, i.e., the load is causing destruction of the generated boundary layer.

The term α_i, which is correlated to CSL, assumes forms that drive towards the TIP, and therefore towards a more general outline.

R. C. Dante found [8] that Eq. (3.2) formally resembles the relation between entropy production σ, the force X, and the corresponding flux J_x,

$$\sigma = J_x X \tag{3.15}$$

The relation of α_i with σ is proposed to be the following:

$$\sigma_{c\tau} = z\, d\alpha_i/dt, \tag{3.16}$$

where the sub-indexes c and τ stay for chemical reaction and shear stress, respectively; while z is a proportionality factor. Since there is a relation between the chemical structure of lubricant and entropy; for analogy, it was assumed that α_i is directly linked to the entropy related to chemical reactions induced by friction.

3.2.4 Empirical Verification of α_i Model and Its Derivative Equation Concerning Heterogeneous Catalysis

3.2.4.1 Verification of α_i model in tribological tests

The α_i model was verified in relation to tribological processes during model tribological tests using the four-ball apparatus. The number of gear, transmission and hydraulic oils, containing different lubricating additives were tested using those tests. Verification of α_i model needed calculation of values of α_i coefficient. Based on the conclusions from α_i model and in their relation to tribological process of boundary lubrication it was found that coefficient is a measure of the influence of the given additive on boundary layer durability. The α_i should be related to triboreactions that take place during boundary layer creation or during boundary layer destruction. The criteria used in tribological tests are related to destruction of boundary layer–seizure load.

The criteria used in tribological tests determine the meaning of α_i: it can be a measure of additive ability to the boundary layer formation or additive influence on the boundary layer durability. In the latter case, reactions before the boundary layer destruction are the most important. Figure 3.12 demonstrates schematically major steps of triboreactions from the beginning of tribological process to the seizure or weld. At the α_i model protective layer destruction is described by critical applied load (seizure or weld load) P and there is no parameters describing protective layer formation. The value of critical load P depends on durability of protective layer and durability of protective layer in given tribological process depends on initial stage of the process of protective layer formation.

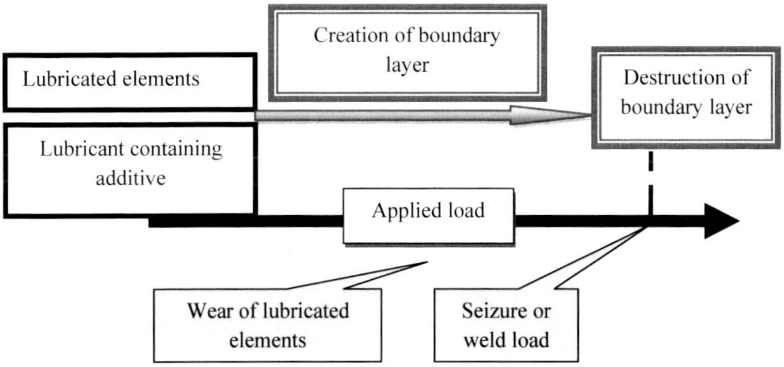

Figure 3.12 Steps of boundary layer destruction leading to the seizure or weld processes.

For more details see Chain 1: Additive in lubricant + surface of lubricated element → reaction 1_1 → reaction 2_1 ... → reaction n_1 → reaction of boundary layer destruction m_1 and Chain 2: Additive in lubricant + surface of lubricated element → reaction 1_2 → reaction 2_2 ... → reaction n_2 → reaction of boundary layer destruction m_2.

The changes of initial state of the tribological system can be one of the possible reasons of changes of the mechanism of protective layer creation. During a chosen tribological test beginning from the same initial state of tribological system, each active lubricating additive, under test condition undergoes all reactions of chain 1. In this case, the mechanism of boundary layer formation by all tested additives is the same. Consequently looking on the α_i model value of C should be constant for all tested additives under given conditions. On the other hand, the influence of given additive on boundary layer durability is constant for a given range of initial states of tribological system. Consequently, α_i value for the given additive should be constant in this range of initial states.

Concluding, α_i value for each tested additive can be determined by measuring the influence of the change of initial state of tribological system on the variation of the load of the boundary layer destruction, for example seizure load P_t (see Eq. 3.17). The changes of initial state of tribological system were caused by duration of initial period of test carried out under constant load $P_0 = 900$ N.

$$\alpha_i = d[C(L - L_0)]/d\,(\text{initial state of tribological system}), \quad (3.17)$$

where the initial state of the tribological system can be expressed by t_0—duration of initial period of test carried out under constant load P_0 = 900 N. In the range of initial-state changes that do not cause changes of the type of protective layer, α_i does not depend on the changes of initial state of tribological system. There is the border value of initial state, which changes the mechanism of protective layer creation and consequently the value of α_i changes as well.

Experimental determination of α_i needs measure of $f(y)$ and $\phi(y)$ values for each tested lubricant. $f(y)$ can be work done on tribological system L, which is a function of applied load, friction coefficient and sliding speed, all parameters are usually measured during tribological tests. One of the applied four-ball test is conducted under uniformly increasing load up to seizure. This test can be useful for the measure of protective layer durability in case the initial-state changes. For this purpose, the standard test with four-ball apparatus was modified and the beginning of the test was under fixed load of 900 N (P_0) for the duration of t_0; after t_0, the load begins to uniformly increase to seizure, similar to the standard test. For this purpose the standard test on four-ball apparatus was modified (new test) and the beginning of test was under fixed load of 900 N (P_0) during t_0 time; after t_0 load begins uniformly increase to seizure like in standard test. During first part of the test, initial state of tribological system was changed proportionally to time t_0 and energy L_0 and was introduced into this system.

$$L_0 = \mu P_0 t_0 V_s, \tag{3.18}$$

where μ is the friction coefficient, P_0 applied load 900 N, t_0 time of this part of test duration (s), and V_s rotation speed of upper ball equal to 470 rpm.

Value of L_0 was changed by the change of time t_0, stepwise from 0 to 60 s. Introduced into tribological system energy L_0 was partly accumulated in tribological system before boundary layer was created under uniformly increasing load. Total energy introduced into the system during test, up to the seizure, was

$$f(y) = L_t = \mu P_0 t_0 V_s + ½\mu(P_t^2 - P_0^2)V_p^{-1}V_s. \tag{3.19}$$

The main idea of experimental determination of α_i values for additives of different structure is to compare its effectiveness after different duration of initial stage of test (different values of t_0).

$$F(y)_1 = f(b)_{t02} - f(b)_{t01} - f(a)_{t02} + f(a)_{t01}$$

$$\Phi(y)_1 = \phi(b)_{t02} - \phi(b)_{t01} - \phi(a)_{t02} + \phi(a)_{t01}$$

$$F(y)_2 = f(b)_{t03} - f(b)_{t02} - f(a)_{t03} + f(a)_{t02}$$

$$\Phi(y)_2 = \phi(b)_{t03} - \phi(b)_{t02} - \phi(a)_{t03} + \phi(a)_{t02}$$

The character of $\phi(y)$ was based on Eq. (3.4). It was assumed that this dependence is a function of additive chemical change (the rate of additive reaction) leading to products of the formed protective layer. Consequently, $\phi(y)$ is a function of initial additive concentration [Add], activation energy E_a, critical value of y – seizure load $b = \mathrm{P_t}$ and trigonometric function of y.

$$\phi(y) = \phi\{[\mathrm{Add}], \exp(E_a/y), \cos(k_2 y), y\}$$

For base oil, $\phi(a) = 0$ because [Add] = 0.

$B = [\mathrm{Add}]\, e^{Ea/b} \cos(k_2 b)$ is nearly constant in the range of $b = P_t$ of similar values.

Consequently $\Phi(y)_1 = B[(b)_{t02} - (b)_{t01}]$ and $\Phi(y)_2 = B[(b)_{t03} - (b)_{t02}]$.

In order to calculate α_i values for chosen lubricating additives above described test was carried on using various mineral oils of different viscosity without additives (base oils) as lubricants, and lubricants containing additives of different chemical structure added to the base oils at various concentrations. The effect of additive action can be seen when test results for lubricating oil (containing additive) will be compared with base oil (without additive). In this case the difference between values of functions describing lubricating oil $f[(b)]$, $\phi[(b)]$ and base oil $[f(a)]$ and $\phi[(a)]$ were introduced into Eq. (3.1).

$$\alpha_i = [F(y)/\Phi(y)]\, d[\Phi(y)]/d[F(y)], \tag{3.20}$$

where $F(y) = f(b) - f(a)$; $\Phi(y) = \phi(b) - \phi(a)$; value $y = b$ is related to oil with additive and value $y = a$ is related to base oil (without additive).

$$[1/F(y)]\, d[F(y)] = (1/\alpha_i)\, [1/\Phi(y)]\, d[\Phi(y)] \tag{3.21}$$

The changes of initial state of tribological system resulted from the changes of initial period of test t_0 (see Fig. 3.13) are the reason of changes of b (P_t for lubricating oil) and a (P_t for base oil) values. Integrating of Eq. (3.21) in the limits related to $F(y)$ values for t_0 to t_{01} the following dependence was found:

$$\ln [F(y)_2/F(y)_1] = [1/\alpha_i] \ln [\Phi(y)_2/F(y)_1]$$

$F(y)$ values were determined experimentally according to Eqs. 3.19 and 3.20. It was concluded that when $F(y) < 0$ the influence of changes of initial state of tribological system on boundary layer formation by oil with additive and by base oil are on the same way. These results were related to Anti-Wear (AW) type of boundary layer formation. Oil with lubricating additive creates the EP boundary layer in different way as compared to the base oil (without additive). Those for EP layer $F(y) > 0$.

Based on the above dependences and experimentally obtained test results the α_i values were calculated for a number of gear, hydraulic and transmission oils of viscosity at 40°C in the range between 32 and 220 mm^2/s.

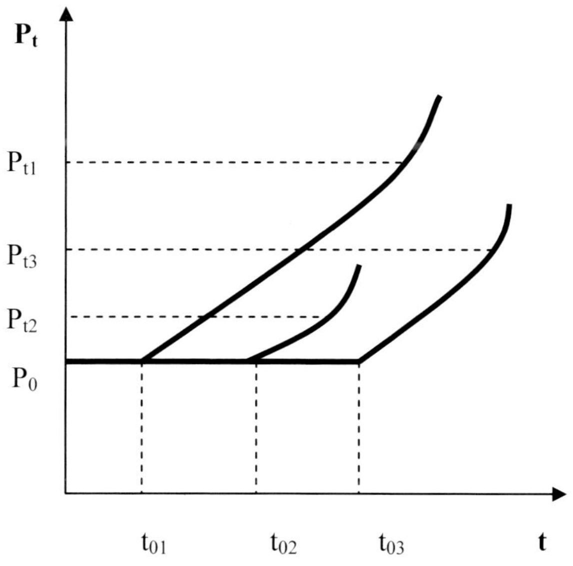

Figure 3.13 The principles of test used for experimental determination of α_i values for lubricants containing various lubricating additives.

The α_i values assigned to those lubricants were related to the results of tribological standard tests. Therefore, it was necessary to find out tribological experimental methods able to provide test results concerning lubricant's ability to create both AW and EP types of the boundary layer. In earlier work [5] there were two selected different tests carried out using a four-ball machine. The first test was used to determine seizure load P_t under the following operating conditions:

- Rotating speed of the upper ball was 470 +/− 20 rpm.
- Load was increased continuously from 0 to the seizure load.
- Seizure load was detected by friction coefficient measurement.
- Experimentally, the following dependence between α_{AW}, determined by the procedure described above, and P_t values was found:

$$\alpha_{AW} = (0.000086\nu_{40} - 0.01)\ 0.5\ \mu\ P_t^2\ \nu_s\ \nu_p^{-1} + 0.2 - 0.00073\nu_{40}, \tag{3.22}$$

where ν_{40} is the kinematic viscosity of tested oil at 40°C, v_s the sliding velocity (0.18 m/s), v_p the speed of load increase (45 N/s), and μ the friction coefficient.

Dimensions of constants in Eq. (3.22) are chosen to be dimensionless α_{AW}.

This empirical relationship confirms assumptions described in Section 3.2.2 of this chapter. Based on the Gibbs phase rule and taking into account above assumptions, the number of phases $f = 1$, the number of reagents $k_r = 1$, and the number of independent variables = $2 + k_r f = 3$.

Relationship (3.22) contains three variables: α_{AW}, P_t and ν_{40}. This dependence points out that the mechanism of AW type of protective layer formation depends on base oil viscosity.

The α_{EP} values determined using above procedure were related to second four-ball test, where welding load P_w was that in the test result. This test was carried out under the following conditions:

- Rotating speed of the upper ball was 1470 +/– 30 rpm.
- Stepwise load was increased according to Polish Standard PN-76/C-04147 (ASTM D2783: Standard Test Method for Measurement of EP Properties of Lubricating Fluids, Four-Ball Method).
- Test duration on each step was 10 s.

It was found that P_w values depend on durability of EP type of the boundary layer and was experimentally found the linear dependence between α_{EP} and P_w:

$$\alpha_{EP} = 0.48 - 0.00013\, P_w \tag{3.23}$$

Dimensions of constants in Eq. (3.23) are chosen to be dimensionless α_{EP}.

This empirical relationship confirms also assumptions discussed in Section 3.2.2 of this chapter. Based on the Gibbs phase rule and taking into account above assumptions: the number of phases $f = 1$, the number of independent variables = $2 + k_r f = 2$, relationship (3.23) contains 2 variables: α_{EP} and P_w. The number of reagents under reaction in lubricant phase should be $k_r = 0$ ($2 + k_r f = 2$). EP layer is formed by additive under reaction of this additive or product of its transformation with the surface of lubricated element. Taking into account the assumption that reactions described by Eq. (3.1) are in lubricant phase the reaction leading to EP layer formation is out of system (lubricant phase). Consequently the number of reagents under reaction in lubricant phase is $k_r = 0$.

In that context, the mechanism of boundary lubrication holds two different types of the boundary protective film beginning from (i) pure AW up, through (ii) mixed AW and EP to (iii) pure EP layer. Lubricants ability to create AW layer determines α_{AW} coefficient, and EP layer determines α_{EP} coefficient. Consequently it was concluded that every mechanism of boundary lubrication can be described by α_i coefficient, defined as the sum of α_{AW} and α_{EP} coefficient:

$$\alpha_i = p_{AW}\,\alpha_{AW} + (1 - p_{AW})\,\alpha_{EP}, \tag{3.24}$$

where p_{AW} is a coefficient that determines the ratio of AW mechanism taking place in the protective layer formation. Equation (3.24) can be used to calculate α_i values if adequate experimental method to determine values of α_{AW} and α_{EP} coefficients is applied.

3.2.4.2 Verification of derivative equation concerning heterogeneous catalysis; electron and photon emission anisotropy

Equations (3.1) and (3.8) lead to conclusion that the emission of energy (electrons and photons) from the surface of solid body is anisotropic one. Exhaustive literature review revealed that there is no data on the angular distribution of triboemitted electrons, but that existing research works on the emission of electrons from cathodes consistently report anisotropic distributions [10–13]. Highly anisotropic distributions with a maximum in the direction normal to the emitting surface were measured.

Figure 3.14 shows an example of anisotropic electron emission from sandwich cathodes. At the emission temperature of 300 K the measured electrons present a quasi-isotropic characteristic. However, at a lower temperature of 80 K, the isotropic component vanishes.

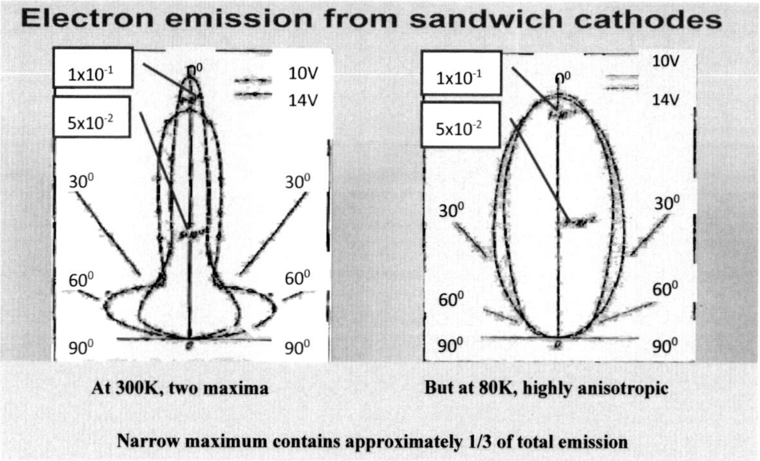

Figure 3.14 Examples of EE from sandwich cathodes presenting two maxima with a maximum in direction normal to the emitting surface, and highly anisotropic EE: based on Hrach research results. Reproduced from R. Hrach, J. May, The angular distribution of electrons emitted from thin-film MIM structures at various temperatures, *Phys. Stat. Sol. (a)*, 4 (1971) 637–642, with permission from John Wiley and Sons.

Works [10–11] allowed to make an estimate that at room temperature the narrow maximum around the direction normal to the emitting surface contained approximately 1/3 of the total emission. Further work of Hrach [12–13] was of particular significance as he measured the energy characteristics of emitted electrons at different angles by means of a hemispherical collector and the retarding-grid technique. He found that at room temperature (300 K) anisotropic energy spectra of emitted electrons were in the range of 0 and 7 eV, but for emission angles closer to the normal to the surface the measured energy was between zero and 4 eV. Figure 3.15 depicts some of these findings.

A more recent study was presented by Fitting et al. [14] on the angular distribution of thermostimulated exoelectron emission (TSEE) from alpha-Al_2O_3. They found anisotropic angular distributions strongly directed normally to the surface; the energy of the emitted electrons was in the range of 0 to 4 eV. Thus, the literature review revealed strong evidence that low energy-electrons are typically emitted for highly anisotropic distributions with maxima in the direction normal to the surface. Interestingly, model calculations are consistent with experimental data. Kortov and Zolnikov's work [15] predicted that the angular distribution from an Auger-type emission would not be influenced by roughness in homogeneous rough samples.

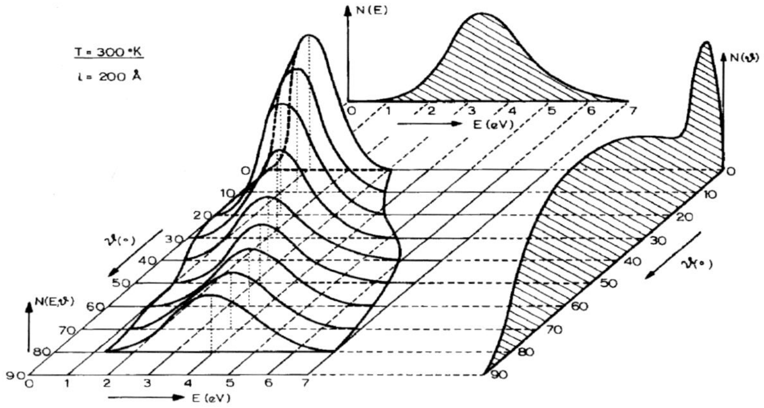

Figure 3.15 Typical experimental angular and energy distribution of electrons emitted from Al–Al_2O_3–Au sandwich cathode structures. Temperature: 300 K, applied voltage: 10 V. Reprinted from *Thin Solid Films*, 15, H. Hrach, Energy angular distribution of electrons emitted from MIM systems, 65–69, Copyright (1973), with permission from Elsevier.

The measurements of angular distribution of electron emission are limited in the available literature to cathode emission [11–13], TSEE [14] and backscattered emission [16] from electron bombardment of the emitting surfaces. Angular distribution of photoelectrons and emitted photons were also extensively studied; Jablonski and Zemek [17] found highly anisotropic distribution similar to one of electron-type for the photon emission from X-ray-irradiated thin polycrystalline aluminum foils. Hucek et al. [18] found similar anisotropy for irradiated Cu, Ag, and Au foils, and that isotropic component is due to the elastic collisions that photoelectrons suffer on their way out of the target. Anisotropy also occurs for different spectral lines.

3.2.5 Hypothesis on E_a of Heterogeneous Catalysis and Tribocatalysis

Chemical reactivity and the mechanical work done on surfaces seem to be clearly connected to the material surface states. These reactions under friction (known as tribochemical reactions) proceed much faster than thermochemical ones for the same reactants. In recent work Kajdas and Kulczycki [4] introduced the novel concept that the action of solid heterogeneous catalysts could be studied by comparing it to the action of mechanically activated solids in tribocatalysis. They proposed that for catalyzed heterogeneous reactions the same E_a (real activation energy for the non-catalyzed reaction) is needed to initiate the process, and that the difference between E_a and a reduced "apparent" activation energy (E_{app}), can be explained by an input of "catalytic reaction" activation energy (E_c).

The work of Kajdas and Kulczycki [21] hypothesized that for standard and catalyzed heterogeneous reactions the same E_a (real activation energy) is needed to initiate reaction processes. It is proposed that energy introduced into tribological system as mechanical work, done on the solid material, is accumulated in this material and then emitted as electrons and/or photons to the space in which reactions take place. The developed model includes the specific γ angle at which the stream of energy e_γ in the range of 3–5 eV, is emitted and the reaction can be initiated. This idea results from α_i model but should be confirmed experimentally. As it resulted from work of Prevenslik [41], during tribological processes energy introduced to tribological system is stored in molecules—harmonic oscillators, and then emitted by UV photons. This phenomenon can take place in liquid phase (lubricant) or in solid phase on the surface of lubricated elements or/and in the material of solid in dislocations. The possibilities and mechanism of energy storage is the key problem for tribochemistry, mechanochemistry, and catalysis.

The distribution of the energy emitted by the solid body (e.g., by the catalyst) is controlled by the angle of emission,

$$e_\gamma = e_0 \cos \gamma,$$

where γ is a function of P (loading) and depends on the stream of energy introduced into the tribological system, e_0 is the density of the energy flux in the perpendicular direction to the solid body surface [21]. Figure 3.16 shows the idea of the process by which the energy emitted by surface as impulses can reach value of E_a and the heterogeneous catalytic process starts.

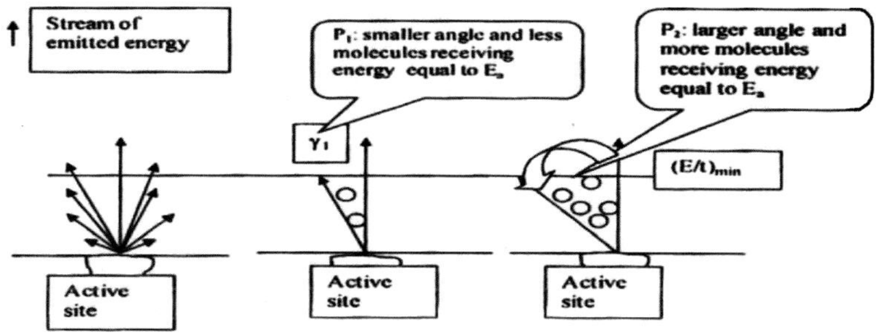

Figure 3.16 Illustration of the main concept of the model; stream of the energy emitted by the catalyst depends on the angle γ, hence the number of molecules receiving energy equal to E_a depends on γ angle.

The emission of electrons, ions, neutral particles, photons, and acoustic emission under conditions of tribological contact and damage is called triboemission [22]. Triboemitted electrons, which make the major part of the charged-triboparticles emission stream from insulators and semiconductors, are known to be related to exoemission (including chemical exoemission and fractoemission), and triboemitted electrons have been postulated as key factors in the initiation and control of tribochemical reactions [23].

Molina et al. [23] have recently proposed that electron triboemission can be the result of non-adiabatic surface reactions according to the Kasemo model [24]. This model is based on an Auger decay process in which an impingent gas particle provides an empty level that is filled by an electron from the solid, while energy is transferred to another emitted electron. The probability of such event is very low, of about 10^{-4} per impinging molecule, but consistent with observed exoemission processes.

Chemical reactions may dissipate part of their energy to electron–hole pairs in the surface region: part of the impinging molecule energy may be transferred to electronic states which are lower by several eVs from the ground state potential of the surface, and a low-energy electron would then be emitted in a non-adiabatic process; this phenomenon of low-energy "chemi-emission" has been extensively observed. Although the classic modeling of surface chemistry processes (i.e., quantum or first principles descriptions of the atomic interactions) assumes that such deviations from adiabatic behavior are small and of no practical importance, recent experimental works [25–27] have shown that exoelectron emission is just a fraction of the total electronic excitation on the surface. The majority of the non-adiabatic excitation would not be typical exoemission but being detectable as "chemicurrents," which are due to chemically generated hot carriers and stay in the solid. Gergen et al. [27] have detected such chemicurrents by using thin film electronic devices that allow non-emitted excited electrons to traverse their Schottky barrier (of only about 0.7 eV) as "internal exoemission." By this method, they estimated that, for instance, for the oxidation of Mg surfaces approximately 10% of the chemical reaction energy is dissipated by non-adiabatic electronic excitation.

Boettcher et al. [28] reported the evolution of electron work function (WF) and of the intensity of exoemission upon exposure of Cs to O_2. They found that no electron emission was detected upon clean surface exposure to increasing O_2 until about a pressure of 10^{-6} Torr, i.e., that partial oxidation would be needed for this chemiemission to occur. A decrease of the WF also was observed, but its minimum was reached earlier, for lower O_2 exposure than that of the electronic peak. This result suggests that the electronic excitation through chemical reaction may govern exoemission, rather than the lowering of WF being the dominant factor. The mechanism for this emission can be interpreted by the Kasemo et al. model [24]. Bottcher et al. [28] also demonstrated that this probability of emission by this model monotonically decreases upon approaching completion of oxide layer formation. XPS (X-ray-photoelectron spectroscopy) is in regular use for surface analysis and recently the knowledge of photoelectron angular distribution has been presented and discussed by Y. F. Chen [29].

All the above reported results are in agreement with the hypothesis that the solid catalytic effect can be connected with energy emission from the solid surface.

3.2.6 How Do We Understand the Basic Arrhenius Equation?

As already noticed in Section 3.2.1, (a) catalyst is a material that increases the rate at which a chemical system approaches equilibrium, without being consumed in the process and, (b) catalysis is the phenomenon of a catalyst in action while the catalyst lowers the activation energy. In chemical kinetics, E_a is the height of the potential barrier separating the products and reactants. It determines the temperature dependence of the reaction rate. When regular solid catalysts, as those applied in industry (for instance, nanodispersed metal on amorphous aluminosilicates), are considered, the effect of this type of material might be compared with the action of mechanically activated solids as in tribocatalysis [30]. Tribocatalytic effect is also due to the synthesis reaction of carbon dioxide [31]. Heterogeneous catalysis accelerates the movement of atoms on the passage through the transition state; it provides the link between reactants and products on a reaction pathway that involves simultaneous motion of several to very many atoms [2]. Friction processes enhance energy of the mating elements by, for instance, dislocation defects. Analysis of the α_i model, described above, in relation to empirical results of tribological processes leads to conclusions towards mechanisms of catalytic processes. According to the model, it is possible to say that: (i) E_a is constant for given type of reaction and the reaction start depends not only on the quantity of energy delivered to substrates (E_a), but also on density of the introduced energy stream; (ii) catalyst collects energy introduced into the system and emits it as impulses of high density flux of energy, thereby triggering the reaction or increasing its rate, (iii) catalyst emits impulses of energy flux of high density and at angle γ the value of energy emitted in short time is equal to activation energy E_a.

The model can be applied to tribological systems in which the mechanical work done can be treated as input and dissipated energy—predominantly heat—as output. In this, the functional input–output relation may also concern tribochemical reactions which are known to proceed much faster than thermochemical ones. One further extension of the model compares tribochemical reactions with heterogeneous catalytic reactions [4, 32]. The accumulated energy in the solid increases enthalpy of the system. Tkacova et al. [33] shows that the dislocation concentrations with the specific energy of defect structures are of particular meaning for the energetic condition of the energy contributions. It relates to changes of both structure and enthalpy and it is in line with the statement that tribochemical reactions are mostly stimulated by triboelectrons emitted under friction conditions. Thereby, this work proposes that both heterogeneous catalytic and tribochemical processes are controlled by triboelectrons [34]. Direct correlation between catalytic activity and the EEE intensity for different deposited and promoted catalysts was clearly shown by Sato and Seo [3]. Biernacki et al. [35] demonstrated the temperature dependence of the intensity of photostimulated EEE from Raney nickel, and cobalt alloys.

While theoretical friction dissipation models and computer simulations have significantly improved in sophistication over the past decade, their applicability is hampered by lack of empirical data; most of these models either are of phenomenological nature only or discuss friction solely at the skin of rigid corrugated surface potentials.

The fact that friction also involves intrinsic activation modes, e.g., relaxation modes in organic solid materials has only been recognized in few isolated studies. Knorr et al. [36] provides direct insight into the basic material intrinsic mechanism for frictional dissipation on the submolecular scale of amorphous organic systems. They demonstrated that at low pressures and temperatures below the glass transition point, the phenyl pendant side groups of polystyrene, known for their preferential orientation at the free surface, were noticed to be the primary channel for dissipation of kinetic sliding-energy. This process was found to be truly enthalpic with activation energy of 33 kJ/mol. The energy dissipation was shown to possess both enthalpic and cooperative entropic contributions above the loading capacity of the surface phenyl groups. They also found apparent Arrhenius activation energies of frictional dissipation of 92 and 376 kJ/mol, respectively, and cooperative contributions up to 80% [36]. As such, that study highlighted issues critical to organic lubricant design, i.e., the intrinsic enthalpic activation barriers of mobile linker groups, the evaluation of cooperative mobility phenomena, and critical tribological parameters to access or avoid coupling between shear disturbances and molecular actuators.

Concluding, Arrhenius equation in classic form is not suitable for catalytic reactions. In Arrhenius equation the rate of reaction depends on the value of activation energy E_a and ambient temperature of reaction mixture. The result of analysis of α_i model lead to conclusion that the rate of catalytic and tribocatalytic reactions depends not only on E_a and ambient temperature, but on energy emitted by active centers on the surface of solid body in angle γ. Consequently Arrhenius equation can be modified as shown below:

$$k = \phi\{A_k, [\exp(B - E_a/RT)], e_a, \cos\gamma\}, \tag{3.25}$$

where e_a is the average density of the introduced energy flux into tribological system.

3.2.7 Practical Significance of the New Approach Concerning Activation Energy (E_a)

The new approach concerning activation energy (E_a) can be applied to make more understanding tribochemistry and mechanochemistry. This better understanding results from the use of α_i coefficient to describe quantitatively lubricants influence on protective layer durability and from interpretation of C in Eq. (3.2) as a function of activation energy of tribchemical process E_a. Empirically found linear dependence between α_i and given tribological test results obtained for a number of lubricants of different chemical structure of additives and base oil (the slope of determined straight line is C in Eq. (3.2)) points that for each of these lubricants in this tribological test E_a is of the same value and consequently triboreactions proceed according to the same mechanism.

3.2.7.1 Application to tribochemistry

This new approach can be used as a bridge between the results of various tribological tests, lubricated by the same lubricants. In this model, there are two independent variables: applied load (P) and chemical structure of lubricant. The chemical structure of lubricant is a qualitative notion, but it is necessary to express it as quantitative variable, if tribological process will be described by mathematical dependence. In this model chemical structure of lubricants is expressed by α_i. If the mechanism of protective layer formation is fixed for a number of tested lubricants, value of α_i is a linear function of critical load P—load under which takes place destruction of boundary layer. As shown above, linear function of α_{AW} and seizure load P_t—the result of described above four-ball tests, were found. This test was chosen for group of tested lubricants as a reference for their antiwear (AW) properties determination.

The ranking list of tested lubricants for test I is described by the following dependence:

$\alpha_i = C_I\,(P_{I\,critical\,state} - P_0)$, where in the case of test I;

$\alpha_i = \alpha_{AW}$ (see Fig. 3.17).

Comparison between test I and another tribological test II

All lubricants (1–6) were tested in test II (for instance FZG) and $P_{II\,critical\,state}$ were determined experimentally. The α_i values determined in test I were related to experimental results obtained in test II—$P_{II\,critical\,state}$.

Values of $C_{II\,1}$ and $C_{II\,2}$ resulted from assumed values of α_i in test I and experimentally determined values of $P_{II\,critical\,state}$. For example, for lubricants 1, 3, 5 and 6 one linear dependence was obtained $\alpha_i = (P_{II\,critical\,state} - P_0)C_{II\,1}$ and for lubricants 2 and 4 another one $\alpha_i = (P_{II\,critical\,state} - P_0)C_{II\,2}$. In consequence the same value of E_a of reactions of protective layer formation should be found for lubricants 1, 3, 5 and 6 and different one for lubricants 2 and 4. In this case, for the analysis of the mechanism of protective layer creation in test II, the third test (test III) should be used. This third, reference test is carried on four-ball apparatus and weld load (P_w) is the criteria of EP properties of lubricant. Based on Eq. (3.23), the α_{EP} values can be determined for each tested lubricant. Having for each of tested lubricant values of α_{AW}, α_{EP}, and $P_{II\,critical\,state}$ the following linear dependence is found:

$$p_{AW}\alpha_{AW} + (1 - p_{AW})\,\alpha_{EP} = C_{II3}\,P_{II critical\,state} \qquad (3.26)$$

Parameter p_{AW} is equal to participation of AW mechanism in protective layer formation.

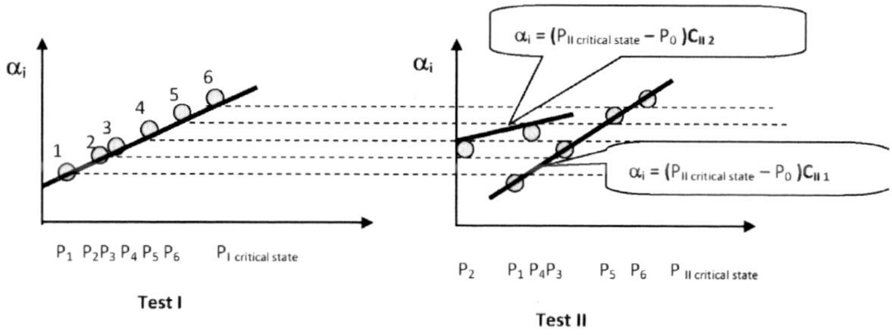

Figure 3.17 The comparison of the α_i and results of two different tribological tests.

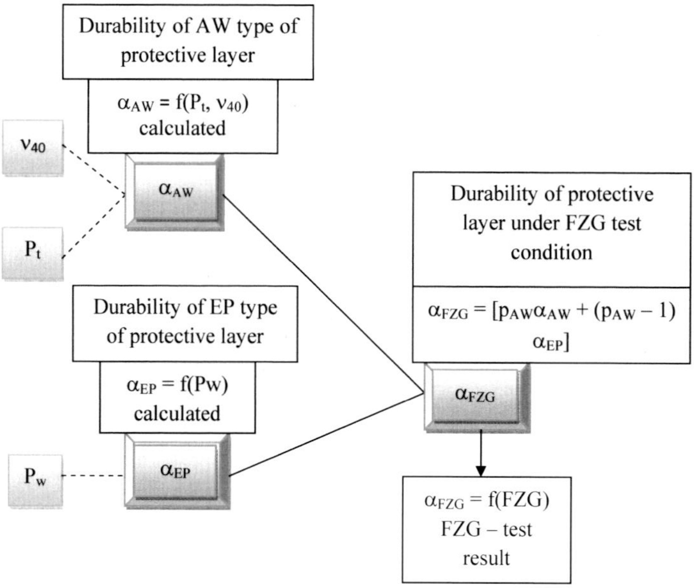

Figure 3.18 Approach to analysis of the protective layer structure and results of tests prediction based on the results of α_{AW} and α_{EP} calculation.

Using described above procedure the following dependence was found for FZG (model gear) test and transmission, gear and hydraulic oils:

$$0.75\alpha_{AW} + 0.25\alpha_{EP} = -0.01\ \text{FZG} + 0.23, \tag{3.27}$$

where FZG is the number of loading stage, under which seizure occurs for tested oil.

Figure 3.18 illustrates the main idea of the analysis of protective layer structure and prediction of complex tribological tests results on the basis of results of α_{AW} and α_{EP} calculation.

3.2.7.2 Application to mechanochemistry

Mechanochemistry is the coupling of mechanical and chemical phenomena on a molecular scale and includes mechanical breakage, chemical behavior of mechanically stressed solids (e.g., stress-corrosion cracking), tribology, polymer degradation under shear, cavitation-related phenomena (e.g., sonochemistry and sonoluminescence), shockwave chemistry and physics, and even the burgeoning field of molecular machines. Mechanochemistry can be seen as an interface between chemistry and mechanical engineering. A smart method was proposed recently, in order to measure the energy involved during mechanical transformations. Displacement reactions between a metal oxide and a more reactive metal can be induced by ball milling. In some cases, the reaction progresses gradually and a metal/metal-oxide nanocomposite is formed. Ball milling may also initiate a self-propagating combustive reaction. The information available about these processes is reviewed. It is argued that the gradual or combustive nature of the reaction depends on thermodynamic parameters, the microstructure of the reaction mixture, and the way they develop during the milling process.

L. Takacs [37] developed reactions between a metal oxide and a more reactive metal, which can be induced by high-energy ball milling. The reaction may progress gradually, producing a nanocomposite powder. The mechanical agitation may also initiate combustion in highly exothermic systems, melting the reaction mixture and destroying the ultrafine microstructure. In order to avoid this problem, reaction couples with a smaller driving force have been investigated. The role of intermediate phases in understanding the mechanism of these mechanochemical processes is emphasized. The reduction of Cr_2O_3 by aluminum or zinc and the reduction of Fe_3O_4 by zinc are identified as promising candidates for further investigations.

Takacs et al. [38] described the reduction of magnetite by Al and Mg induced by mechanical alloying. Milling initiates self-propagating thermal "explosion" in these systems after some incubation period. Moessbauer spectroscopy and x-ray diffraction have been used to investigate the reaction products. A mixture of α-Fe, $FeAl_2O_4$, α-Al_2O_3, γ-Al_2O_3 and a small percentage of the starting materials have been found when the stoichiometric reaction was stopped seconds after explosion.

Peter Baláž et al. [39] investigated the mechanochemical treatment of solids, which led to a positive influence on the solid–liquid kinetics. They used Arrhenius equation for activation energy analysis. The breaking of bonds in the crystalline lattice of solids brings about a decrease (ΔE^*) in the activation energy and an increase in the rate of leaching

$$(\Delta E^*) = E - E^* \tag{3.28}$$

$$k^* = k \exp(\Delta E^*/RT), \tag{3.29}$$

where E is the apparent activation energy of the non-disordered solid, E^* is the apparent activation energy of the disordered solid, k, R, and T stand for the rate constant of leaching for the non-disordered solid, (the pre-exponential factor) gas constant and reaction temperature, respectively; k^* is the rate constant of leaching for the disordered solid. If $E > E^*$, then $\exp(\Delta E^*/RT) > 1$ and thus it follows from Eq. (3.29) that $k^* > k$, i.e., the rate of leaching of a disordered solid is greater than that of an ordered mineral.

M. A. Grinfeld et al. [40] studied the current state-of-the-art of thermodynamic methods in the theory of phase transformations and fracture in solids. Their opinion is that this state-of-the-art can be characterized by a single word: "crisis." It might look surprising when somebody talks about crisis in the era when thousands of papers appear in scientific journals every year. In those sometimes sophisticated and well-elaborated papers, it is claimed that theory is in remarkable harmony with the experiment. The crisis is in the predictive ability of the current thermodynamic theories of heterogeneous systems with solid phases. The often-demonstrated agreement of the theory and experiment should not be exaggerated. It is basically achieved by pumping the models with sufficiently many parameters. However, try to make an unbiased prediction based on those allegedly universal theories. In doing so, one will find that those predictions do not have much in common with reality. The problem is not only with the number of tuning parameters per se. More importantly, those parameters require permanent and essential adjustments even when dealing with very close experimental settings.

The unsatisfactory state-of-the-art in the theories of phase transformations and fracture is the reason of the current (and future) study. To make some progress in understanding the existing difficulties, Grinfeld proposes exploring the strong features and weaknesses of the energetic (thermodynamic) methods by making theoretical and experimental studies of macroscopic adhesive systems.

From the standpoint of thermodynamics, all these systems include, as an essential feature, the energy exchange between two storages of the accumulated energy—the energy accumulated due to reversible elastic deformations and the "chemical" energy stored in the chemical bonds. It is for these reasons that the systems require almost identical thermodynamic methods.

Thermodynamic methods are essentially macroscopic by origin and nature. They appear in the analysis of macroscopic engineering systems. They have been reliably validated in numerous macroscopic experiments and observations. As a result, there can presumably be found areas that permit analysis of mechanochemical systems by means of relatively simple thermodynamic methods. From a purely thermodynamic point of view, the central problem of mechanochemistry is the exchange of energy between the (long-range) elastic energy and the (short-range) energy accumulated in individual bonds.

As shown above, there is no theory that should be adopted to mechanochemistry. The α_{i} model worked out for tribochemical applications can also be adopted to mechanochemistry. This model can be helpful in general dependences formulation, related to kinetics of mechanochemical reactions and to mechanical forces used for reactions activation. The theory based on α_{i} model assumes that mechanical energy introduced into solid body—reagent or catalyst, is accumulated in it and then emitted as low energy electrons or photons of energy equal or higher than activation energy of the reaction. The general Eq. (3.10) can be used to determine quantitatively relationship between mechanical stress (L), the possibility of solid body to accumulate and then emit energy ($e_0/e_{\mathrm{a}} \cos\gamma$) and kinetics of mechanochemical reaction.

3.2.8 Concluding Remarks

The model put forward in this paper attempts to correlate mechanical work performed on a solid with its catalytic activity. The tribological system contains fluid reagents and a solid (a specific material) which is the border of the system. The material plays the role of catalyst (tribocatalyst). The mechanical work done on the system is transformed to internal energy increase. The internal energy is distributed into a liquid/fluid phase bringing about ambient temperature increase (*Tb*), and energy of electrons/photons emitted by solid surface to liquid phase. The energy accumulated in the solid is emitted from its surface as pulses of electrons and/or photons. There is a specific angle γ at which the reaction can reach a critical rate or may be initiated, e.g., when e_γ equals some 3–5 eV. The energy accumulated in the liquid phase is insufficient to overcome the threshold of the real activation energy E_a. The energy emitted from surface as pulses of 3–5 eV can reach the value of E_a and the triboreaction process starts to proceed or reaches the critical rate. Based on the discussion concerning the α_i model, thermionic emission, and the NIRAM approach, it was concluded that for both thermochemical heterogeneous reactions and catalyzed heterogeneous processes, the same activation energy (E_a value) is needed to initiate the reaction process. The model discussed in this paper provides a new approach to the activation energy and leads to better understanding of heterogeneous catalysis as well as mechanochemistry. Although the discussed anisotropic EE relates to metal–insulator–metal thin-film sandwich structures, this fact is considered as a first evidence for the proposed model, which predicts such triboelectron distribution. The fact that triboreactions take place and the critical rate of triboreactions is reached under different conditions (load, temperature) for different reactants cannot be explained by decreasing of activation energy (E_a) by the solid catalyst because E_a is not a function of *P*.

In summary, it should be emphasized that the activation energy is constant for a given type of reaction and the reaction onset depends not only on the energy quantity added to reactants but on density of the introduced energy stream. The catalyst emits pulses of energy flux of high density and at a given angle the value of emitted energy is equal to real activation energy E_a.

References

1. G. C. Bond, *Heterogeneous Catalysis: Principles and Applications*, 2nd ed., Clarendon Press, Oxford, (1987).
2. K. Klier, The transition state in heterogeneous catalysis, *Topics Catal.*, 18(3–4) (2002) 141–156.
3. N. Sato, M. Seo, Chemically stimulated exoelectron emission from silver catalyst during partial oxidation of ethylene, *J. Catal.*, 24 (1972) 224–232.
4. C. K. Kajdas, A. Kulczycki, A new idea of the influence of solid materials on kinetics of chemical reactions, *Mater. Sci. Poland*, 26 (2008) 787–796.
5. A. Kulczycki, The correlation between results of different model friction tests in terms of an energy analysis of friction and lubrication; *Wear*, 103 (1985) 67–75.
6. C. Kajdas, Importance of ionic reactive intermediates for lubricant component reactions with friction surfaces, *Lubr. Sci.*, 6 (1994) 203–228.
7. C. Kajdas, *Proceedings of the International Tribology Conference*, Yokohama 1995, Satellite Forum on Tribochemistry, JST, Tokyo, (1995) 31–35.
8. R. C. Dante, C. K. Kajdas, A. Kulczycki, Theoretical advances in the kinetics of tribochemical reactions, *React. Kinet. Mech. Catal.*, 99 (2010) 37–46.
9. A. Kulczycki, C. Kajdas, The influence of base oil properties on the efficiency of action of ZDDP-type additives, *Lubr. Sci.*, 6 (1994) 161–180.
10. R. Hrach, J. May, The angular distribution of electrons emitted from thin-film MIM structures at various temperatures, *Phys. Stat. Sol. (a)*, 4 (1971) 637–642.
11. R. D. Gould, C. A. Hogarth, Angular distribution measurements of electrons emitted from thin film Au–SiO_x–Au diode and triode structures, *Phys. Stat. Sol. (a)*, 41 (1977) 439–442.
12. R. Hrach, Emission of electrons from MIM systems: Disscussion of processes in the cathode, Czech. *J.f Phys. B*, 23 (1973) 234–242.
13. H. Hrach, Energy angular distribution of electrons emitted from MIM systems, *Thin Solid Films*, 15 (1973) 65–69.
14. H.-J. Fitting, H. Glaefeke, W. Wild, J. Lang, Energy and angular distribution of exoelectrons, *Phys. Stat. Sol., (a)*, 42 (1977) K75–K77.

15. V. S. Kortov, P. P. Zolnikov, Computation of the energy and angular distribution of exoelectrons by the Monte-Carlo method, *Phys. Stat. Sol. (a)*, 31 (1975) 331–339.

16. A. Darlinski, Measurements of angular distribution of the backscattered electrons, *Phys. Stat. Sol. (a)*, 63 (1981) 629–633.

17. A. Jablonski, J. Zemek, Angular distribution of photoemission from amorphous and polycrystalline solids, *Phys. Rev. B*, 48 (1993) 4799–4805.

18. S. Hucek, I. S. Tilinin, J. Zemek, Angular distribution of photoelectrons emitted from solids, *J. Electron Spectrosc. Relat. Phenomena*, 85 (1997) 263–268.

19. S. R. De Groot, *Thermodynamics of Irreversible Processes.* North Holland Publishing Company, Amsterdam (1952).

20. L. Onsager, Reciprocal relations in irreversible processes, I., *Phys. Rev.*, 37 (1931) 405–426.

21. C. K. Kajdas, A. Kulczycki, K. J. Kurzydłowski, G. J. Molina, Activation energy of tribochemical and heterogeneous catalytic reactions, *Mater. Sci. Poland*, 28, (2010) 523–533.

22. C. Kajdas, M. J. Furey, A. L. Ritter, G. J. Molina, Triboemission as a basic part of the boundary friction regime, *Lubrication Sci.*, 14 (2002) 223–252.

23. G. J. Molina, C. Kajdas, M. J. Furey, Charged triboemission from sliding contacts: possible mechanisms and its relation with surface chemical processes, *Proceedings of the Adhesion Society Meeting*, Savannah, Georgia, USA, February (2009) pp. 327–329.

24. B. Kasemo, E. Tornqvist, J. K. Nordskold, B. I. Lundqvist, Photon and electron emission as indicators of intermediate states in surface reactions, *Surf. Sci.*, 89 (1979) 554–565.

25. H. Nienhaus, S. Glass, Probing the distribution of hot charge carriers generated in Mg surfaces by oxidation, *Surf. Sci.*, 600 (2006) 4285–4289.

26. S. Glass, H. Nienhaus, Continuous monitoring of Mg oxidation by internal exoemission, *Phys. Rev. Lett.*, 93 (2004) 168302.

27. B. Gergen, H. Nienhaus, W. H. Weiberger, E. W. McFarland, Chemically induced electronic excitations at metal surfaces, *Science*, 294 (2001) 2521–2523.

28. A. Boettcher, R. Imbeck, A. Morgante, G. Ertl, Nonadiabatic surface reaction: Mechanism of electron emission in the $Cs+O_2$ system, *Phys. Rev. Lett.*, 65 (1990) 2035–2037.

29. Y. F. Chen, Surface effects on angular distribution in X-ray-photoelectron spectroscopy, *Surf. Sci.*, 519 (2002) 115–124.
30. K. Hiratsuka, C. Kajdas, Wear and chemical reactions, *Tribology Science and Application*, Proceedings of the Review Conference on the Scientific Cooperation between Austria, Poland and their Worldwide Partners, editors F. Franek and C. Kajdas. Scientific Centre of the Polish Academy of Sciences, Vienna, Austria (2004), pp. 71–90.
31. K. Hiratsuka, C. Kajdas, M. Yoshida Tribo-catalysis in the synthesis reaction of carbon dioxide, *Tribol. Trans.*, 47 (2004) 86–93.
32. C. Kajdas, Tribochemistry, *Proceedings of the 2nd World Tribology Congress Tribology 2001*, (F. Franek, W. J. Bartz, A. Pauschitz, eds.), The Austrian Tribology Society, Vienna (2001) 39–41.
33. K. Tkacova, H. Heegn, N. Stevulova, Energy transfer and conversation during comminution and mechanical activation, *Int. J. Miner. Process.*, 40 (1993) 17–31.
34. C. K. Kajdas, Importance of the triboemission process for tribochemical reaction, *Tribol. Int.*, 38 (2005) 337–353.
35. L. Biernacki, C. Gorecki, T. Goreceki, Exoelectron emission from skeletal (Raney) catalysts, *Jap. J. Appl. Phys.*, 24 (1985) 116–117.
36. D. B. Knorr, Jr., T. O. Gray, R. M. Overney, Cooperative and submolecular dissipation mechanisms of sliding friction in complex organic systems, *J. Chem. Phys.*, 129, 074504 (2008); http://dx.doi.org/10.1063/1.2968548 (6 pp.).
37. L. Takacs, Metal-metal oxide systems for nanocomposite formation by reaction milling, *Nanostruct. Mater.*, 2 (1993) 241–249.
38. L. Takacs, H. Pollak, H. Dlamini, Effect of composition on the solid state reaction of magnetite with Al and Mg induced by mechanical alloying, in *Conference Proceedings vol. 50, ICAME-95* (I. Ortalli, ed.), SIF, Bologna (1996) 149–152.
39. P. Balaż, W. S. Choi, M. Fabian, E. Godocikova, Mechanochemistry in the preparation of advanced materials, *Acta Montanistica Slovaca*, 11 (2006) 122–129.
40. M. A. Grinfeld, S. B. Segletes, Towards mechanochemistry of fracture and cohesion: General introduction and the simplest model of Velcro, ARL-TR-5309 September (2010).
41. T. V. Prevenslik, Nanocatalysts by quantum electrodynamics induced electromagnetic, *Chin. J. Catal.*, 29(11) (2008) 1073–1078.
42. M. Hebda, A. Wachal, *Trybologia*, WNT Warsaw, (1980).

3.3 Theoretical Advances in Tribochemical Reaction Kinetics Based on Thermodynamics of Irreversible Processes

Roberto C. Dante[a,b]

[a]*Department of Rheology and Complex Materials, Instituto de Investigaciones en Materiales, Universidad Nacional Autónoma de México, Mexico*
[b]*Laboratory of Advanced Materials ETSIIA, University of Valladolid, Spain*

rcdante@yahoo.com

3.3.1 Introduction

Thermodynamics is widely used by chemists to define the equilibrium state of chemical reactions, and their interest is mainly focused on enthalpy and Gibbs free energy. This kind of thermodynamics describes reversible chemical processes rather than irreversible processes, which are those of Nature. However, it is a powerful tool in chemical engineering, widely used in fuel industry, etc. Nevertheless, there were significant advancements towards a non-equilibrium thermodynamics, mostly obtained around the half of the 20th century, which allowed us to get a better description of the complex irreversible phenomena of physics, chemistry, and especially biology. Onsager, Prigogine, Glansdorf, and De Groot [1–4] were among the scientists who mostly contributed to the main achievements of non-equilibrium thermodynamics.

These efforts were directed to explain the local negative production entropy of some complex systems as manifested by living systems, and several chemical reactions (e.g., Belousov–Zhabotinsky reaction) [5]. Many advances of thermodynamics of irreversible processes (TIP) were linked to the study of chemical reactions, and Onsager's relations can be considered one of the most important achievements of TIP [6]. However, most of the theorems and relations are valid only in proximity of equilibrium.

Chemical reaction rate and catalysis (both homogeneous and heterogeneous) were often studied through mechanistic approaches of direct kinetic methods which failed into provide a description of complex chemical transformations often with many arbitrary assumptions. Nevertheless, there were certain attempts that already achieved a high grade of maturity to provide a general TIP-based framework and description of catalysis, which is a typical non-equilibrium phenomenon. The work of Parmon is one of the best examples of building a general catalysis theory based on TIP [7]. The chemical transformations are initiated by the thermodynamic driving forces, which change along with transformations' advancement. The prediction of the system evolution can be achieved through the analysis of these driving forces even if the intimate mechanism of the analyzed process is not fully understood.

Moreover, Kuzharov et al. demonstrated through TIP that friction generates so high chemical affinities (the driving force of chemical reactions) in the friction interface that chemical reactions would be speeded up in a significant way [8].

This approach can be extended to tribocatalysis giving a new insight into the observed rate enhancement of many triboreactions, which is compatible and fits with negative ion-radical action mechanism (NIRAM) theory [9–10]. TIP implies that a flow of energy through specific carriers such as triboelectrons, is directed towards reagents and can be able to enhance reaction rate in a significant manner, as explained by Dante et al. [11]. These results in conditions near to equilibrium can be extended also to situations far from equilibrium. NIRAM, TIP, and α_{i} model converge to provide a thorough theory to describe tribocatalysis and tribochemical activated reactions.

3.3.2 Thermodynamics Applied to Chemistry

The focal point of thermodynamics is not that it provides any theoretical insight into the structure of matter but rather that it describes accurately the state or condition of substances and how they behave in chemical or physical changes. Thermodynamics also deals with the movement of energy from one system to another, from one form to another, from one chemical species to another.

Thermodynamics is used in chemistry to describe the transition from an equilibrium state to another one, to determine both the heat and spontaneity of reactions. The purpose of this section is to provide a concise but complete presentation of the foundations of thermodynamics through an axiomatic approach.

The approach of Tisza [12] and Callen, who *postulate* the existence of entropy with certain properties, permits a more precise analysis. Callen proposed the following postulates for classical thermodynamics, which provide an axiomatic framework [13]:

Postulate I. There exist particular states (called equilibrium states) of simple systems that, macroscopically, are characterized completely by the internal energy *U*, the volume *V*, and the mole numbers $n_1, n_2, \ldots, n_r$ of the chemical components.

Postulate II. There exists a function (called the entropy *S*) of the extensive parameters of any composite system, defined for equilibrium states and having the following property: The values assumed by the extensive parameters in the absence of an internal constraint are those that maximize the entropy over the manifold of constrained equilibrium states.

Postulate III. The entropy of a composite system is additive over the constituent subsystems. The entropy is a continuous and differentiable function and is a monotonically increasing function of the energy.

Postulate IV. The entropy of any system vanishes in the state for which

$$\left(\frac{\partial U}{\partial S}\right)_{V,n_i} = 0. \tag{3.30}$$

It is easy to show that the Clausius entropy fulfills the requirements of the postulates.

Thermodynamic state functions depends only on the initial and final states, in other words they are independent of the transformation pathway. A state function *f* is differentiable at all points of an open domain *D*; then at any point of the domain, the total differential can be expressed in the following way:

$$df(x,y) = \left(\frac{df}{dx}\right)_y dx + \left(\frac{df}{dy}\right)_x dy. \tag{3.31}$$

Moreover, also the following condition must be satisfied:

$$\left(\frac{d^2 f}{dydx}\right)=\left(\frac{d^2 f}{dxdy}\right). \qquad (3.32)$$

Among these functions we can account internal energy, enthalpy, entropy, Gibbs free energy, chemical affinity, etc. On the other hand, work and heat are process parameters since they depend on the pathway from one equilibrium state to another one. In thermodynamics, different state functions are chosen depending on the most convenient variables to describe the system under study, but it is noteworthy to point out that they are Legendre transformations of each other. This makes possible to find a way to face a thermodynamic problem according to the available variables without affecting conclusions.

We shall define the Legendre transform. Let f: $\mathfrak{R}^n \rightarrow \mathfrak{R}$ be a convex function, i.e., $\Delta^2 f > 0$. The Legendre transform can be expressed in the following way:

$$L(f)(m)=\max_x(m\cdot x-f(x))=g(m), \qquad (3.33)$$

where $g(m)$ is the transform function. So if $x \in \mathfrak{R}$, essentially we are measuring the maximal distance between the line $y = mx$ and $f(x)$. Note that the maximum occurs when

$$\nabla_x(m\cdot x-f(x))=0, \qquad (3.34)$$

so that

$$m = \Delta_x f(x). \qquad (3.35)$$

It should be noted that these equations determine *x as a function of m* (x and m are called conjugate variables). The convexity condition ensures that there is only one of these critical points, and that is a maximum. Another important relation that can be easily obtained from previous statements is the following:

$$f(x) + g(m) = xm, \qquad (3.36)$$

satisfied under the conditions of Eq. 3.33. Taking internal energy $U(T, V)$ as starting thermodynamic function, which satisfies the Legendre conditions, the others can be obtained as Legendre transform such as enthalpy H, Gibbs and Helmoltz free energies, G and F, respectively:

$$-H(S, p) = L(U)(-p), \tag{3.37}$$

$$-G(p, T) = L(H)(T), \tag{3.38}$$

$$-F(T, V) = L(U)(T), \tag{3.39}$$

where p is pressure, V volume, and T the absolute temperature.

According to Eq. 3.36 and Eqs. 3.37–3.39, we can write the following equations:

$$U - H = -pV, \tag{3.40}$$

$$H - G = TS, \tag{3.41}$$

$$U - F = TS, \tag{3.42}$$

The right hand side of Eqs. 3.40–3.42 shows the product of the conjugate variables: p and V, T and S. According to Eq. 3.35, the following parameters can be expressed in terms of first derivative of thermodynamic functions:

$$p = -\left(\frac{\partial U}{\partial V}\right)_S, \tag{3.43}$$

$$T = \left(\frac{\partial H}{\partial S}\right)_p, \tag{3.44}$$

$$T = \left(\frac{\partial U}{\partial S}\right)_V, \tag{3.45}$$

from which it is possible to show that the exact differential dU is given:

$$dU = \left(\frac{\partial U}{\partial S}\right)_V dS + \left(\frac{\partial U}{\partial V}\right)_S dV = TdS - pdV, \tag{3.46}$$

which depends on the extensive variables S and V.

Applying Eq. 3.32 to Eq. 3.46, e.g., the following equality of second derivatives of thermodynamic functions can be written:

$$\left(\frac{\partial T}{\partial V}\right)_S = -\left(\frac{\partial p}{\partial S}\right)_V. \tag{3.47}$$

When a chemical reaction occurs, we can write Eq. 3.46 in the following form:

$$dU = TdS - pdV + \sum_i \mu_i dn_i, \tag{3.48}$$

$$\mu_i = \left(\frac{\partial U}{\partial n_i}\right)_{S,V,n_{j\neq i}}, \tag{3.49}$$

where μ_i is the chemical potential of the i-th component (reagents and products) and n_i the moles of the i-th component. It should be noted that dn of consuming reagents will be negative.

Now, the absolute temperature T can be expressed in the following general way:

$$T = \left(\frac{\partial U}{\partial S}\right)_{V,n_i}, \tag{3.50}$$

and this clarifies the meaning of the Postulate IV, which thus implies that entropy S vanishes at $T \to 0$.

Summing Eq. 3.40 and 3.41, a new equation is obtained:

$$U - G = TS - pV, \tag{3.51}$$

and in its differential form it is

$$\begin{aligned} dG &= -d(TS) + d(pV) + dU \\ &= -SdT - TdS + pdV + Vdp + TdS - pdV + \sum_i \mu_i dn_i \\ &= -SdT + Vdp + \sum_i \mu_i dn_i \end{aligned} \tag{3.52}$$

and since G can be written also in this form,

$$\begin{aligned} G &= H - TS + \sum_i \mu_i N_i \\ &= U + pV - TS + \sum_i \mu_i n_i \\ &= TS - pV + pV - TS + \sum_i \mu_i n_i = \sum_i \mu_i n_i. \end{aligned} \tag{3.53}$$

Therefore, in brief the chemical potential μ is an intensive variable obtained by division of two extensive variables: the Gibbs free energy G and the number of mole n.

Finally, the differential dG becomes

$$dG = \sum_i \mu_i dn_i + \sum_i n_i d\mu_i, \tag{3.54}$$

equaling the final result of Eqs. 3.52 and 3.54, the Gibbs–Duehm equation is obtained:

$$\sum_i n_i d\mu_i = -SdT + Vdp, \tag{3.55}$$

which shows that the intensive variables that characterize the system μ_i, T and P are not independent, and at least one is fully determined by the others.

3.3.3 Basic Concepts and Achievements of Thermodynamics of Irreversible Processes

The extent of reaction is a fundamental parameter and it is defined in the following way:

$$d\xi = \frac{dn_i}{\nu_i}, \tag{3.56}$$

where dn_i and ν_i are, respectively the moles' variation and the stoichiometric coefficient of the i-th reaction component, which is taken as negative in case of reagents.

The reaction rate that is the thermodynamic flux for chemical reactions

$$\frac{d\xi}{dt} = \frac{1}{\nu_i}\frac{dn_i}{dt} = \dot{\xi}, \tag{3.57}$$

and for a chemical reaction at stationary T and p, dG can be written in following form:

$$dG = \sum_i \mu_i dn_i = \sum_i \nu_i \mu_i d\xi, \tag{3.58}$$

$$A \equiv -\frac{dG}{d\xi} = -\sum_i \nu_i \mu_i, \tag{3.59}$$

where A is called chemical affinity and is the driving force of chemical reactions [4]. In an open system, entropy may change by internal irreversible processes such as for chemical reactions and for exchange with surroundings.

Therefore the change of entropy is given by the contribution of two components:

$$dS = dS_{in} + dS_e, \tag{3.60}$$

where dS_{in} and dS_e are the internal entropy and the external entropy, respectively. For reversible processes $dS_{in} = 0$, otherwise $dS_{in} > 0$, according to the Postulate II of Callen. In isolated systems, $dS_e = 0$; thus Eq. 3.60 is reduced to

$$dS = dS_{in} \geq 0. \tag{3.61}$$

The rate of entropy production is so defined:

$$\sigma = \frac{dS}{dt} = \frac{dS_{in}}{dt} + \frac{dS_{ex}}{dt} = \sigma_{in} + \sigma_{ex}, \tag{3.62}$$

where σ_{in} and σ_{ex} are the rates internal and external entropy productions, respectively. In the presence of any kind of irreversible process in the system,

$$\sigma_{in} > 0, \tag{3.63}$$

but it is noteworthy to point out that σ_{ex} may be negative. This allows open system to evolve and organize themselves with a local decrease of entropy such as for living organisms.

In case of closed systems, only heat and work can be exchanged with the environment, so that the internal entropy increment can be related to chemical reactions

$$dS_{in} = dS - dS_e = dS - \frac{\delta Q}{T}, \tag{3.64}$$

remembering that the heat δQ is given by the following equation,

$$\delta Q = dU + pdV, \tag{3.65}$$

and substituting δQ of Eq. 3.65 in Eq. 3.64, it is found

$$dS_{in} = dS - \frac{1}{T}(dU + pdV) = -\frac{1}{T}(dU - TdS + pdV), \tag{3.66}$$

and at constant pressure and temperature, Eq. 3.66 becomes

$$dS_{in} = -\frac{1}{T}d(U - TS + pV) = -\frac{1}{T}dG > 0. \tag{3.67}$$

It is possible to show that this equation can be extended also to open systems. Therefore, the rate of entropy production is given by

$$\sigma_{in} = -\frac{1}{T}\frac{dG}{dt} > 0. \tag{3.68}$$

Equation 3.68 can be rewritten in terms of the chemical affinity A:

$$\sigma_{in} = -\frac{1}{T}\frac{d\xi}{dt}\frac{dG}{d\xi} = \frac{1}{T}\dot{\xi}A > 0. \tag{3.69}$$

From this result and other observations, it is possible to argue that the rate of entropy production can be expressed by the following general equation [6]:

$$\sigma_{in} = \frac{1}{T}\sum_i \frac{d\alpha_i}{dt}\left(\frac{\partial S_{in}}{\partial \alpha_i}\right) > 0, \tag{3.70}$$

where $d\alpha_i/dt$ is the flux J_i and $\partial S_{in}/\partial \alpha_i$ the conjugated thermodynamic force X_i, and in our case,

$$\alpha_i = \xi, \tag{3.71}$$

$$\left(\frac{\partial S_{in}}{\partial \xi}\right) = A. \tag{3.72}$$

Equation 3.70 is of general validity and written in terms of fluxes and forces exhibits the following form:

$$\sigma_{in} = \frac{1}{T}\sum_i J_i X_i > 0, \tag{3.73}$$

$$J_i = \left(\frac{\partial \alpha_i}{dt}\right), \tag{3.74}$$

$$X_i = \left(\frac{\partial S_{in}}{d\alpha_i}\right), \tag{3.75}$$

In our case, J_i is the reaction rate $\dot{\xi}$ and X_i is the chemical affinity A.

The thermodynamic fluxes J_i are generally functions of thermodynamic forces X_i that induces fluxes. When the chemical affinity $RT >> A$, i.e., the thermal energy is much higher than the chemical affinity, the reaction is close to equilibrium. Near equilibrium fluxes and forces are very small, and fluxes can be approximated by means of a Taylor series truncated at the second term:

$$J_i(\bar{X}) = J(0) + \sum_j \left(\frac{\partial J_i}{\partial X_j}\right)\Bigg|_{\bar{X}=0} X_j + \Theta(\bar{X}) \approx \sum_j L_{ij} X_j, \tag{3.76}$$

$$L_{ij} = \left(\frac{\partial J_i}{\partial X_j}\right)\Bigg|_{\bar{X}=0} \tag{3.77}$$

$$J_i = \sum_j L_{ij} X_j \tag{3.78}$$

where $\Theta(\bar{X})$ includes all the terms of greater orders of a set of forces $\bar{X}$ being flux $J_i(0)$ null at equilibrium, and coefficients L_{ij} independent of $\bar{X}$.

Therefore, near equilibrium Eq. 3.73 becomes

$$\sigma_{in} = \frac{1}{T}\sum_i \sum_j L_{ji} X_j X_i > 0, \tag{3.79}$$

which is known as the Rayleigh–Onsager equation. Moreover, according to the Onsager's reciprocal relations [6]:

$$L_{ij} = L_{ji}. \tag{3.80}$$

Onsager achieved this result assuming microscopic reversibility, e.g., the principle of "detailed balancing." For a chemical reaction, at equilibrium the reciprocal relations correspond to the basic assumption that the direct ($K_{AB}n_A$) and the reverse ($K_{BA}n_B$) reaction rates are equal on the basis of microscopic reversibility.

$$K_{AB}n_A = K_{BA}n_B, \tag{3.81}$$

where n_A and n_B are the equilibrium moles of the reagent and the product, respectively.

The Onsager relations are very useful also when the coefficients are empirically determined to analyze some properties of catalytic systems, etc.

The following relations are also valid for Onsager coefficients [7]:

$$L_{ii} \geq 0; \tag{3.82}$$

$$L_{ij} > -\frac{1}{2}\frac{L_{ii}X_i^2 + L_{jj}X_j^2}{X_iX_j}. \tag{3.83}$$

The rate of entropy production in stationary states is related to the Prigogine theorem of the evolution for systems that are close to their thermodynamic equilibrium. In summary, if we consider an open system with N isothermal irreversible processes, the rate of entropy production is given by Eq. 3.73, and next to equilibrium can be expressed through Eq. 3.79.

The steady state is determined by the following equation:

$$\frac{dP}{dt} = \frac{d}{dt}\sum_i J_iX_i = \sum_i\left(\frac{dJ_i}{dt}X_i + J_i\frac{dX_i}{dt}\right) = 0, \tag{3.84}$$

where P is called rate of energy dissipation

$$P = T\sigma_{in}, \tag{3.85}$$

which is satisfied when both fluxes J_i and forces X_i are constant. Near equilibrium Eq. 3.84 becomes

$$\frac{dP}{dt} = \frac{d}{dt}\sum_i J_i X_i = \sum_i \left(\frac{dJ_i}{dt} X_i + J_i \frac{dX_i}{dt} \right)$$

$$= \sum_i \left(\sum_j L_{ij} \frac{dX_j}{dt} X_i + L_{ij} X_j \frac{dX_i}{dt} \right) = \sum_i J_i \frac{dX_i}{dt} = 0. \tag{3.86}$$

It should be noted that that the spontaneous evolution towards the steady state is characterized by $d\sigma_{in}/dt < 0$. Taking the partial derivative of the rate of energy dissipation with respect to X_i at constant X_j, the following expression is obtained by applying the Onsager reciprocal relations:

$$\left(\frac{\partial P}{\partial X_i} \right)_{X_{j \neq i}} = 2\sum_j L_{ij} X_j = 2J_i. \tag{3.87}$$

The flux J_i, which depends on an internal parameter α_i, will be null because this parameter will be constant during the steady state. This will correspond to an extreme point:

$$\left(\frac{\partial P}{\partial X_i} \right)_{X_{j \neq i}} = 0. \tag{3.88}$$

Remembering that in stationary state Eq. 3.79 is valid, so that is positive and quadratic in its domain, and null at equilibrium. Equation 3.88 states that for an isothermal open system near equilibrium the stationary state corresponds to the minimal rate of entropy production.

To be a minimum the analytic requirement is that the second derivative must be positive

$$\left(\frac{\partial^2 P}{\partial X_j \partial X_i} \right)_{X_{k \neq i,j}} > 0. \tag{3.89}$$

Any deviation from the stationary state will cause an increment of the rate of entropy production, and the system will tend spontaneously to the minimal rate of production entropy of the stationary state. The spontaneous evolution of the system to its stationary state of the system can be accompanied by only a monotonous decrease of P, that is

$$\frac{dP}{dt} < 0. \tag{3.90}$$

The stability of the stationary state can be showed supposing that a disturbance of the steady state causes a change of the forces X_j at equilibrium by quantities δX_j.

Equation 34.3 will be written in the following way:

$$J_i = \sum_j L_{ij}(\hat{X}_j + \delta X_j) = \sum_j L_{ij}\hat{X}_j + \sum_j L_{ij}\delta X_j, \tag{3.91}$$

where $\hat{X}_j$ is the value of the force in the stationary state, remembering that the first term corresponds to the flux in the stationary state and that is null, follows that

$$J_i = \sum_j L_{ij}\delta X_j, \tag{3.92}$$

i.e., fluxes are proportional to the forces' fluctuations. Applying this result to the Rayleigh–Onsager equation, the fluctuation of the rate of energy dissipation must satisfy the following condition:

$$\delta P = \sum_i \sum_j L_{ij}\delta X_j \delta X_i > 0. \tag{3.93}$$

This condition is equivalent to those of Eqs. 3.89 and 3.90, which are well proven, and shows that any fluctuation of an isothermal stationary state characterized by the minimum rate of entropy production causes an increment of this rate. Fluxes caused by fluctuations will act to vanish the perturbation until returning to the minimum rate of entropy production. Therefore, steady states are the stable states for isothermal systems out of equilibrium and are characterized by the minimum rate of entropy production.

In a steady state far from equilibrium Eq. 3.84 is still valid in the stationary state but no longer Eq. 3.85. In general, the time derivative of the rate of energy dissipation can be separated in two terms:

$$\frac{dP}{dt} = \frac{d}{dt}\sum_i J_i X_i = \sum_i \left(\frac{dJ_i}{dt} X_i + J_i \frac{dX_i}{dt} \right) = \frac{d_J P}{dt} + \frac{d_X P}{dt}, \tag{3.94}$$

In this case the linear Onsager relations are not satisfied, so that the differentials d_JP and d_XP are no longer total differentials. Hence, variations of the parameter P depend on the transition route. However, Glansdorf and Progogine demonstrated that for a system far from equilibrium evolving towards the steady state

$$\frac{d_XP}{dt} \leq 0, \tag{3.95}$$

which reflects the monotonous decrement of the forces X approaching the steady state [3, 7]. It should be noticed that near equilibrium is valid the following equation:

$$\frac{d_JP}{dt} = \frac{d_XP}{dt} = \frac{1}{2}\frac{dP}{dt}. \tag{3.96}$$

The evolution of systems far from equilibrium is definitely more complex. However, it remains a fact that the stationary states in TIP have the same role of equilibrium states in reversible thermodynamics. When irreversible processes occur, the system tends spontaneously to stationary states.

3.3.4 Reaction Kinetics and TIP

The reaction kinetic equations belong completely to the field of TIP. The reaction rate ξ is expressed in the following way according to the mass action law:

$$\dot{\xi} = k_f \Pi_i c_i^{\nu_i} - k_r \Pi_j c_j^{\nu_j}, \tag{3.97}$$

where k_f is the rate constant of the forward reaction, and k_r is that of the reverse reaction. Sub-indexes i and j correspond to the i-th reagent and j-th product, respectively. For instance, c and ν are the molar concentration and the stoichiometric coefficient, respectively.

The rate constant of forward reaction depends on the activation energy, which is the difference between the standard chemical potential of the transition state μ_0^* (i.e., of the activated complex), and that of reagents

$$\Delta G_f^* = \mu_0^* - \sum_i \nu_i \mu_i. \tag{3.98}$$

The rate constant of the forward reaction can be thus expressed in terms of the activation energy ΔG_f^*:

$$k_f = \chi_f \frac{RT}{h\gamma^*} \exp\left(-\frac{\Delta G_f^*}{RT}\right), \tag{3.99}$$

where χ_f is the permeability of the activation barrier, γ^* and h are the activity coefficient of the hypothetical the transition state and the Planck's constant, respectively.

Equation 3.97 can be written in this form:

$$\dot{\xi} = k_f \Pi_i c_i^{\nu_i} \left(1 - \frac{k_r}{k_f} \frac{\Pi_j c_j^{\nu_j}}{\Pi_i c_i^{\nu_i}}\right) = k_f \Pi_i c_i^{\nu_i} \left(1 - k_{\mathrm{rf}} \frac{\Pi_j c_j^{\nu_j}}{\Pi_i c_i^{\nu_i}}\right), \tag{3.100}$$

where k_{rf} is the ratio between the reverse reaction rate k_r and the forward one k_f.

The relationship between chemical potential and concentration is expressed as follows:

$$\mu_l = \mu_l^0 + RT \ln \gamma_l c_l, \tag{3.101}$$

being μ_l the chemical potential of the reaction component l (either reagent or product), μ_l^0 the standard potential, and γ_l is the activity coefficient.

From Eq. 3.101 can be obtained easily the molar concentration as function of the chemical potential:

$$c_l = \frac{1}{\gamma_l} \exp(\mu_l/RT) \exp(-\mu_l^0/RT). \tag{3.102}$$

On the basis of these results, Eq. 3.100 can be written in the following form:

$$\dot{\xi} = k_f \Pi_i c_i^{\nu_i} \left[1 - k_{rf} \frac{\Pi_i \gamma_i}{\Pi_j \gamma_j} \exp\left(-\frac{1}{RT} \sum_l \nu_l \mu_l^0\right) \exp\left(\frac{1}{RT} \sum_l \nu_l \mu_l\right)\right], \tag{3.103}$$

It should be noted that stoichiometric coefficients of reagents are negative. Since the equilibrium constant is given by

$$K_{\mathrm{eq}} = \exp\left[-\left(\frac{1}{RT} \sum_l \nu_l \mu_l^0\right)\right], \tag{3.104}$$

and observing that

$$k_{rf} \frac{\Pi_i \gamma_i}{\Pi_j \gamma_j} = K_{\text{eq}}^{-1}, \tag{3.105}$$

Equation 3.101 can be simplified, and the reaction rate $\dot{\xi}$ can be expressed in terms of the chemical affinity A, remembering the definition of A given in Eq. 3.59.

$$\dot{\xi} = k_f \Pi_i c_i^{\nu_i} \left(1 - e^{-\frac{A}{RT}} \right). \tag{3.106}$$

Indicating $\dot{\xi}$ as the flux J_c, and the first factor of Eq. 3.104 as the forward reaction rate J_{cf}, the Eq. 3.106 can be written in the following way:

$$J_c = J_{cf} \left(1 - e^{-\frac{A}{RT}} \right). \tag{3.107}$$

Since near equilibrium $\dot{\xi} \to 0$, this requires that $A << RT$ as previously asserted, and developing the exponential function of Eq. 3.107 as a Taylor series cut at the second term, flux J_c exhibits a linear dependence on the chemical affinity A:

$$J_c = L \frac{A}{RT}, \tag{3.108}$$

$$L = J_{cf}. \tag{3.109}$$

In any condition, the rate of entropy production σ_{in} is given by this equation:

$$\sigma_{in} = J_c \frac{A}{T} = J_{cf} \left(1 - e^{-\frac{A}{RT}} \right) \frac{A}{T}. \tag{3.110}$$

It is noteworthy to point out that if $A >> RT$ this relation is linear respect to A and depends only on the forward rate, i.e., very far from equilibrium:

$$\sigma_{in} \approx J_{cf} \frac{A}{T}. \tag{3.111}$$

3.3.5 TIP Description of Catalysis

Considering that Eq. 3.99 contains in the exponential term the chemical potential of the transition state added to those of reagents, Eq. 3.107 can be expressed with a rate constant J_{ij}^{*}, which depends only on the chemical potential of transition state, taking the following form [14]:

$$\sigma_c = J_{ij}^{*}\left(e^{-\frac{A_i}{RT}} - e^{-\frac{A_j}{RT}}\right)\frac{A_{ij}}{T}, \tag{3.112}$$

$$J_{ij}^{*} = \frac{RT}{h}\exp\left(-\frac{\mu^{*}}{RT}\right), \tag{3.113}$$

where A_i is the chemical affinity of reagents and A_j that of products, and A_{ij} as the chemical affinity of the reaction from the pool of reagents i to the pool of products j. Observing the new form of the rate constant J_{ij}^{*}, we realize that it is independent of the reaction direction:

$$J_{ij}^{*} = J_{ji}^{*}, \tag{3.114}$$

$$\mu^{*} = \mu_0 + RT\ln(\gamma^{*}/\chi_f). \tag{3.115}$$

Hereafter, the two exponential terms of Eq. 3.112 will be indicated as the dimensionless parameters $\tilde{c}_i$ and $\tilde{c}_j$, linked to the concentrations of reagents and products, respectively. According to these definitions, Eq. 3.112 can be written in the following form:

$$\sigma_c = J_{ij}^{*}(\tilde{c}_i - \tilde{c}_j)\frac{A_{ij}}{T}, \tag{3.116}$$

$$\tilde{c}_i = e^{\frac{\mu_i}{RT}}, \tag{3.117}$$

$$\tilde{c}_j = e^{\frac{\mu_j}{RT}}. \tag{3.118}$$

It should be noticed that μ_i and μ_j are the summations of the chemical potentials of reagents and products, respectively. Since many reactions exhibit several steps in the steady-state conditions, the rates of each step are equal:

$$\text{R} + \text{K} \rightarrow \text{X}_1 \rightarrow \dots \text{X}_{n-1} \rightarrow \text{P} + \text{K}, \tag{3.119}$$

$$J_{\text{RK}} = J_{\text{X}_1} = \dots = J_{\text{X}_{n-1}}, \tag{3.120}$$

$$\sigma_{\text{c}} = J^*_{\text{RX}_1}(\tilde{c}_{\text{K}}\tilde{c}_{\text{R}} - \tilde{c}_{\text{X}_1})\frac{A_{\text{RP}}}{T}. \tag{3.121}$$

In a catalyzed reaction R are the reagents and P the products, the catalyst K is regenerated:

$$\text{R} + \text{K} \rightarrow \text{P} + \text{K}. \tag{3.122}$$

Equation 3.116 takes the following form for a one-step reaction:

$$\sigma_{\text{c}} = J^*_{\text{RP}}\tilde{c}_{\text{K}}(\tilde{c}_{\text{R}} - \tilde{c}_{\text{P}})\frac{A_{\text{RP}}}{T}. \tag{3.123}$$

Showing that the rate of entropy production depends also on the catalyst concentration $\tilde{c}_{\text{K}}$.

Catalyzed reactions often lead to the formation of intermediates (indicated as [RK] in our case) between reagents and catalyst.

$$\text{R} + \text{K} \rightarrow [\text{RK}] \rightarrow \text{P} + \text{K} \tag{3.124}$$

In the steady state, we can write the following equalities:

$$J^*_{\text{RK}}(\tilde{c}_{\text{K}}\tilde{c}_{\text{R}} - \tilde{c}_{\text{RK}}) = J^*_{[\text{RK}]}(\tilde{c}_{\text{RK}} - \tilde{c}_{\text{K}}\tilde{c}_{\text{P}}). \tag{3.125}$$

From this equation through algebraic manipulations, the following is obtained:

$$\sigma_{\text{c}} = \frac{J^*_{\text{RK}}J^*_{[\text{RK}]}}{J^*_{\text{RK}} + J^*_{[\text{RK}]}}\tilde{c}_{\text{K}}(\tilde{c}_{\text{R}} - \tilde{c}_{\text{P}})\frac{A_{\text{RP}}}{T}. \tag{3.126}$$

This result shows that the rate constant J^*_{RP} of Eq. 3.123 may contain an apparent activation energy of a virtual transitory state, which is composed by the activation energies of several energy barriers. In fact, J^*_{RP} is given by the following equation:

$$J^*_{\mathrm{RP}} = \frac{J^*_{\mathrm{RK}} J^*_{[\mathrm{RK}]}}{J^*_{\mathrm{RK}} + J^*_{[\mathrm{RK}]}}. \tag{3.127}$$

Therefore, the apparent activation energy μ^*_{RP} will be

$$\mu^*_{\mathrm{RP}} = \mu^*_{\mathrm{RK}} + \mu^*_{[\mathrm{RK}]} - RT\ln(J^*_{\mathrm{RK}} + J^*_{[\mathrm{RK}]}). \tag{3.128}$$

This kind of approach to catalysis is very flexible and emphasizes the role of the catalyst in stabilizing the intermediates. This means that (non-equilibrium) steady state stabilizes the formation of an intermediate involving the reagent R and the catalyst K speeding up the reaction extent.

3.3.6 TIP Approach to Tribochemistry and Tribocatalysis

Hiratsuka et al. found that the gas-gas reactions were enhanced by friction and called this phenomenon "friction catalysis" or "tribocatalysis" [15, 16]. For example, the synthesis reactions listed below were enhanced by the rubbing of platinum or palladium: (i) water synthesis from hydrogen and oxygen [15], (ii) carbon dioxide synthesis from carbon monoxide and oxygen [16]. In another paper [17], Hiratsuka et al. showed that the synthesis of CO_2 from CO and O_2 was enhanced in a considerable manner by friction between palladium and alumina surfaces. They estimated activation energy of 2.49 kJmol^{-1} against values reported around 100 kJmol^{-1} [18–20], found for the same thermally activated catalytic reaction. The authors guess that the enhanced reaction rate can be due to triboemitted electrons, which should favor the formation of very reactive radicals and ions. Triboemission of electrons is a specific case of the general phenomenon of exoelectron emission [21–23], which is observed when a material surface is perturbed, and whose origin is still unclear. Triboemitted electrons are thought to play a significant role in tribochemical reactions under boundary lubrication conditions. Triboemission of electrons has been observed from both metals and non-metals and there is strong evidence that oxides or other non-metallic surface layers are needed for triboemission of electrons to occur. Important research in the field of electron triboemission was carried out by Nakayama et al. [24] using a Faraday-cup type assembly to detect bursts of electric current (either charged particle emission or surface-charge variation) from scratching diamonds on metals and insulators. Kim et al. [25] investigated photon emission, electron emission intensities, and electron kinetic-energy from reciprocating scratching of MgO with diamond.

The work of Molina et al. [26–28] characterized triboemission of electrons from diamond-on-alumina, diamond-on-sapphire, alumina-on-alumina, and diamond-on-aluminum. The three ceramic-materials consistently showed burst-type negatively charged triboemission during contact at constant load and speed, while the aluminum system produced no significant emission. For the cases of diamond-on-alumina and diamond-on-sapphire, energy spectrometry showed that a large fraction of the triboemitted negative charges were of low-energy (1–5 eV, i.e., 262–1310 kJ mol^{-1}). This finding was of significant importance because in the negative ion–radical action mechanism (NIRAM) approach, it was inferred that the energy level of triboelectrons to initiate tribochemical reaction should be 1–4 eV (262–1048 kJ mol^{-1}) [29]. A book chapter by Kajdas [30] shows that in both tribochemical reactions and some heterogeneous catalytic reactions, the intermediate reactive species are formed by a mechanism mostly governed by the NIRAM approach.

To understand the difference between chemically and mechanically catalyzed hydrolysis some detailed information is given. The ester base-promoted hydrolysis process shows the following leading to the ester substrate: the carboxylic acid in the form of anion [$RCOO^-$] and alcohol (R_1OH):

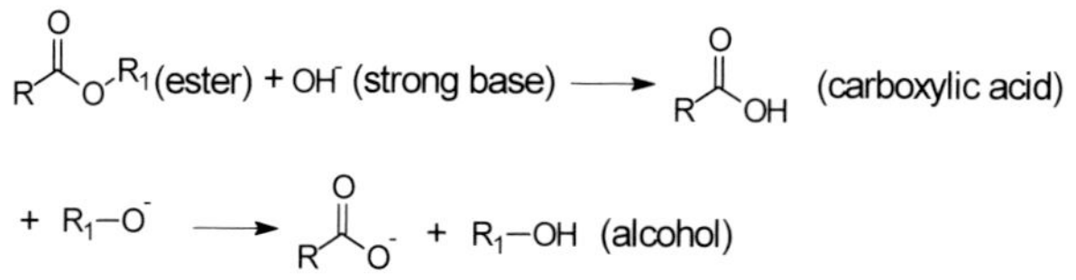

(3.129)

However, the pathway of a tribochemical reaction can be different and exoelectron can act as catalysts themselves. Actually, it is necessary to bear in mind another, and even more important process, namely, the exoelectron interaction with an ester molecule, producing directly two types of negative ions $RCOO^-$ and RO^-.

The ester reactive intermediates produced via the dissociative electron attachment, showing two types of C–O bond cleavage, are presented in Fig. 3.19.

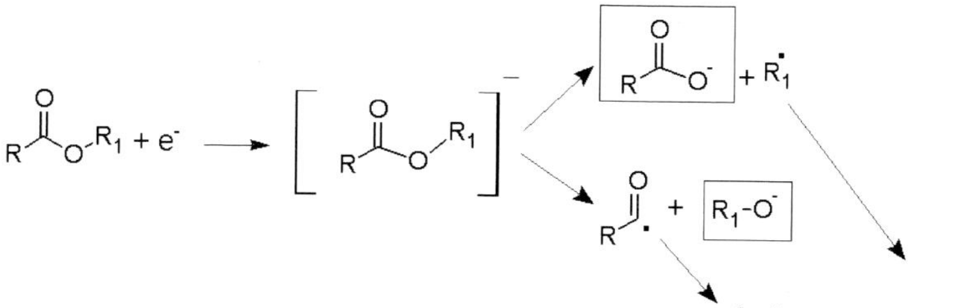

Figure 3.19 Ester reactive intermediates produced via dissociative electron attachment. Reprinted from *Tribol. Int.*, 38, C. K. Kajdas, Importance of the triboemission process for tribochemical reaction, 337–353, Copyright (2005), with permission from Elsevier.

The first bond cleavage (left side in Fig. 3.19) produces carboxylate anion ($RCOO^-$) and the free radical R_1. The second bond cleavage (right side in Fig. 3.19) generates the alkoxide anion (RO^-) and the R–•C=O free radical that undergoes further reactions.

To produce free radicals and thereby initiate the free radical chain reaction process either heat or catalyst is needed. The electron attachment in this case acts as a catalyst. The exoelectron dissociative attachment to an ester molecule yielding two types of negative ions is well evidenced by the electron attachment mass spectrographic results [31]. The carboxylate anion $RCOO^-$ interacts with a positively charged friction solid surface site to produce a salt (soap). Alkoxide ions ($R–CH_2–O^-$) are known to split off two hydrogen atoms [32]. Therefore, it is conceivable to say that the NIRAM approach is responsible for direct triggering tribochemical reactions of esters.

At this point, it is necessary to emphasize that real triboreaction kinetics can be approached by TIP. This fact provides a good base to follow up better understanding of tribochemical heterogeneous catalysis, as well as to plan crucial experiments to determine the actual role of triboemitted electrons in the enhancement of triboreaction rate.

3.3.7 TIP Interpretation of Enhanced Rate of Triboreactions

Kulczycki and Kajdas developed a model to describe the observed enhanced tribochemical reactions' kinetics. This model was described in detail in Chapter 3.2, and can be called the α_i *model*, which hypothesizes that for tribocatalyzed heterogeneous processes the catalyst emits impulses of energy flux of high density, and at the specific angle γ the value of emitted energy is equal to real activation energy E_a (or μ^*). Detailed consideration of the developed model in relation to empirical results of tribological processes allows to propose a new approach aiming at better understanding tribochemical catalytic processes. In this model there are two independent variables: applied load L and chemical structure of lubricant (CSL), i.e., the reagent in the case of the report by Kajdas and Kulczycki [33]. The latter variable was given often qualitatively, but it was necessary to express it as a quantitative variable, if the tribochemical process was intended to be described by mathematical dependence. In this model it was assumed that CSL is expressed by α_i, which is a measure of lubricants ability to boundary layer creation. It was assumed, that value of α_i is a linear function of critical load L, i.e., the load is causing destruction of the generated boundary layer. It should be noticed that this α_i is different from that of equations 3.70–3.75, in which it is used to indicate a series of generic variables (this notation is kept in both cases only because it is strictly related to the most relevant literature on the corresponding subject). The term α_i correlated to CSL assumes forms that drive towards the TIP, and therefore towards a more general outline. In fact, Kajdas et al. [33] proposed that α_i can also be expressed as

$$\alpha_i = \frac{w - w_0}{u - u_0}\left(\frac{\partial u}{\partial w}\right)_{L,T} = \frac{\Delta w}{\Delta u}\left(\frac{\partial u}{\partial w}\right)_{L,T}, \qquad (3.130)$$

where u is the internal energy and w is the work done on the tribological system, which is function of the applied load L. Kajdas et al. rewrite Eq. 3.130 in the following form to be used to fit experimental data

$$\alpha_i = (w - w_0)\left[\frac{(\partial u/\partial w)_{L,T}}{u - u_0}\right] = (w - w_0)\,C. \quad (3.131)$$

The relation of α_i with $\sigma_{c\tau}$ is proposed to be the following:

$$\sigma_{c\tau} = z\frac{d\alpha_i}{dt}, \quad (3.132)$$

where the sub-indexes c and τ stay for *chemical reaction* and *shear stress*, respectively, while z is a proportionality factor. Since a relation there is between the chemical structure of lubricant and entropy; for analogy, we assume that α_i is directly linked to the entropy related to chemical reactions induced by friction. The following equations show the relation between α_i and σ_in more explicitly:

$$\sigma_{c\tau} = zCP_w, \quad (3.133)$$

$$P_w = \left(\frac{dw}{dt}\right)_T, \quad (3.134)$$

where P_w is thus the power dissipated into the system.

If a chemical reaction occurs, according to TIP the rate of internal entropy production can be expressed in the following way:

$$\sigma_\mathrm{in} = J_\tau\frac{\tau}{T} + J_c\frac{A}{T}, \quad (3.135)$$

where σ_in is the rate of internal entropy production, τ is the shear stress due to friction and A is the chemical affinity, J_τ is the flux associated to the shear stress, and J_c is the reaction rate. The nature of J_τ is still not well defined, Kuzharov et al. [8] proposed that it is the ratio between sliding velocity v and absolute temperature T, where the contact area A is introduced as a proportionality factor

$$J_\tau = \frac{Av}{T}. \quad (3.136)$$

It should be noted that we deal only with scalars because sliding velocity and shear stress vectors are parallel but with opposite direction.

Equation 3.135 takes the following form close to equilibrium [11]:

$$\sigma_{\text{in}} = c_{11}\left(\frac{\tau}{T}\right)^2 + c_{12}\frac{\tau}{T}\frac{A}{T} + c_{21}\frac{A}{T}\frac{\tau}{T} + c_{22}\left(\frac{A}{T}\right)^2, \tag{3.137}$$

applying the Onsager's reciprocal relations [23], the following is obtained:

$$\sigma_{\text{in}} = c_{11}\left(\frac{\tau}{T}\right)^2 + 2c_{12}\frac{\tau}{T}\frac{A}{T} + c_{22}\left(\frac{A}{T}\right)^2. \tag{3.138}$$

The interaction term may correspond to that of Eq. 3.133, and represents the contribution given to the reaction rate by the interaction between friction and the chemical reaction

$$\sigma_{c\tau} = 2c_{12}\frac{A}{T}\frac{\tau}{T}, \tag{3.139}$$

c_{12} should be zero if there is no energy flow from triboelectrons to boost the reaction rate. Thus, in first approximation we can write

$$c_{12} = bJ_u, \tag{3.140}$$

$$U_c^* = \frac{J_u}{J_c}. \tag{3.141}$$

In Eq. 3.140, J_u is the energy flow towards the reagents due to triboelectrons, b is a constant, where in Eq. 3.141, U_c^* is the so-called *energy of chemical transfer* [4].

Moreover, taking into account Eq. 3.140, Eq. 3.139 can be written in the following ways:

$$\sigma_{c\tau} = 2\frac{bJ_u}{J_c J_\tau} J_c \frac{A}{T} J_\tau \frac{\tau}{T}, \tag{3.142}$$

$$\sigma_{c\tau} = 2\frac{bJ_u}{J_c J_\tau}\sigma_c \sigma_\tau. \tag{3.143}$$

Consequently comparing Eq. 3.133 with Eq. 3.142, and assuming that the dissipated power P_w is given by

$$P_w = TJ_\tau \tau, \tag{3.144}$$

the *zC* term can be recognized as

$$zC = b\frac{J_u J_c}{J_c J_\tau}\frac{A}{T}\frac{1}{T} = \frac{bU_c^* \sigma_c}{TJ_\tau}, \tag{3.145}$$

which resembles the equation for C reported by Kulczycki and Kajdas [34].

Entropy production takes the following form:

$$\sigma_{in} = \sigma_\tau + 2b\frac{J_u}{J_c J_\tau}\sigma_\tau \sigma_c + \sigma_c = \sigma_\tau + 2b\frac{U_c^*}{J_\tau}\sigma_\tau \sigma_c + \sigma_c. \tag{3.146}$$

Equation 3.146 shows clearly that if U_c^* is null, no interaction there is between friction phenomena and reaction. Rearranging Eq. 3.146 the following expressions can be found:

$$\sigma_{in} = \left(1 + b\frac{J_u \sigma_c}{J_c J_\tau}\right)\sigma_\tau + \left(1 + b\frac{J_u \sigma_\tau}{J_c J_\tau}\right)\sigma_c, \tag{3.147}$$

$$\sigma_{in} = \left(J_\tau + bJ_u\frac{A}{T}\right)\frac{\tau}{T} + \left(J_{c1} + bJ_u\frac{\tau}{T}\right)\frac{A}{T}. \tag{3.148}$$

Thus

$$J_c = J_{c1} + bJ_u\frac{\tau}{T} = L_{22}\frac{A}{T} + L_{21}\frac{\tau}{T} = J_{c1} + J_{c2}, \tag{3.149}$$

$$L_{21} = bJ_u. \tag{3.150}$$

J_{c1} (Eq. 3.149) is the thermal activated part of the reaction rate, which depends on the activation energy μ^*, and the speed-up term of the reaction J_{c2} (Eq. 3.149) depends not only on J_u but also on the rate of entropy production due to friction work. In conditions close to equilibrium, J_{c1} takes the following form in accordance with Eq. 3.149:

$$J_{c1} = J_{cf}\left(1 - e^{-\frac{A}{RT}}\right) = \frac{RT}{h}\exp\left(-\frac{\mu^*}{RT}\right)\prod_i \tilde{c}_i^{\nu_i}\left(1 - \exp\left(-\frac{A}{RT}\right)\right)$$
$$\approx \frac{RT}{h}\exp\left(-\frac{\mu^*}{RT}\right)\prod_i \tilde{c}_i^{\nu_i}\frac{A}{T}, \tag{3.151}$$

$$L_{22} = \frac{RT}{h}\exp\left(-\frac{\mu^*}{RT}\right)\prod_i \tilde{c}_i^{\nu_i}. \tag{3.152}$$

The term J_{c2} also should be an approximation next to equilibrium of a more complex function.

It is possible to express J_{c2} as a Taylor series of τ/T, which can be cut off at the first order, according to the condition to be next to equilibrium

$$J_{c2} \approx J_{c2}(0) + J'_{c2}(0)\frac{\tau}{T}. \tag{3.153}$$

Reminding that $J_{c2}(0)$ is zero at equilibrium, we obtain the following equation to be coherent with the coefficient L_{21} of Eq. 3.150:

$$J_{c2} \approx J_{c2}(0) + J'_{c2}(0)\frac{\tau}{T}. \tag{3.154}$$

$$J_{c2} = k\left[1 - \exp\left(-J_u ct\frac{\tau}{T}\right)\right]\prod_i \tilde{c}_i^{\nu_i}. \tag{3.155}$$

This derivative and $J_{c2}(0) = 0$ are consistent with the following equation for J_{c2}:

$$J_{c2} = k\left[1 - \exp\left(-J_u ct\frac{\tau}{T}\right)\right]\prod_i \tilde{c}_i^{\nu_i}. \tag{3.156}$$

Making the derivative of Eq. 3.64 and calculating it for $\tau \to 0$, b is obtained as

$$b = k\theta t\prod_i \tilde{c}_i^{\nu_i}, \tag{3.157}$$

where k and θ are other constants, and t is time. Therefore, the whole reaction rate can be expressed as

$$J_c = \left[\frac{RT}{h}e^{-\frac{\mu^*}{RT}}\left(1 - e^{-\frac{A}{RT}}\right) + k\left(1 - e^{-J_u\theta t\frac{\tau}{T}}\right)\right]\prod_i \tilde{c}_i^{\nu_i}. \tag{3.158}$$

Equation 3.158 is valid far from equilibrium [11]. It is noteworthy to point out the similitude between Eq. 3.158 and Eq. 3.128, where there are to positive terms with fixed values and other two negative terms that reduce the activation barrier and vanish at equilibrium (Eq. 3.158). This equation is compatible with a reaction mechanism in which the low energy tribolelectrons cause the formation of excited radical ions $A^{\bullet -}$ (or excited radical $R^{\bullet}$ and the excited anion A^{-}) which react with the reagent B to form intermediates (in the intermediate of sketch Eq. 3.159, S is the catalyst, A and B are the reagents, $\delta+$ and $\delta-$ are the partial positive and negative charges, respectively).

Figure 3.20 shows how in the steady state the triboelectrons can stabilize the formation of an intermediate enhancing the reaction rate. In brief, triboelectrons can provide the formation of an intermediate much more stable than that of the thermal activated reaction, although the heterogeneous catalyst (the substrate) is the same.

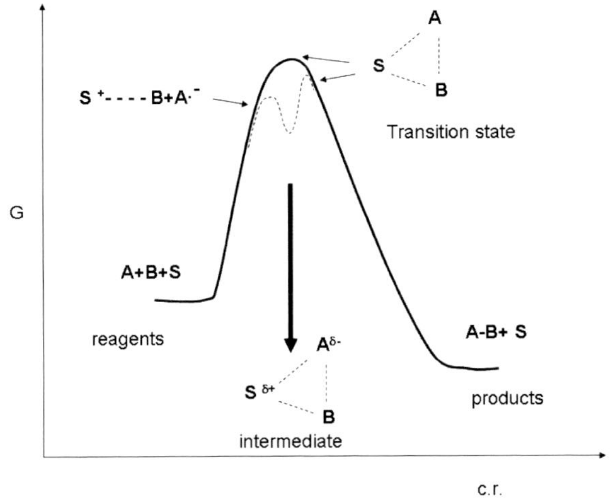

Figure 3.20 Stabilization of the intermediate formed by the reagents A, B, and catalyst S by means of the tribolectrons (dashed curve). $A^{\bullet -}$ is the ion radical formed by the tribolectron and the reagent A. The y axis corresponds to the Gibbs free energy G, the x axis corresponds to the coordinate of reaction $c.r.$ The bold curve corresponds to the thermal activated reaction (the intermediate is not evidenced since the stabilization is much lower than that of the tribocatalytic reaction).

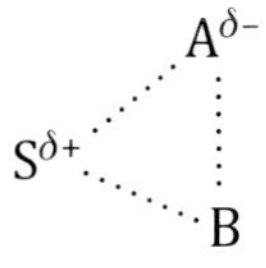

(3.159)

When the system will tend to equilibrium the pathway to the product will approach the bold curve of the Gibbs free energy and the activation barrier will be only the thermal.

Moreover, Eq. 3.145 becomes

$$zC = b\frac{J_u J_c}{J_c J_\tau}\frac{A}{T}\frac{1}{T} = \frac{J_u k\theta t \prod_i \tilde{c}_i^{\nu_i} \sigma_c}{J_c J_\tau T}$$
$$= \frac{J_u t(k'\sigma_c)}{J_c J_\tau T} = \frac{U_c^* t(k''\sigma_c)}{T}. \quad (3.160)$$

This equation for C is equivalent to that obtained by Kajdas and Kulczicki [34].

The reaction rate curve of triboreaction as function of time may appear like that shown in Fig. 3.21, which exhibits an initial peak, and, for comparison, the curve shape of the normal thermal reaction is also shown.

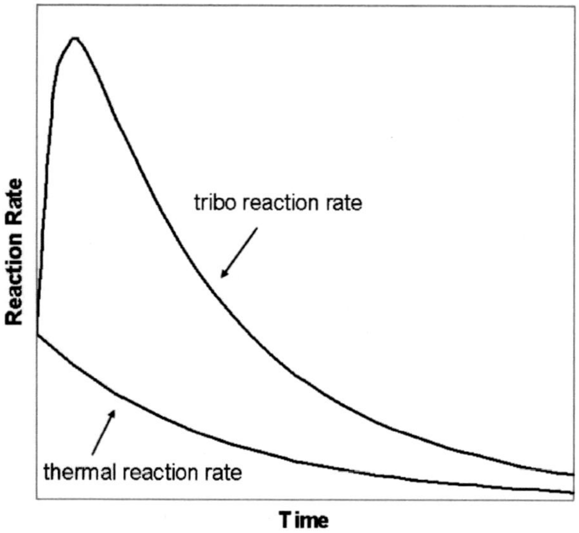

Figure 3.21 Comparison between reaction rate curves of triboreactions and the rate of normal thermal reactions.

In addition, the general mechanism proposed in Fig. 3.20 is somehow comparable with the tribocatalytic CO oxidation mechanism proposed by Hiratsuka et al. [17], which is described in a re-elaborated praiseworthy way in Figs. 3.22a,b, while Fig. 3.22c shows a plausible mechanism step of the thermal activated reaction. Step 2, assumed to be the fastest, is shown in Fig. 3.22b and it is the same for both tribocatalytic and thermal reactions. Step 1 (Fig. 3.22a: tribocatalytic reaction, and Fig. 3.22c: thermal activated reaction) is the slower one, and determines the reaction rate.

It is noteworthy to point out that the kinetics results for CO_2 synthesis, obtained by Hiratsuka et al. [17], shown in Fig. 3.23, have a similar behavior to those proposed in Fig. 3.21, both exhibiting the initial overshoot of production rate in support of the herein described TIP-α_i derived model. The overshoot, correlated to exposure time to gases, seems also to be correlated to the reaction step where triboelectrons intervene on chemisorbed reagents, as proposed in Fig. 3.22a.

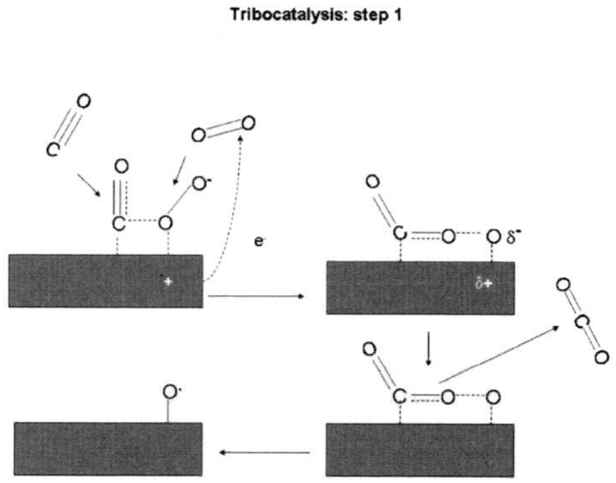

Figure 3.22 (a) Step 1 of the tribocatalytic CO oxidation.

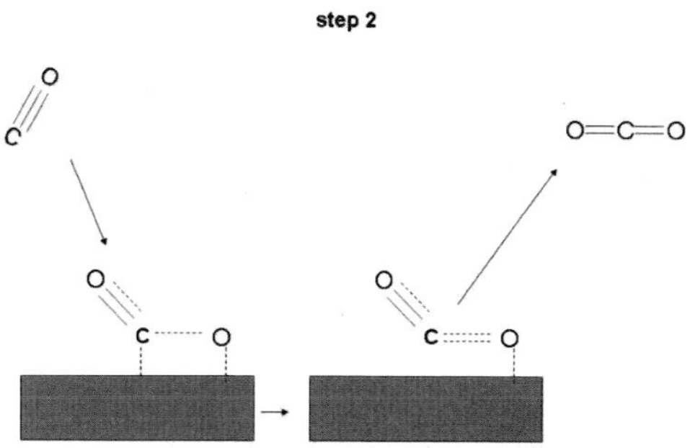

Figure 3.22 (b) Step 2 of both tribocatalytic and thermal activated CO oxidation.

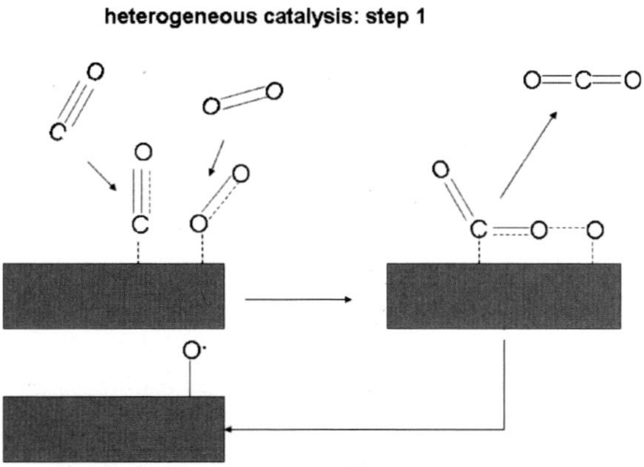

Figure 3.22 (c) Step 1 of the thermal activated CO oxidation.

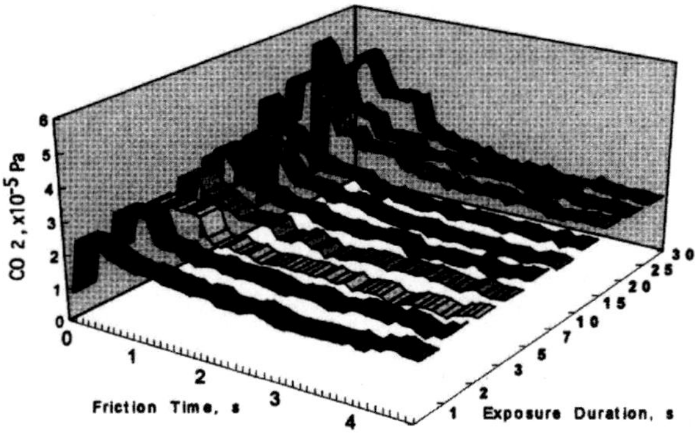

Figure 3.23 Effect of exposure duration before friction on CO_2 pressure (CO and O_2: 5.0×10^{-5} Pa), (sliding velocity = 93 mm/s). Reprinted from K. Hiratsuka, C. Kajdas, M. Yoshida, Tribocatalysis in the synthesis reaction of carbon dioxide, *Tribol. Trans.*, 47 (2004) 86–93, by permission of Taylor & Francis Ltd.

3.3.8 TIP Approach to Tribocatalysis and NIRAM Theory

NIRAM approach explains the role of exoelectrons in some relevant tribochemical reactions, in particular in tribopolymerization of addition type monomers. NIRAM comprises the following major stages as exposed by Kajdas [9, 10]:

(i) Low-energy electron emission and creation of positively charged spots, generally on the top of asperities.

(ii) Action of the emitted electrons with the lubricant molecules L causing the formation of negative ions and radicals on the rubbing surfaces.

(iii) Reaction of negative ions with metal surfaces, and other reactions such as free radical reactions, forming an organometallic (or inorganic) film, which protects the rubbing surfaces from wear.

(iv) If the shear strength is high, chemical bonds of organometallic compounds are cleaved resulting in producing inorganic films and further radicals.

(v) Eventual destruction of protective layer caused by wear, followed by electron emission and subsequent formation of a new protective film according to stages (i)–(iv).

The initiation reaction process is due to triboemitted low-energy electrons combined with flash temperature. The latter might also be considered in electronic form (thermionic emission). NIRAM stages are compatible with the mechanism proposed in Section 3.3.7 and can be considered a typical case of the TIP-α_i model.

The described NIRAM stages can be interpreted through the stabilization due to the formation of an intermediate as for TIP model. Figure 3.24 shows the activation energy lowering due the reaction intermediates' formation between a lubricant molecule L and a surface site S, so that the NIRAM mechanism can be assimilated to that depicted in Fig. 3.20.

In spite of the difficulty to detect the ion intermediate formation between sliding surfaces, a direct evidence that such a mechanism can involve excited ions-radicals came from an experiment where lubricant molecules of A20H are bound to the DLC surface of a head disk by UV-185 nm (corresponding to photons of 6.5 eV) irradiation for 10–20 s. The work function of carbon is 5 eV. This lubricant molecule, which is a Fomblin Z type perfluoropolyether with the hydroxyl end-group, $-O-CF_2-CH_2-OH$, at one end, and the cyclo-triphosphazene end-group, $R_5(PN)_3-O-$, at the other end (where R is an m-trifluoromethyl-phenoxy group) interacts with surface graphitic regions. Low energy photoelectrons favor the formation of large ionic fragments, which were detected by time-of-flight secondary ion mass spectroscopy (TOF-SIMS). These large ions react anchoring to the surface carbon coat [34].

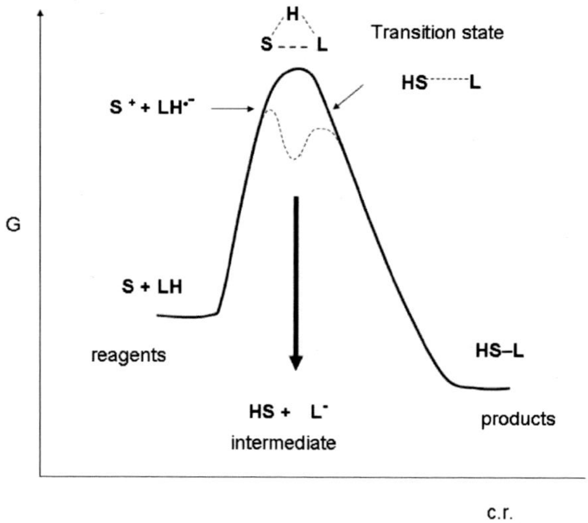

Figure 3.24 Stabilization of the intermediate by means of tribolectrons (dashed curve) according to the joint NIRAM and TIP interpretation. *S* is the surface site, and *L* is the lubricant molecule, $LH^{\bullet -}$ is the radical ion formed, and L^{-} is the ion intermediate. In the *y* axis corresponds to the Gibbs free energy *G*, *x* axis corresponds to the coordinate of reaction *c.r.* The bold curve corresponds to the thermal activated reaction.

This mechanism can work when there is a source of low energy exoelectrons such as triboemitted electrons. Experimental results of head-disk friction show that fluorocarbon fragments are generated from Z-Dol and X-IP lubricants during sliding [35].

Tribochemical transformations characterize the behavior of many lubricating systems of practical interest. For example, lubrication of silicon nitride by means of alcohols can be understood in terms of alkoxylation of silicon, through the radicals and ions activated by friction on the silicon nitride surface [36]. TIP can provide a general framework to the observed enhanced rate in comparison to thermal reactions. The formation of ammonia by grinding of silicon nitride in air [37], as well as the irreversible water absorption, indicates that certain energy barriers are easily overcome by the formation of different intermediates. Nevertheless, until now the lack of thorough energetic and kinetic results makes difficult the quantitative TIP analysis in this field.

References

1. I. Prigogine, *Introduction to Thermodynamics of Irreversible Processes*, Wiley-Interscience, New York (1967).
2. P. Glansdorff, I. Progogine, *Thermodynamic Theory of Structure, Stability and Fluctuations*, Wiley-Interscience, Chichester (1971).
3. I. Prigogine, Time, structure, and fluctuations, *Science*, 201/4358 (1978) 777–785.
4. S. R. De Groot, *Thermodynamics of Irreversible Processes*, North-Holland Publishing Company, Amsterdam (1952).
5. I. R. Epstein, K. Showalter, Nonlinear chemical dynamics: Oscillations, patterns, and chaos, *J. Phys. Chem.*, 100 (1996) 13132–13147.
6. L. Onsager, Reciprocal relations in irreversible processes. I., *Phys. Rev.*, 37 (1931) 405–426.
7. V. Parmon, *Thermodynamics of Non-Equilibrium Processes for Chemists with Particular Application to Catalysis*, Elsevier, Oxford UK (2010).
8. A. S. Kuzharov, S. B. Bulgarevich, V. E. Burlakova, A. A. Kuzharov, E. E. Akimova, Molecular mechanisms of self-organization at friction. Part VI. Analysis of thermodynamic features of tribochemical reactions, *J. Friction Wear*, 28/2 (2007) 218–223.
9. C. Kajdas, K. Hiratsuka, Tribochemistry, tribocatalysis, and the negative-ion-radical action mechanism, *Proc. IMechE, Part J J. Eng. Tribol.*, 223 (2009) 827–848.
10. C. K. Kajdas, Importance of the triboemission process for tribochemical reaction, *Tribol. Int.*, 38 (2005) 337–353.
11. R. C. Dante, C. Kajdas, A. Kulczycki, Theoretical advances in the kinetics of tribochemical reactions, *Reac. Kinet. Mech. Cat.*, 99 (2010) 37–46.
12. L. Tisza, *Generalized Thermodynamics*, The MIT Press, Cambridge MA (1978).
13. H. B. Callen, *Thermodynamics and Introduction to Thermostatics*, 2 ed. John Wiley & Sons, New York (1985).
14. V. N. Parmon, Catalysis and non-equilibrium thermodynamics: Modern in situ studies and new theoretical approach, *Catal. Today*, 51 (1999) 435–456.
15. K. Hiratsuka, M. Kuzuya, T. Sasada, Friction Catalysis in the synthesis reaction of H_2O during adhesive wear, *Proc. 33rd Japan Congress Mater. Res.*, 33 (1990) 191–195.

16. K. Hiratsuka, M. Kuzuya, T. Sasada, Friction catalysis in the synthesis reaction of CO_2 during adhesive wear, *Proc. 34th Japan Congress Mater. Res.*, 34 (1991) 119–123.
17. K. Hiratsuka, C. Kajdas, M. Yoshida, Tribocatalysis in the synthesis reaction of carbon dioxide, *Tribol. Trans.*, 47 (2004) 86–93.
18. G. W. Graham, A. D. Logan, M. Shelef, Oxidation of carbon monoxide by oxygen, nitric oxide and mixtures of O_2 and NO over palladium (100), *J. Phys. Chem. Lett.*, 97/21 (1993) 5445–5446.
19. A. W. Adamson, A. P. Gast, *Physical Chemistry of Surfaces*, 6th ed. John Wiley, New York (1997).
20. R. P. H. Gasser, *An Introduction to Chemisorption and Catalysis by Metals*, Clarendon Press Oxford (1985).
21. T. M. Roberts, D. B. Hibbert, Field stimulated exoelectron emission from borosilicate glass, *Nature*, 297 (06 May 1982) 42–43.
22. H. Shimada, K. Nakajima, Thermally stimulated exo-electron emission in copper, zinc and their alloys, *Surf. Sci.*, 88 (1979) 751–759.
23. S. A. Hoenig, M. G. Utter, Observations of exoelectron emission associated with heterogeneous catalysis, *J. Catal.*, 47 (1977) 210–213.
24. K. Nakayama, N. Susuki, H. Hashimoto, Triboemission of charged particles and photons from solid surfaces during frictional damage, *J. Phys. D Appl. Phys.*, 25 (1992) 303–308.
25. M. Kim, S. C. Langford, J. T. Dickinson, Electron and photon emission accompanying the abrasion of MgO with diamond, *Tribol. Lett.*, 1 (1995) 147–157.
26. G. J. Molina, M. J. Furey, A. L. Ritter, C. Kajdas, Triboemission from alumina, single crystal sapphire, and aluminium, *Wear*, 249 (2001) 214–219.
27. G. J. Molina, M. J. Furey, B. Vick, A. L. Ritter and C. Kajdas, Triboemission from the sliding contact of alumina systems, *Proceedings of the 2nd World Tribology Congress*, Vienna, Austria, CD ROM (2001).
28. G. J. Molina, Triboemission from ceramics: Charge intensity and energy distribution characterizations. Ph.D. Dissertation, Department of Mechanical Engineering, Virginia Polytechnic Institute and State University, Blacksburg VA, USA (2000).
29. C. Kajdas, On a negative-ion concept of EP action of organo-sulfur compounds, *Tribol. Trans.*, 28 (1985) 21–30.
30. C. Kajdas, Hydrolysis, in: *Surface Modification and Mechanisms* (G. E. Totten, H. Liang, eds.), Marcel Dekker Inc., New York, Basel, 2004, 203–241 (2004).

31. M. von Ardenne, K. Steinfelder and, R. Tuemmler, *Elektronenanlagerungs-Massenspektrographie Organischer Substanzen*, Springer Verlag, Berlin (1971).
32. C. E. Melton, P. S. Rudolf, Negative ion mass spectra of hydrocarbons and alcohols, *J. Chem. Phys.*, 31 (1959) 1485–1488.
33. C. K. Kajdas, A. Kulczycki, A new idea of the influence of solid materials on kinetics of chemical reactions, *Mater. Sci. Poland*, 26/3 (2008), 787–796.
34. P. H. Kasai, A. Wakabayashi, Disk lubricant additives, A20H and C2: Characteristics and chemistry in the disk environment, *Tribol. Lett.*, 31 (2008) 25–35.
35. X. Zhao, B. Bhushan, Lubrication studies of head-disk interfaces in a controlled environment Part 1: Effects of disk texture, lubricant thermal treatment and lubricant additive on the durability of the head-disk interface, *Proc. Inst. Mech. Eng. Part J.*, 214 (2000) 535–546.
36. R. S. Gates, S. M. Hsu, Boundary lubrication of silicon nitride, *Natl. Inst. Stand. Technol. Spec. Pub. 876*, Washington D.C. (1995).
37. M. Volante, B. Fubini, E. Giamello, V. Bolis, Reactivity induced by grinding in silicon nitride, *J. Mater. Sci. Lett.*, 8 (1989) 1076–1078.

Chapter 4

Tribocorrosion

Czesław Kajdas,[a] Ken'ichi Hiratsuka,[b] Feng Gao,[c] Sukbae Joo,[c] and Hong Liang[c]

[a]*Institute of Chemistry in Płock, Warsaw University of Technology, Poland*
[b]*Department of Innovative Mechanical and Electronic Engineering, Chiba Institute of Technology, Japan*
[c]*Department of Mechanical Engineering, Texas A&M University, USA*

4.1 Tribocorrosion

Czesław Kajdas[a] and Ken'ichi Hiratsuka[b]

[a]*Automotive Industry Institute in Warsaw, Poland*
[b]*Department of Innovative Mechanical and Electronic Engineering, Chiba Institute of Technology, Japan*

khiratsuka@it-chiba.ac.jp

4.1.1 Introduction

4.1.1.1 General overview and the present state of the art

Considering Landolt's work [1], tribocorrosion might be defined as a complex mechanochemical process proceeding under friction in the presence of any corrosive environment. Since the mechanochemical mechanisms of tribocorrosion are still incompletely understood, the primary goal of this chapter is focused on better understanding mechanochemistry (mechanochemical mechanisms) along with tribochemistry (chemomechanical mechanisms).

Tribocatalysis, Tribochemistry, and Tribocorrosion
Edited by Czesław Kajdas and Ken'ichi Hiratsuka

ISBN 978-981-4316-95-8 (Hardcover), 978-1-315-36498-8 (eBook)
www.panstanford.com

In Chapter 1, we defined the mechanochemistry process as a coupling of mechanical and chemical phenomena on a molecular scale leading to initiation of heterogeneous chemical reactions due to specific mechanical bond breakage of the molecules. The activation energies of such reactions are lower than those of regular thermochemical processes. Consequently, these reaction types should also be considered as catalyzed ones. In other words, the mechanical interaction of a solid–solid–chemical environment system generates catalyzed processes, and the catalyst is defined as a powerful substance/agent that enables a chemical reaction to be initiated by lower activation energy and/or proceed at a faster rate than thermochemical reactions. It is of note that the tribocorrosion process also involves chemoelectrochemical interactions between surfaces in relative motion in the presence of any corrosive atmosphere.

Detailed description of tribochemistry is given in Section 2.1. By and large, we consider tribochemistry as a subfield of mechanochemistry. The same is due to mechanocatalysis, mechanosynthesis and tribocatalysis. A common denominator for all these specific, mechano-activated processes is the fact that they are governed by lower activation energies than those of regular thermochemical or catalytic processes. Thus, mechanochemical reactions are clearly distinct from those of thermochemical ones. Processes initiated by mechanical action are also widely known as tribochemical reactions. The book [2] provides a good evidence for that.

Research on very specific mechanochemical reactions (mechanochemistry) is growing rapidly and, the number of published papers/books is regularly increasing. However, it was also mentioned that the universally agreed definition of the mechanochemistry term is not yet widely and explicitly used. Works [3–4] demonstrate detailed discussion of the use of the term mechanochemistry. Work [4] clearly presents distinction between the mechanical breaking of covalent bonds, where mechanical energy is used for bond breaking as the first step in chemical transformations.

4.1.1.2 Practical importance of tribocorrosion processes

Tribocorrosion processes and/or tribocorrosion phenomena are considered in technological areas where they cause damage to installations, machines, and devices. Sometimes tribocorrosion damages are problems for both safety and human health. In specific applications focused on manufacturing, tribocorrosion phenomena are put to good use. Electrochemical methods are widely used for the study of tribocorrosion reactions. To get information about synergistic and antagonistic mechanisms they must be applied in situ under strictly controlled mechanical conditions, using materials with well-characterized surface properties [1]. Recent progress in modeling and understanding of tribocorrosion systems is discussed and some challenges and opportunities for future research are identified [4].

Tribocorrosion processes are very widely spread, including among others: automotive, aeronautic, and naval industries, mining plus metallurgy, chemical and crude oil industries, medical prostheses and micronanotechnologies. Works [5–6] present more detailed information/data on tribocorrosion. Paper [6] defines tribocorrosion as the electrochemical and mechanical process leading to wear of metallic materials immersed in a corrosive environment under sliding or rolling contact conditions. Due to synergy effects between mechanical and electrochemical phenomena, the material loss in tribocorrosion usually is very higher. Early paper by Landolt et al. [7] reviewed and discussed the synergy effects. The friction process significantly affects the corrosion resistance of metallic surfaces. On the other hand, the corrosion process influences the composition of material surfaces by enhancement of adsorbed layers and corrosion oxides, and thereby changes their mechanical properties. Significance of tribocorrosion in biomedical applications has recently been overviewed by Mathew et al. [8].

4.1.1.3 Scientific approach to tribocorrosion processes

Another significant work considers tribocorrosion fundamentals along with materials and their applications. It describes the science of surface transformations resulting from the interaction of mechanical loading and chemical reactions that occur between elements of a tribosystem exposed to corrosive environments [9]. Thus, it combines the mechanical and chemical interactions of the contacting bodies/materials in the given tribosystem, taking into account activated chemical and electrochemical reactions. However, the tribocorrosion mechanisms are still not adequately known.

In order to benefit from tribocorrosion phenomena it is crucial to control them as well as the nature of rubbed or worn surfaces, which have undergone tribological stresses. Fischer and Mischler propose that reasons for this stem not only from the complex interactions of chemical and mechanical factors but also from the experimental difficulty of characterizing mechanically and/or thermally induced surface phenomena occurring inside small contacts [9]. For example, due to the fact that the duration of a single asperity contact is short and that it takes place directly at the interface of contacting bodies, even modern surface analytical techniques may fail. Therefore, in situ techniques should be more suitable for tribocorrosion investigations.

Since the tribocorrosion basic research had been performed for more than 40 years, the special cluster issue within the *Journal of Physics D: Applied Physics* considers state-of-the-art information on tribocorrosion [10].

4.1.1.4 Significance of mechanical and chemical effects

Generally, tribocorrosion causes degradation of the materials and negatively influences the performance of a contact; however, in some applications tribocorrosion can be useful and it relates just to CMP. For instance, CMP helps for multilevel metallization in advanced integrated circuit device technology, where simultaneous chemical and mechanical material removal is used for surface planarization and smoothing. Materials degradation due to simultaneous chemical and mechanical effects may occur under a variety of conditions as schematically depicted in Fig. 4.1.

Either two-body or three-body contacts between sliding surfaces contribute to the tribocorrosion major cause. The relative motion of the surfaces can be unidirectional such as for example in a pin-on-disk wear test apparatus or any fretting apparatus. Tribocorrosion is also observed in regular ball bearings. Particle impact can result in a combined mechanical and chemical attack of the material; a well-known example is erosion corrosion of pumps and pipes carrying slurries [7]. Considering physical point of view, tribocorrosion includes a wide range of mechanical and chemical degradation processes such as wear accelerated corrosion, erosion corrosion, corrosive wear, oxidative wear, etc. On the other hand, the term tribocorrosion is mostly used in much more restricted sense, referring to sliding contacts.

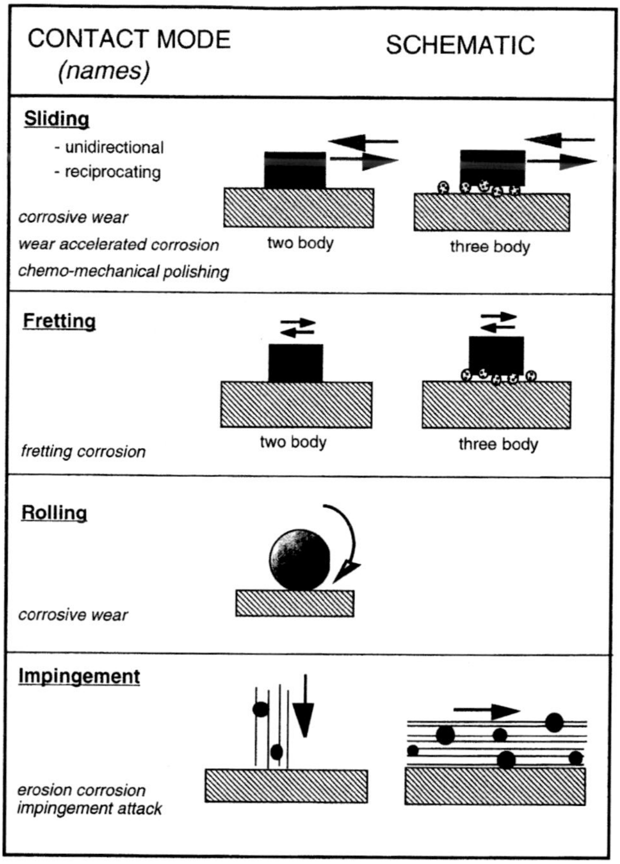

Figure 4.1 Schematic of different simultaneous tribological contact modes of mechanical and chemical effects. Reprinted from *Electrochim. Acta*, 46, D. Landolt, S. Mischler, M. Stemp, Electrochemical methods in tribocorrosion: A critical appraisal, 3913–3929, Copyright (2001), with permission from Elsevier.

4.1.2 Tribocorrosion vs. Tribochemistry

4.1.2.1 Importance of electrochemistry in tribocorrosion processes

Electrochemical aspects of tribocorrosion are widely presented and discussed by Landolt [1] with the focus on potential measuring techniques. By changing the applied potential, corrosion conditions can be adequately controlled. Further, Landolt in his paper discusses different electrochemical techniques that can be applied in tribocorrosion studies as shown in Fig. 4.2.

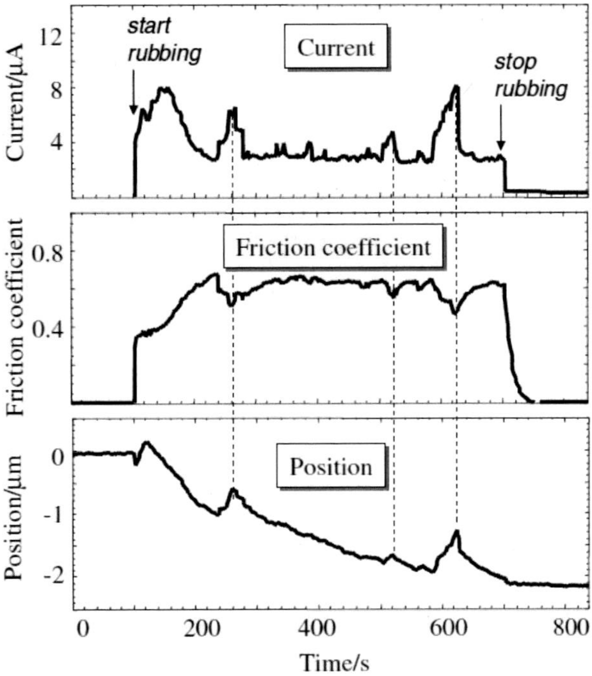

Figure 4.2 Correlation between measured anodic current, coefficient of friction and vertical pin position observed during rubbing of a titanium alloy (Ti_6Al_4V) against alumina in 0.15 M NaCl using an alternating motion electrochemical tribometer. Applied anodic potential: 0.5 V versus Ag/AgCl. Load 10 N, oscillation frequency 1 Hz. Average values over each stroke are shown. Reprinted from *J. Phys. D Appl. Phys.*, 39, D. Landolt, Electrochemical and materials aspects of tribocorrosion systems, 3121–3127, Copyright (2006), with permission from Elsevier.

They include (i) the application of a constant potential by means of a potentiostat and monitoring the variation of current with time, (ii) wear particles form discontinuously from the titanium alloy, as evidenced by the maxima and minima in the curves representing the advance of the pin and the friction coefficient, respectively; (iii) detachment of a particle locally exposes the bare alloy surface to the electrolyte and its re-passivation leads to an anodic current peak; (iv) after some time, the wear particle is ejected from the contact or crushed, and as a consequence the pin position, the coefficient of friction and the current return to their previous values; (v) application of a small amplitude ac signal at open circuit or at constant imposed potential and measurement of the electrochemical impedance, as the function of frequency (EIS, electrochemical impedance spectroscopy).

Azzi and Szpunar [11] define tribocorrosion as the term describing the synergy between tribological and electrochemical processes focused on biomaterials. To study the tribocorrosion behavior of biomaterials, electrochemical setup with three electrodes was used for controlling the potential of the surface of a conducting material subjected to classical wear testing. It was shown that applying this equipment, it is possible to carry out friction and wear tests in electrolytic solution under well-defined electrochemical conditions determined by the applied electrode potential. Tests of deposited TiN on pure Ti for corrosion and tribocorrosion behavior under simulated body fluid were conducted. Azzi and Szpunar found that the presence of TiN layer on the surface of Ti has increased the open circuit potential. Interestingly, the charge transfer resistance (Rct) determined using EIS was higher for the nitrided surfaces than for the Ti substrates; after the wear test, Rct was reduced as the protective layer was damaged [11].

Kok et al. [12] used the scanning reference electrode technique (SRET) to monitor the galvanic currents in the electrolyte during rubbing under open circuit conditions. A conventional SRET has successfully been modified to provide a novel method for characterizing the real-time localized tribocorrosion behavior of uncoated and coated samples. The influence of various experiments involving both (i) contact load and (ii) and test environments was presented, and their effect on the following parameters was found [12]:

- Local anodic current activity
- The number of active anodic areas
- Sample recovery (re-passivation) time
- Free corrosion potential
- Amount of wear

- Wear coefficient
- Wear morphology

In dry air, PVDSS sample (PVD coated stainless steel) exhibited 2–3 orders of magnitude improved wear resistance compared to the uncoated stainless steel.

According to Landolt et al. [7] the term tribocorrosion designates changes and/or processes of irreversible transformation of a material in the tribological contact. Such changes result from simultaneous physicochemical and mechanical surface interactions. tribocorrosion has been a point of basic research for almost 50 years and seems to be as such in the near future [1].

Simultaneously acting the mechanical and physicochemical aspects are strongly interrelated; synergistic or antagonistic effects are observed; these effects may be deleterious or beneficial for the performance of the tribological contact. Thus, the study of tribocorrosion should be based on a systems approach taking into account the combined effects of all significant parameters. Since the wear-corrosion synergy is very complex, this problem resembles the tribochemistry complexity. Such problems appear when one tries to understand a detailed reaction mechanism by which the tribocorrosive and tribochemical films are generated.

Work by Borodich et al. [19] is focused on physical and tribochemical effects in sliding contact between the DLC coated slider and the counterpart. Paper [20] considers wear and corrosion as two-material degradation phenomena that involve a number of mechanical and electrochemical mechanisms. Another work [21] deals with the activation process by mechanical action and the tribochemical process.

4.1.2.2 Knowns and unknowns of tribocorrosion and tribochemistry

Tribocorrosion behavior depends on properties of the contacting materials, mechanics of the tribological contact and the environmental conditions. The tribocorrosion mechanisms are still not well understood and the same is due to tribochemistry. This section mostly aims at looking for similarities and differences concerning both tribocorrosion and tribochemistry mechanisms from the viewpoint of mechanochemistry.

Tribochemistry defined as the heterogeneous chemical reaction process initiated either by friction of solids and/or by any other mechanical energy input, well corresponds also to tribocorrosion. By and large chemical reactions initiated by friction processes relate to tribochemistry that can be defined as "a branch of chemistry dealing with the chemical changes of solids due to the influence of mechanical activation" [22]. The same reference book defines mechanochemistry as "a branch of chemistry dealing with the chemical and physicochemical changes of substances of all aggregation states due to the influence of mechanical energy."

Todres, in his book on organic mechanochemistry and its practical applications [2], emphasizes that mechanically induced generation of organic free radicals is one of the most illustrative result of the mechanical action and, in Chapter 4 of the book, provides principal examples of mechanically induced organic reactions leading to tribochemistry. However, a better understanding of reaction mechanisms is needed. At this point one might ask the following question: "How far tribocatalytic reaction mechanisms are contributing to mechanisms of tribocorrosion processes"? Catalysts are known to cause reactions to proceed faster than thermochemical ones, but how do catalysts work?

There is no simple answer to that question. A general approach can be presented as follows: "A catalyst provides a new and easier pathway for reactant molecules to be converted into product molecules." At this moment, another question should be asked: "Might we consider 'a catalyst' as a part of the corrosive environment"?

Most surface catalytic reactions and the production surface intermediates involve charge transfer, either an electron transfer or a proton transfer; these processes are often viewed as modified acid base reactions [23]. The electron transfer capability of a catalyst is expressed according to the Lewis definition. Considering now the NIRAM [24] approach and the NIRAM-HSAB (Hard and Soft Acids and Bases) concept [25], it is possible to look for similarities and differences between tribochemistry and catalysis. Taking into account these similar processes it seems to be possible to compare tribochemical reactions and some heterogeneous catalytic reactions, because we assume that the intermediate reactive species for both processes are formed by the same mechanism governed by the NIRAM approach.

Test results obtained under sliding (ball-on-disk tester) and reciprocating sliding (HFRR test rig) can be considered tribocorrosive. For example, palmitic acid (C_{16}) solutions in hexadecane (cetane) or with other fatty acids can be considered as tribocorrosive. Figure 4.3 illustrates this phenomenon and shows that in the range of 50–500 ppm palmitic acid concentration the ball wear slightly increases, but at the 750 ppm concentration there is very significant wear increase! It is suggested that this wear enhancement relates to a tribocorrosion process. The ball relative wear for pure cetane is much lower than for the 750 ppm concentration.

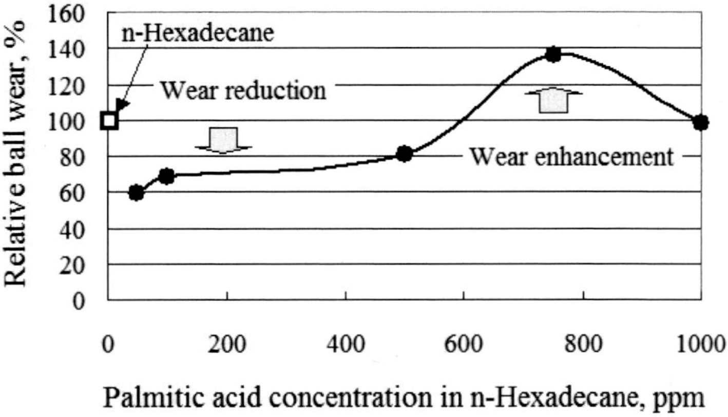

Figure 4.3 Tribological wear in steel–steel sliding in ball-on-disk machine system, lubricated with 50–1000 ppm solutions of palmitic acid in n-hexadecane; to draw the figure, experimental data from papers [26, 27] were used.

Corrosive/abrasive wear combined with oxidation/corrosion between moving parts are discussed in detail in [28]. Corrosion is here defined in terms of chemical attack on a metallic material by reaction with the environment, and it is emphasized that corrosion reactions strongly interact with mechanically induced local deformation of material. Thus, it strictly concerns tribocorrosion. The corrosion process comprises an anodic and cathodic reaction as demonstrated in Fig. 4.4.

The anodic reaction is a metal (M) dissolution by oxidation to the metal ions (M^{2+}). The right hand part in Fig. 4.4 is the reduction of oxygen and H_2O by the cathodic reaction. By and large, the electrons released in the anode are conducted through the metal to the cathode where they are consumed. The system is closed by ion conduction via the electrolyte; metal ions are conducted to meet OH^- ions, producing the metal hydroxide. Since wet corrosion is characterized by an electric current across the metal-electrolyte interface at the electrodes, the corrosion rate can be considered related to current. As the result, the local corrosion current density on the anode is the anodic current divided by the anodic area [28]. The potential of electrodes can be only measured relative to the potential of the reference standard hydrogen electrode (SHE), with reaction $2H^+ + 2e^- \rightarrow H_2$. Equilibrium potentials of different reactions form the basis for determination of possible directions of the electrode reactions of selected electrode combinations. Actually, they show if the dissolution of a metal is possible or not. Detailed SHE potentials are available in adequate references, e.g., [29].

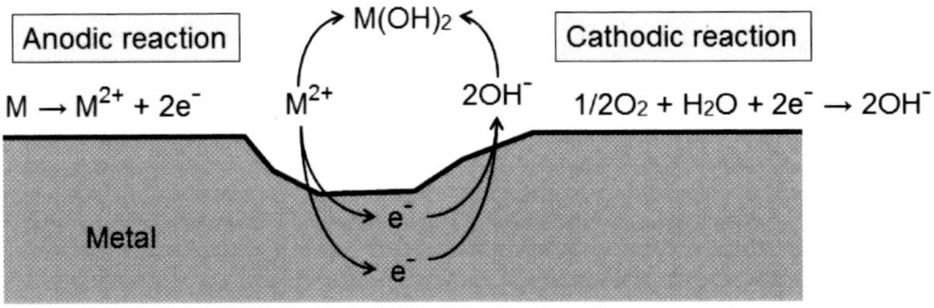

Figure 4.4 Aqueous corrosion of a divalent metal M in an electrolyte containing oxygen. Copyright (2004). From Corrosion and its Impact on wear processes, in: *Surface Modification and Mechanisms* (G. E. Totten, H. Liang, eds.; Marcel Dekker, New York, Basel) by E. Bardal, A. Bardal. Reproduced by permission of Taylor and Francis Group, LLC, a division of Informa plc.

To understand better the tribocorrosion process it is necessary to combine the regular corrosion process with a mechanical action or just friction in this case. Figure 4.5 depicts the apparatus used for electrochemical and CMP research tests. Reminding now that CMP tests should also relate to tribocorrosion processes.

This novel system consists of an 8 in. platen and a 3 in. carrier; the platen is bonded with slurry holder to avoid splash of slurry during the polishing process; the platen rotating speed is fixed at 100 rpm. The down force between the specimen carrier and the platen may be varied from 3 to 9 psi. A Rodel Politex polish pad was affixed to the platen for polishing [30]. The complexity of this apparatus recalls very much the complexity of tribocorrosion system combining mechanical/electrochemical action with characteristics of the pin and plate materials and the anodic metal solubilization reaction, as depicted in Fig. 4.4.

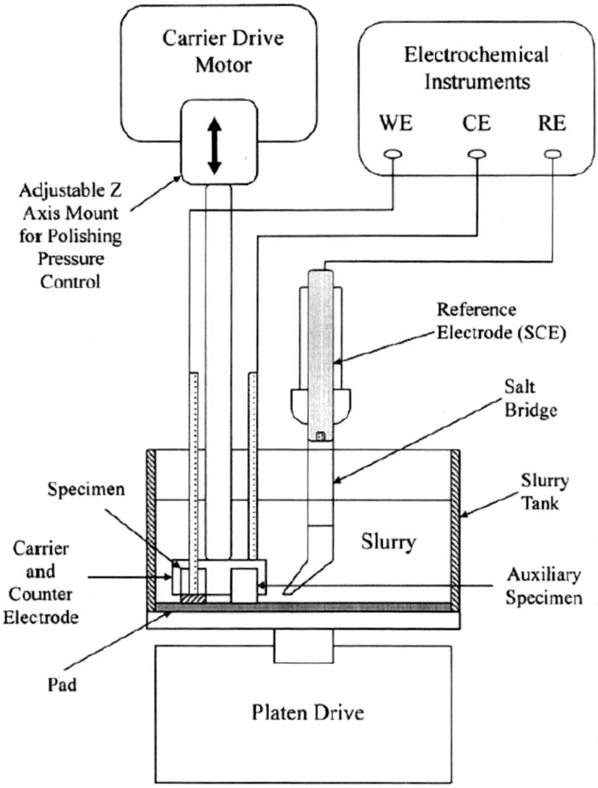

Figure 4.5 Diagram presenting the apparatus used for electrochemical and CMP tests. From H.-S. Kuo, W.-T. Tsai, Effect of applied potential on the chemical mechanical polishing of aluminum in phosphoric acid base slurry, *J. Electrochem. Soc.*, 147(6) (2000) 2136–2142. Reproduced by permission of The Electrochemical Society.

Figure 4.6 demonstrates the behavior of a sliding contact (tribocorrosion system) under electrochemical control.

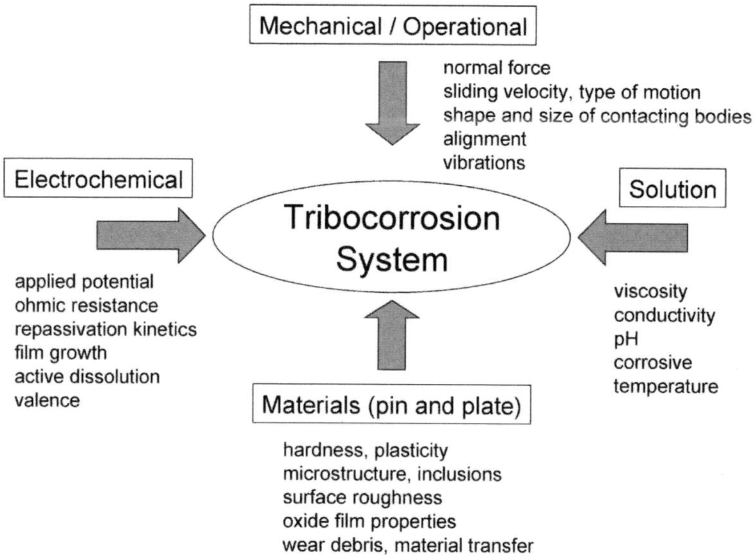

Figure 4.6 Schematic showing four types of parameters that affect the tribocorrosion behavior of a sliding contact (tribocorrosion system) under electrochemical control. Reprinted from *Electrochim. Acta*, 46, D. Landolt, S. Mischler, M. Stemp, Electrochemical methods in tribocorrosion: A critical appraisal, 3913–3929, Copyright (2001), with permission from Elsevier.

Work [30] presents/describes another apparatus for the investigation of fretting–corrosion (tribocorrosion) behavior of metallic materials used in prosthetic implants. In the apparatus well-defined micromovements are accomplished via a rigid overall structure and the precise alignment of the components.

The mobile beam supporting the elastic bearing is visible in the center of the structure with the piezo-actuator fixed on its left side: (1) main structure, (2) elastic bearing, (3) mobile beam, (4) screw to move the mobile beam up or down, (5) piezo-actuator, (6) force sensor (hatched part), (7) specimens.

The surface chemistry of the metal in contact is controlled by special electrochemical cell of the anodic oxidation. Performing experiments, the most relevant mechanical and electrochemical parameters are continuously monitored using a computer-based data acquisition system [30]; therefore, the apparatus allows perform in vitro experiments under well-controlled mechanical and electrochemical conditions. Preliminary results demonstrated that the prevailing corrosion conditions critically affect the overall fretting behavior of Ti_6Al_4V.

Figure 4.7 illustrates the electrochemical cell for triboelectrochemical measurements. The cell is fixed on the sample holder, which is in turn fixed on the elastic bearing. An alumina disk is used to electrically insulate the lower sample holder from the main frame [30].

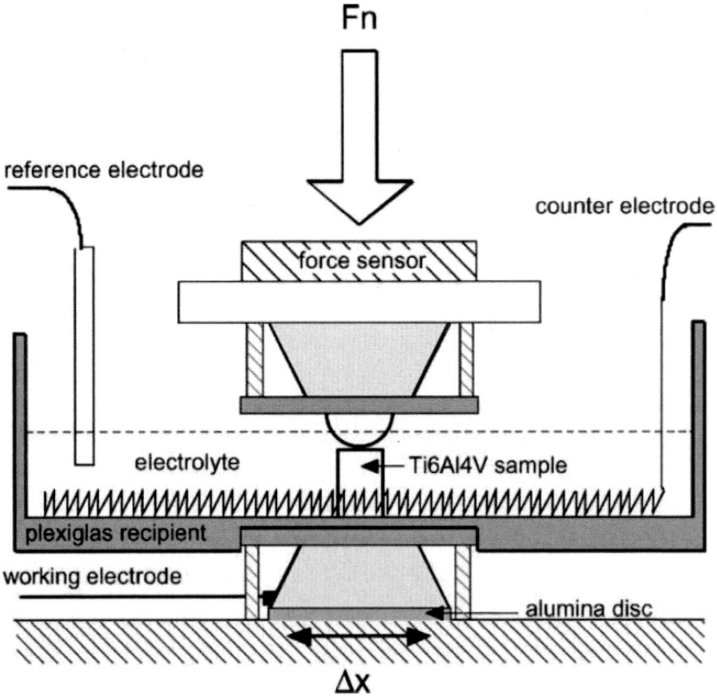

Figure 4.7 Electrochemical cell for triboelectrochemical measurements. Reprinted from *Wear*, 252, S. Barril, N. Debaud, S. Mischler, D. Landolt, A tribo-electrochemical apparatus for in vitro investigation of fretting-corrosion of metallic implant materials, 744–754, Copyright (2002), with permission from Elsevier.

As to the fretting corrosion behavior of Ti_6Al_4V sample in contact with a smooth alumina ball, Barril et al. [31, 32] demonstrated that the real wear volume was in proportion to the anodic volume, which was estimated by the current needed for the oxidation of titanium. They also found out that the wear volume evolved linearly with the cumulated energy dissipated in the contact: Comparison with the other data meant that the mechanical energy needed to initiate wear was drastically decreased by the application of anodic potential.

Recently very important aspects of tribocorrosion are reviewed by Mathew et al. [8]. It presents basic methodology and steps in the present tribology study.

4.1.2.3 Similarities and differences between tribocorrosion and tribochemistry

We hypothesize that in both tribochemical reactions and the tribocorrosion processes the intermediate reactive species are formed by the same mechanism mostly governed by the NIRAM approach. Thereby we can look for some similarities between tribochemistry and tribocorrosion as in the case of tribocatalysis and tribochemistry, as evidenced in [3]. Thus, in both tribochemical reactions and some heterogeneous catalytic reactions the intermediate reactive species ("activated complex" or "transition state") are formed by the same mechanism mostly governed by the NIRAM approach.

The emphasis was on the role of thermal electrons (thermionic emission) on catalysis and both thermal electrons and exoelectrons on tribocatalysis and tribochemistry. Usually, it is accepted that the anti-wear property of fatty acids is controlled by carboxylates produced under a friction process; the longer the alkyl chain the better wear reduction effect. The same is due to lubricity (oiliness); however, the wear/friction correlation is not simple. Accordingly, it was suggested [26–27] to consider the "prowear" behavior of fatty acids and esters, appearing at some concentrations in hydrocarbon solvents, as a tribocorrosion effect.

That is why both TRIBOCHEMISTRY and TRIBOCORROSION are so complex! The term tribocorrosion designates the process of irreversible transformation of a material, which results from simultaneous physicochemical and mechanical surface interactions and takes place in a tribological contact [7]. Landolt [1] emphasizes that tribocorrosion behavior depends on the (a) properties of the contacting materials, (b) mechanics of the tribological contact, and (c) physico-chemical properties of the environment.

It is of note here that tribocorrosion had been a point of basic research for over 40 years and, this science of surface transformations would remain as such in the near future [1]. Special cluster issue edited by Fischer and Mischler [9], provides state-of-the-art information on tribocorrosion. The editors stressed that this research area is very large and complex and therefore the included papers are not representative of the entire field.

Synergistic or antagonistic effects are observed and they might be deleterious or beneficial for the performance of the tribological contact. The study of tribocorrosion therefore should be based on a systems approach taking into account the combined effects of all significant parameters.

Wear-corrosion involving a sliding contact is of particular importance. Landolt et al. [33] describe and discuss in detail the third-body effects and material fluxes in tribocorrosion systems involving a sliding contact. An illustrative approach of such effects is shown in Fig. 4.8.

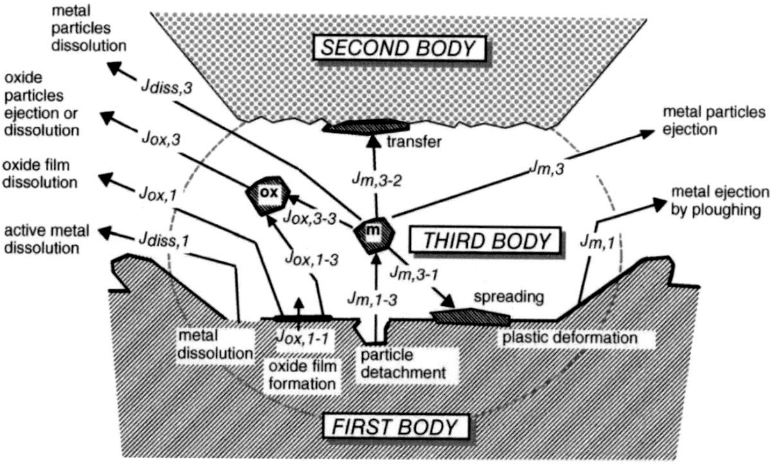

Figure 4.8 Schematic illustration of a tribocorrosion contact. Reprinted from *Wear*, 256, D. Landolt, S. Mischler, M. Stemp, S. Barril, Third body effects and material fluxes in tribocorrosion systems involving a sliding contact, 517–524, Copyright (2004), with permission from Elsevier.

Material loss from the first body, as shown in Fig. 4.8, is not only due to formation of wear particles but also occurs by electrochemical oxidation of the metal, which leads to the generation of dissolved ions or of a surface oxide film, and the oxide film may undergo chemical dissolution evolving metal ions or producing third-body particles [33].

As already mentioned, this book chapter also aims at looking for similarities and differences concerning both tribocorrosion and tribochemistry mechanisms from the viewpoint of mechanochemistry. Defining tribochemistry as the heterogeneous chemical reaction process initiated either by friction of solids and/or by any other mechanical energy input (activation of a solid by impact, elastic/plastic deformation, crack propagation, milling, etc.) well corresponds to tribocorrosion.

In the present author's opinion, negative-ion-radical action mechanism (NIRAM) approach, which was developed specifically to account for tribochemical discoveries, might be here helpful in enhancing our knowledge concerning mechanochemical tribocorrosion reaction mechanisms. By simply taking the similarities between tribochemistry and tribocorrosion, it is apparent that both processes involve electron exchange (donation and acceptance) and surface modification (protective or deteriorative).

The similarities and differences between tribocorrosion and tribochemistry are schematically drawn in Fig. 4.9. In tribochemistry, electrons are emitted from the breaking of bonds between a metal and oxygen atom, which results in the metal being positively charged. This is a kind of "anode" in terms of tribocorrosion. The emitted electrons attach to a surrounding molecule, either charging it negatively or dissociating it into anion and radical. These processes are called oxidation/reduction processes. In the second stage, the positively charged spot accepts anions via acid–base reaction.

On the other hand, in corrosion, electrons are extracted from metal atoms, resulting in cations. Then, electrons are conducted to the other surface, which serves as cathode where reduction of oxygen takes place. A cation dissolved in the surrounding solution, react with anions, resulting in e.g., $M(OH)_2$. The dissolution of the formed compound in the surrounding liquid is called corrosion. Tribocorrosion is the process of enhancing corrosion by applying frictional contact. Formation of $M(OH)_2$ is an acid–base reaction.

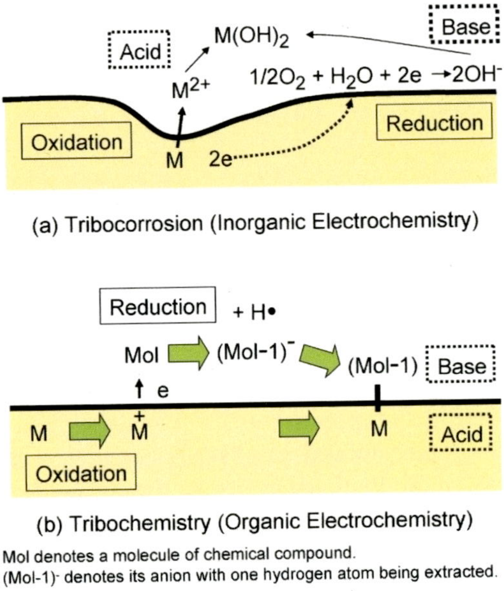

Figure 4.9 Similarities and differences between tribocorrosion and tribochemistry.

Both tribocorrosive and tribochemical reactions produce either a deleterious film or protective one depending on the properties of surrounding liquid as shown in Fig. 4.10. In boundary lubrication, chemisorbed molecules protect the surface from wear. This is due to (1) strong bonds between a molecule and surface, and (2) strong bonds between chemisorbed molecules. When the molecular chain is not long enough to have strong van der Waals interactions between molecules, they can be removed from the surface by the shear force due to friction. In that process, surface atoms can also be eliminated together with the removed molecules. Therefore, the bond between chemisorbed molecules, creating a "network" among them, is essential in boundary lubrication. In general, additives are expected to lubricate tribological systems, i.e., to reduce friction and wear. However, additives contained in some concentrations accelerate wear. This is the tribocorrosion as pointed out in Fig. 4.2.

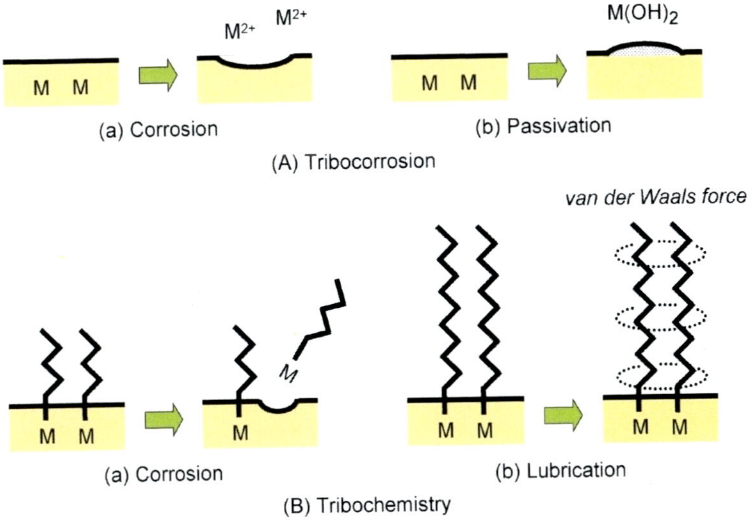

Figure 4.10 Deleterious or protective film formation in tribocorrosion and tribochemistry.

In tribocorrosion, wear is accelerated by corrosive media. Despite this deleterious process, some tribological systems transform into mild wear mode, where protective oxide/hydroxide film is formed. In this case, wear is inhibited by the surface film. This surface film forms a kind of network, same as boundary film in boundary lubrication, to protect surface from wear. To sum up, both tribochemical and tribocorrosive reactions are the combination of redox and acid–base reactions.

4.1.3 Bridging Tribocorrosion with Chemomechanical Polishing

4.1.3.1 Practical importance of the CMP process

The chemomechanical polishing (CMP) technology is applied by the electronics industry for planarization of dielectrics and metallic interconnects in multilevel chips. Several theoretical and applied studies of CMP have been reported with various materials containing an oxidizing agent, usually added to the slurry. Based on the theoretical analysis of the molecular dynamics simulation of sliding contact between a single microparticle and a smooth flat surface, paper [34] proposed a novel dominant mechanism of viscous flow of the amorphous layer on wafer surface for the for the material removal in CMP process.

In paper [35], apart from the primary factors like pressure and velocity, details like pad hardness, pad roughness, abrasive size, and abrasive size distributions are presented/discussed, and also modeled. It is found that with the increase in pressure, materials removal rate (MRR) increases due to increase in the number of active abrasives; the model was validated by comparing the results with experimental results [35]. Work [35] presents the known Preston model of CMP and considers the material removal rate to be proportional to the applied pressure and the sliding velocity but the model does not predict linear pressure dependence. Conversely, the presented equation correlates to some extent the equipment parameters to the polishing rate prediction, and the equation is simple and useful for an engineer to set up a new recipe of a certain combination of the consumables to meet process requirement [36].

4.1.3.2 Modeling and chemical approach to the CMP process

Also Luo and Dornfeld [37] proposed a three-body abrasion model for the mechanical polishing considering the volume removed by individual particles, taking into account their size distribution, and then integrating the volume over the number of particles. The model includes an empirical term that characterizes chemical effects. As already mentioned, Chen et al. [36] proposed a regression model based on dimensional analysis for predicting polishing rate on rotating equipment. Kaufman et al. [38] stressed the importance of passive film formation and removal during CMP of tungsten. The importance of chemical effects in CMP of metals was also pointed by Kuo and Tsai; they found that the removal rate of aluminum in acid slurry depends on the applied potential [29].

Work [30] investigated CMP the behavior of Al in phosphoric acid base slurry and evaluated the applied potential on the CMP metal removal rate; the results indicated that a contact pressure modifies the test system by a decrease in corrosion potential and an increase in passive current density. Additionally, the experimental results also evidenced that the CMP removal rate depends on both the contact pressure and the potential applied

[30]. Interestingly, the behavior of the potential dependence is controlled (i) by the direction of polarization and (ii) by the magnitude of potential used.

Based on the model mechanism proposed by Kaufman et al. [38], CMP involves the repeated processes of passive film formation, removal, and re-passivation of passive film. The removal rate depends on the formation and abrasion of the passive film, thus, the presence of surface film is necessary and essential in the CMP process.

Recently Songbo and Li [39] described a new phenomenon observed in determining the wear-corrosion synergy during a corrosive sliding wear test. Accordingly, the corrosive wear of materials is the very complex surface failure process, which involves both the wear accelerated corrosion and the corrosion-accelerated wear. Separating contributions of these two processes to the total material loss is of significance for gaining better understanding of the corrosion process. Thus, wear and corrosion are two-material degradation phenomena that involve a number of mechanical and electrochemical mechanisms.

4.1.3.3 Importance of the oxidation process for CMP

Researchers of electrochemistry and corrosion primarily are focused on kinetics of metal surfaces re-passivation, activated by scratching, and tribologists, are mainly interested in how surface oxidation during rubbing affects the rate of mechanical wear.

Work on oxidational wear modelling by Quinn [40] deals with measuring the oxidational wear of EN31 steel pins sliding against EN8 steel disks. Experiments were performed at the surface temperatures between 60 and 500°C, and the applied loads varied between about 5 and 150 N. It was demonstrated how one might use some obtained constants to predict the oxidational wear rates of EN31 pins against EN8 steel disks, applying a similar tribometer.

Quinn et al. [40] identified oxidation processes as major mechanisms of wear, and demonstrated that the chemical oxidation of the friction surfaces leads to growth of an oxide film, which periodically is eliminated from the contact due to the effect of mechanical forces.

Another paper [41] reviewed the role of fretting tribological systems subjected to low amplitude vibrations. Early detailed experimental data on the role of corrosion under fretting are included in book [42]. These references also consider the synergy effect between mechanical and chemical metal loss under tribocorrosion processes. Work [43] predicts synergism between corrosion and abrasive wear. Details on tribocorrosion behavior of metallic orthopedic implant materials are presented in [44].

4.1.4 Effect of Water on Wear of Metals

It is generally known that atmospheric oxygen can either increase or decrease wear of metals depending on the type of metal and pressure range of oxygen. In severe wear regions, it increases wear of almost all metals except gold [45]. The effect of oxygen exposure time, on the other hand, is not as widely known as that of oxygen pressure. Figure 4.11 shows an example of copper-to-copper wear with pin-on-disk test rig. The exposure time is expressed as non-friction time to represent rest time in between rubbings. It is apparent that the oxygen pressure and the logarithm of non-friction time similarly increase the wear of copper [46].

The effect of non-friction time on the wear of copper was also demonstrated in laboratory air conditions where the relative humidity was around 55%. The experimental setup also aimed at elucidating the effect of water on wear of metals where water drops were continuously supplied to the friction interface of the periphery of the two identical rings in sliding contact. Before dropping, argon or oxygen gas was provided to water by bubbling to adjust the oxygen levels as low as or as high as possible: the resulting oxygen concentration was 3 or 30 mg per liter, respectively. As shown in Fig. 4.11, wear in all three conditions increased with increasing non-friction time. Wear experiments in laboratory air with the relative humidity of 55% was also conducted as a reference. The slopes of the graphs clearly indicate that the effect of non-friction time on wear is stronger in water than in air. It is also important that the water environment inhibits wear when the non-friction time is 1 s, yet it enhances wear when the interval time is prolonged.

The experimental results mentioned above are explained as follows: Water is a lubricant; interfacial water protects surfaces from wear when the sliding proceeds with short intervals between rubbings. On the other hand, when the time interval between rubbings is extended, water becomes a corrosive media; interfacial water attacks surfaces promoting wear. Dissolved oxygen has little effect on the wear of copper in water.

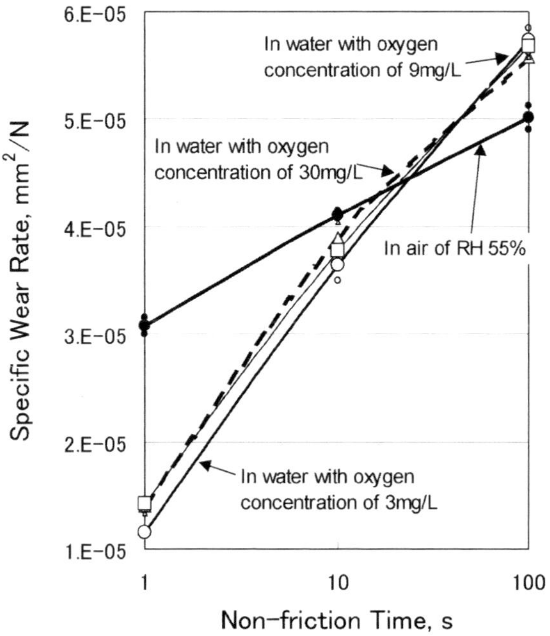

Figure 4.11 Effects of humid air and water with two concentrations of oxygen on wear on Cu-to-Cu reciprocating sliding system.

Compared with copper, aluminum in water has different wear characteristics as shown in Fig. 4.12. The effect of non-friction time on wear in air is almost the same as in copper: it increases wear. On the other hand, water has opposite effect on the wear of aluminum compared with copper: it increases wear for short non-friction time, yet it decreases wear for long non-friction time. Contrary to copper, aluminum strongly takes corrosive attack from water, leading to the increase in wear when water is present nearby. However, when aluminum worn surface is surrounded by water for a long period of time, an oxide film is created on the surface, which protects the surface from being worn out. It is, thus, demonstrated from this experiment that both processes of depassivation and subsequent repassivation of worn surfaces are governed by the duration of non-friction time: Depassivation with short non-friction time increases wear, while repassivation with long non-friction time decreases it. This explains why we often encounter the complex wear phenomena in tribocorrosion. It is also of note that wear is decreased in water with even low oxygen concentration when non-friction time is long. This is because stable mild wear is established in water environment with low concentration of oxygen in it.

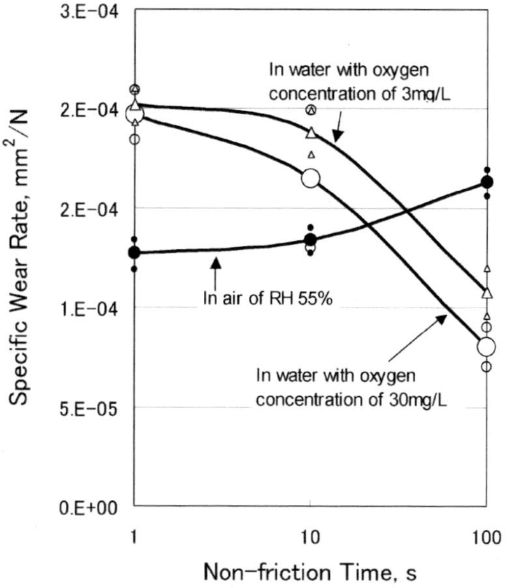

Figure 4.12 Effects of humid air and water with two concentrations of oxygen on wear on Al-to-Al reciprocating sliding system.

4.1.5 Final Discussion and Concluding Remarks

Presently there are only a few works/papers with the term tribocorrosion in the title. They are as follows.

(1) Significance of tribocorrosion in biomedical applications: Overview and current status [8].
(2) Tribocorrosion: Fundamentals, materials and applications [9].
(3) Triboelectrochemical technique for studying tribocorrosion behavior of biomaterials [11].

(4) Tribocorrosion testing of stainless steel (SS) and PVD coated SS using a modified scanning reference electrode technique [12].
(5) Tribocorrosion: Materials behavior under combined conditions of corrosion and mechanical loading [47].
(6) Tribocorrosion of stainless steel: (Review) [48].

The review paper [48] stresses synergistic effects between mechanical, chemical, and electrochemical processes controlling mass removal of metals is not a simple sum of losses caused by mechanical wear, but somewhat exceeds the sum. Thus, the mechanical wear and corrosion should be tested/determined individually. Complexity of tribocorrosion processes has been emphasized, as it is influenced by several parameters (mechanical and corrosion), and also due the lack of an integrated/efficient test system. Additionally, work [48] reviews the recent achievements in the area of biotribocorrosion, especially on orthopedic surgery and dentistry. Paper [49] presents theory and modeling of material removal mechanisms in chemical mechanical polishing. Work [50] relates to third-body effects on friction and wear dry fretting of steel contacts.

The only book on tribocorrosion was recently edited by Landolt & Mischler [51]. Although it is focused on tribocorrosion of passive metals and coatings, it also includes basic knowledge of tribocorrosion fundaments, adsorption layers and passive oxide films on metals. The other subject topics include tribocorrosion mechanisms in sliding contacts, and models plus mechanisms of erosion-corrosion in metals. Part II of the book [51] deals with methods for measurement and prevention of tribocorrosion specifying, for instance, tribocorrosion test protocols for sliding contacts. Part III presents tribocorrosion in engineering and medicine, e.g., tribocorrosion issues in nuclear power generation.

Considering future trends, book [51] concludes that tribocorrosion in the hot temperature metal working is a material deterioration caused by the synergistic effect of wear, dry corrosion, degradation of lubricant or inadequate thickness of the oil film, acting together.

The most recent VIII International Conference on Mechanochemistry & Mechanical Alloying INCOME'2014, adds also some interesting information relating to triboelectrocorrosion processes. A good example considers advances in electromechanochemistry, also presenting and describing a new test method, the device and its possible applications [52].

References

1. D. Landolt, Electrochemical and materials aspects of tribocorrosion systems, *J. Phys. D Appl. Phys.*, 39 (2006) 3121–3127. doi:10.1088/0022-3727/39/15/S01.
2. Z. V. Todres, *On Organic Mechanochemistry and Its Practical Applications*, CRS, Taylor & Francis Group, LLC Boca Raton London New York (2006).
3. C. Kajdas, General approach to mechanochemistry and its relation to tribochemistry, Chapter 11 in *Tribology in Engineering* (ed. H. Pihtili), InTech, published May (2013), Rijeka. http://dx.doi.org/10.5772/50507.
4. G. Kaupp, Mechanochemistry: the varied applications of mechanical bond-breaking, *Cryst. Eng. Commun.*, 11 (2009) 388–403.
5. J.-P. Celis, P. Ponthiaux, F. Wenger, Tribocorrosion of materials: Interplay between chemical, electrochemical, and mechanical reactivity of surfaces, *Wear*, 261(9) (2006) 939–946.
6. V. Vignal, N. Mary, P. Ponthiaux, F. Wenger, Influence of friction on the local mechanical and electrochemical behaviour of duplex stainless steels, *Wear*, 261(9) (2006) 947–953.
7. D. Landolt, S. Mischler, M. Stemp, Electrochemical methods in tribocorrosion: A critical appraisal, *Electrochim. Acta*, 46 (2001) 3913–3929.
8. M. T. Mathew, P. Srinivasa Pai, R. Pourzal, A. Fischer, M. A. Wimmer, Significance of tribocorrosion in biomedical applications: Overview and current status, *Adv. Tribol.* (2009), Article ID 250986, 12 pp.: doi:10.1155/2009/250986.
9. A. Fischer, S. Mischler (Guest eds.) Tribocorrosion: Fundamentals, materials and applications, *J. Phys. D: Appl. Phys.*, 39 (editorial to issue 15) (2006).
10. A. Fischer, S. Mischler, Tribocorrosion: Fundamentals, materials and applications, *J. Phys. D Appl. Phys.*, 39 (2006) http://iopscience.iop.org/0022-3727/39/15/E01.

11. M. Azzi, J. A. Szpunar, Tribo-electrochemical technique for studying tribocorrosion behavior of biomaterials, *Biomol Eng.*, 24(5) (2007) 443–446. Epub 2007 Aug 7.

12. Y. N. Kok, R. Akid, P. E. Hovsepian, Tribocorrosion testing of stainless steel (SS) and PVD coated SS using a modified scanning reference electrode technique, *Wear*, 259 (2005) 1472–1481.

13. S. W. Watson, F. J. Friedersdorf, B. W. Madsen, S. D. Cramer, Methods of measuring wear-corrosion synergism, *Wear*, 181–183 (1995) 476–484.

14. M. M. Stack, N. Pungwiwat, Erosion/corrosion-mapping of iron-in aqueous slurries: Some views on a new rationale for defining the erosion/corrosion interaction, *Wear*, 256 (2004) 565–576.

15. J. Jiang, M. M. Stack, A. Neville, Modelling the tribocorrosion interaction in aqueous sliding conditions, *Tribol. Int.*, 35 (2002) 669–679.

16. M. M. Stack, Bridging the gap between tribology and corrosion: From wear maps to Pourbaix diagrams, *Int. Mater. Rev.*, 50 (2005) 1–17.

17. L. Benea, F. Wenger, P. Ponthiau, J.-P. Celis, Tribocorrosion behaviour of Ni–SiC nano-structured composite coatings obtained by electrodeposition, *Wear*, 266 (2009) 398–405.

18. R. Schumacher, H. Zinke, Tribofragmentation and antiwear behaviour of isogeometric phosphorus compounds, *Tribol. Int.*, 30(3) (1997) 199–208.

19. F. M. Borodich, C. S. Korach, L. M. Keer, Modeling the tribochemical aspects of friction and gradual wear of diamond-like carbon films, *J. Appl. Mech.*, 74(1) (2007) 23–30.

20. H.-M. Chu, J.-L. Chen, K.-C. Hsu, A new phenomenon observed in determining the wear-corrosion synergy during a corrosive sliding wear test, *Tribol. Lett.*, 29 (2008) 45–52.

21. S. B. Bulgarevich, M. V. Boiko, E. N. Tarasova, V. A. Feizova, K. S. Lebedinskii, Kinetics of mechanoactivation of tribochemical processes, *J. Friction Wear*, 33(5) (2012) 345–353.

22. G. Heinicke, *Tribochemistry*, Academie Verlag, Berlin (1984).

23. G. A. Somorjai, *Introduction to Surface Chemistry and Catalysis*, John Wiley & Sons, New York (1994).

24. C. Kajdas, Importance of anionic reactive intermediates for lubricant component reactions with friction surfaces, *Lubrication Sci.*, 6 (1994) 203–228.

25. C. Kajdas, A novel approach to tribochemical reactions: Generalized NIRAM-HSAB action mechanism. *Proceedings of the ITC Yokohama, satellite forum on tribochemistry*, Tokyo Institute of Technology (1998) 31–35.

26. C. Kajdas, M. Majzner, The influence of fatty acids and fatty acids mixtures on the lubricity of low-sulphur diesel fuels, SAE Technical paper series 2001-01-1929 (2001).

27. C. Kajdas, M. Majzner, Boundary lubrication of low-sulphur diesel fuel in the presence of fatty acids, *Lubrication Sci.*, 14 (2001) 83–108.

28. E. Bardal, A. Bardal, Corrosion and its Impact on wear processes, in: *Surface Modification and Mechanisms* (G. E. Totten, H. Liang, eds.), Marcel Dekker, New York, Basel, Chapter 13 (2004).

29. H.-S. Kuo, W.-T. Tsai, Effect of applied potential on the chemical mechanical polishing of aluminum in phosphoric acid base slurry, *J. Electrochem. Soc.*, 147(6) (2000) 2136–2142.

30. S. Barril, N. Debaud, S. Mischler, D. Landolt, A tribo-electrochemical apparatus for in vitro investigation of fretting-corrosion of metallic implant materials, *Wear*, 252 (2002) 744–754.

31. S. Barril, S. Mischler, D. Landolt, Influence of fretting regimes on the tribocorrosion behaviour of Ti_6Al_4V in 0.9% sodium chloride solution, *Wear*, 256 (2004) 963–972.

32. S. Barril, S. Mischler, D. Landolt, Electrochemical effects on the fretting corrosion behaviour of Ti_6Al_4V in 0.9% sodium chloride solution, *Wear*, 259 (2005) 282–291.

33. D. Landolt, S. Mischler, M. Stemp, S. Barril, Third body effects and material fluxes in tribocorrosion systems involving a sliding contact, *Wear*, 256 (2004) 517–524.

34. J.-Z. Jiang, Y.-W. Zhao, J.-B. Luo, A chemical mechanical polishing model based on the viscous flow of the amorphous layer, *Wear*, 265 (2008) 992–998.

35. S. Thagella, A. K. Skder, A. Kumar, Tribological issues and modeling of removal rate of low-k films in CMP, *J. Electrochem. Soc.*, 151 (2004) G205–G215.

36. P. H. Chen, H. C. Shih, B. W. Huang, J. W. Hsu, An engineering approach to predict the polishing rate in CMP with rotational equipment, *J. Electrochem. Soc.*, 151 (2004) G649–G651.

37. J. Luo, D. A. Dornfeld, Material removal mechanism in CMP theory and modeling, *IEEE Trans. Semicond. Manuf.*, 14 (2001) 112–133.

38. F. B. Kaufman, D. B. Thompson, R. E. Broadie, M. A. Jaso, W. L. Guthrie, D. J. Pearson, M. B. Small, Chemical-mechanical polishing for fabricating patterned metal features as chip interconnects, *J. Electrochem. Soc.*, 138(11) (1991) 3460–3465.

39. Y. Songbo, D. Y. Li, A new phenomenon observed in determining the wear-corrosion synergy during a corrosive sliding wear test, *Tribol. Lett.*, 29 (2008) 45–52.

40. T. F. J. Quinn, The oxidational wear of low alloy steels, *Tribol. Int.*, 35 (2002) 691–715.

41. F. H. Stott, The role of oxidation in the wear of alloys, *Tribol. Int.*, 31 (1998) 61–71.

42. R. B. Waterhouse, *Fretting Corrosion*, Pergamon Press, Oxford (1972).

43. A. W. Batchelor, G. W. Stachowiak, Predicting synergism between corrosion and abrasive wear, *Wear*, 123 (1988) 281–291.

44. Y. Yan, Corrosion and tribo-corrosion behaviour of metallic orthopaedic implant materials, PhD Thesis, The University of Leeds, School of Mechanical Engineering, September (2006).

45. H. Mishina, Atmospheric characteristics in friction and wear of metals, *Wear*, 152(1) (1992) 99–110.

46. K. Hiratsuka, Y. Meki, The effects of non-friction time and atmosphere in friction/non-friction areas on the wear of metals, *Wear*, 270(7–8) (2011) 446–454.

47. P. Ponthiaux, F. Wenger, J.-P. Celis, Tribocorrosion: Material Behavior Under Combined Conditions of Corrosion and Mechanical Loading, in: *Corrosion Resistance* (H. Shih, ed.), InTech, Chapter 4 (2012) DOI: 10.5772/1844.

48. V. I. Pokhmurs'kyi, V. M. Dovhunyk, Tribocorrosion of stainless steels (Review), *Mater. Sci.*, 46(1) (2010) 87–96.

49. J. Luo, D. A. Dornfield, Material removal mechanism in CMP: Theory and modeling, *IEEE Trans. Semicond. Manuf.*, 14(2) (2001) 112–133.

50. N. Diomidis, S. Mischler, Third body effects on friction and wear during fretting of steel contacts, *Tribol. Int.*, 44 (2011) 1452–1460.

51. D. Landolt, S. Mischler (ed.), *Tribocorrosion of Passive Metals and Coatings*, Woodhead Pub. Ltd. Oxford, Cambridge, Philadelphia, New Delhi (2011).

52. A. Calka, *SIV-1 in the Income 2014 Book of Abstracts*, Cracow University of Technology, Poland, June (2014) 22–26.

4.2 Tribology in CMP and Case Studies on Tantalum*

Feng Gao, Sukbae Joo, and Hong Liang

Department of Mechanical Engineering, Texas A&M University, USA

hliang@tamu.edu

4.2.1 Fundamentals of Triboelectrochemistry

Triboelectrochemistry includes three components: tribology, corrosion, and electrochemistry. The first component of tribology refers to friction, lubrication, and wear that occur at the presence of mechanical energy. The second component, corrosion, is where the surface reacts in corrosive environments. The third, electrochemistry, affects or assists the reaction of metal during process.

When tribological interactions take place in a corrosive atmosphere, combinatorial phenomenon, so called tribocorrosion, can occur and sometimes cause unexpected results. Tribocorrosion induced damage can occur in many technological areas where two machine parts physically contact and interact with each other leading to failures such as stress corrosion cracking, and corrosion-fatigue under cyclic load in corrosion environment [1]. The metal degradation is caused by two simultaneous removal mechanisms: mechanical wear and corrosion [2]. This synergetic mechanism triggers material removal comparing to the case of individual mechanism. Depending on the environments, the synergism of wear and corrosion can be categorized into two modes: corrosion-induced wear and wear-induced corrosion [3]. Either of them can dominate the material removal mechanism based on a certain condition.

In addition to the degradation of surface as a result of tribocorrosion, passivation film forms as a result of electrochemical reaction between chemical environment and the metal surface

*Partial contents of this chapter have been published and referenced [14–16]. Permission has been granted.

[4–6]. It has been reported that the structure and composition of the passivation film affect the friction and wear behavior during tribocorrosion process [7]. Furthermore, the property of passivation film can be under the control of applied electropotential [8–9]. In many cases, triboelectrochemical analysis has to be applied to diagnose any relevant processes. Electrochemical mechanical planarization (ECMP) process can be one of the candidates. This process is recently developed version of CMP added with electropotential controllability and will be introduced in the following section.

4.2.2 Triboelectrochemistry in Electrochemical Mechanical Planarization

ECMP process has started from electrochemical planarization (ECP). ECP is the material removal process of metal under certain conditions of electrochemical dissolution. The limit of ECP is that this process does not necessarily bring smooth surface and global or near global planarization. On this purpose, newly developed ECP process as a combined form with soft polishing was first introduced in 2001 and regarded as the beginning of ECMP [10]. ECMP process is composed of the electrochemical dissolution of metal and planarization capability of a pad to obtain sub-microscale leveling. Thus, it overcomes the limit of ECP. The advantage of ECMP is efficient removal with low downforce, good planarization, and high controllability [11, 12]. In addition, the process is more eco-friendly because it requires less toxic chemical elements and less consumables and consequently, more than 30% cost reduction over CMP can be achieved [11, 13]. In last 10 years, ECMP has been consistently studied mainly on copper 83–86, and some on tantalum [14–17]. However, there are few studies about ECMP on patterned geometry because the pattern dependency of ECMP is less comparing to conventional CMP [11].

Since ECMP can be regarded as mechano-electro-chemical process, triboelectrochemical characterization is necessary to elucidate the fundamentals of ECMP. The key factors for understanding triboelectrochemistry in ECMP process are the formation of passivation layer on the metal and the removal thereof. Tribological behaviors such as wear and friction strongly depend on the condition of passivation layer [18–20]. The formation kinetics for this type of surface includes anodic dissolution and cathodic passivation resulted from the electrochemical reaction between slurry and metal surface. Many researchers had investigated possible parameters and it is believed that slurry composition [21], temperature [22], pH, and electropotential [18] are all possible candidates for tribological variation.

4.2.3 Material Removal Mechanisms of Tantalum in Electrochemical-Mechanical Polishing of Tantalum

This part of work has been published in [14]. Here we report using the single-frequency EIS to probe the material removal mechanism of tantalum CMP and ECMP in acidic H_2O_2 slurry on a homemade experimental setup. Different polishing parameters were applied, such as the load, rotating speed, potential, and concentration of the abrasives. Through the impedance measurement, it was revealed that there were two competing removal mechanisms in tantalum CMP and ECMP: the formation and removal of the oxide layer on the surface. The thickness of the oxide film was found to be closely related to the mechanical power provided by friction force and rotating speed. The rate of growing and removing the oxide layer was subsequently a function of the mechanical power. The approach we used here can be applied in other metals for polishing and would be beneficial for the optimization of CMP and ECMP processes.

4.2.3.1 Materials

A tantalum rod (Sigma-Aldrich), 6.25 mm in diameter and 99.9 wt% in purity, was used for this study. It was annealed and in a gray color. Its density and resistivity are 16.69 g/cm^2 and 13.5 $\mu\Omega \cdot$cm at 20°C. Politex (Rohm & Haas) is a soft polishing pad made of fiber reinforced polyester. This pad is usually used for final polishing. Hydrogen peroxide (35 wt%) obtained from VWR was used as an oxidant during polishing. Acetic acid (Riedel-de Haën) was added to adjust the pH value of the slurry. Potassium chloride (Fluka

Chemie GmbH) was used to stabilize the conductivity of the slurry. The gamma alumina powder (from ECO) was used as an abrasive. Its average diameter was 50 nm.

4.2.3.2 Experimental configuration

The experimental setup is sketched in Fig. 4.13. The polyvinylchloride (PVC) container was not conductive and did not react with chemicals. The container rotated on a CSM tribometer (CSM Instruments) on which the rotating speed and load are adjustable. A tantalum sample was mounted on a polyethylene (PE) shaft and its bare surface was in contact with the pad with a known load. The ECMP tests were carried out through a three-electrode system with a potentiostat (Reference 600, Gamry). This testing system consisted of three electrodes, namely reference, counter, and working electrode. The tantalum sample is the working electrode, while a saturated calomel electrode the reference and a platinum wire the counter. In tests, every electrode was submerged in the slurry containing H_2O_2, potassium chloride, acetic acid, abrasives, and the deionized water.

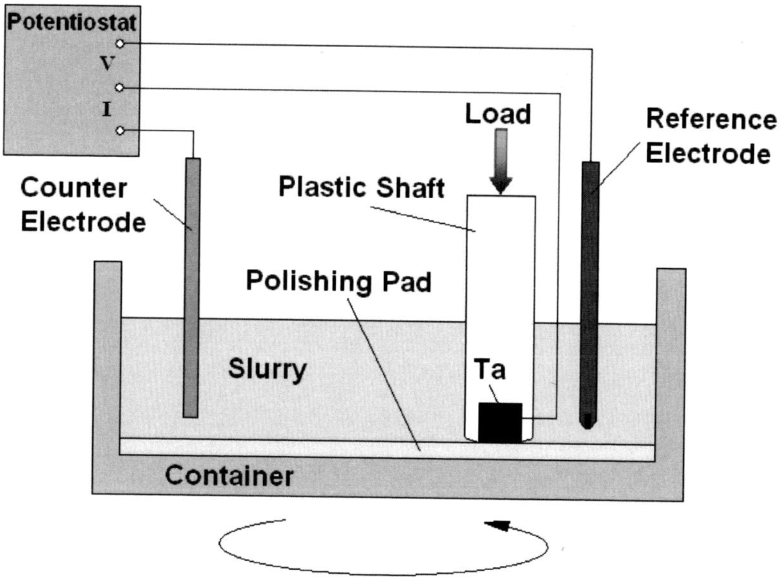

Figure 4.13 Experimental setup for ECMP experiments.

4.2.3.3 Measurement methods

The potentiostatic electrochemical impedance spectroscopy (EIS) is a widely used tool to study surface properties of oxides and coatings on a metal surface. In potentiostatic EIS measurement, an AC potential with small amplitude was applied on the sample and the corresponding current was measured. The impedance was then calculated. Details will be provided in the following. The frequency of the AC potential ranges from high to low values, usually in the range of 100 to 0.01 Hz. Although potentiostatic EIS can reveal surface properties of metals, it has drawbacks. One is its test time. It generally takes at least for several minutes. The potentiostatic EIS is suitable for a static measurement, i.e., the surface conditions are relatively stable. The ECMP is a kinetic process due to the fact that the surface undergoes rapid oxidization and abrasion. It is not feasible to use the potentiostatic EIS to observe the surface during ECMP. In the present work, we applied a technique called single-frequency EIS for real-time measurement.

The principle of single-frequency EIS is similar to that of potentiostatic EIS. In the former, a low value AC potential is applied to the sample with a fixed frequency. The AC potential is applied at an identical time interval to record the impedance change as a function of time. In the single-frequency EIS, the impedance is expressed by using a complex number as shown in Eq. (4.1):

$$Z = a + jb = a + j\frac{1}{\pi fc}, \tag{4.1}$$

where Z is the impedance, a is the real part, and b is the imaginary part. We can denote $b = 1/2\pi fc$, where c is the nominal capacitance, f is the frequency. Then, a and c can be loosely referred to the resistance and capacitance of a coating or an oxide layer [23]. Equation (4.2) shows the magnitude of the impedance Z:

$$|Z| = \sqrt{a^2 + b^2} \tag{4.2}$$

The magnitude of the impedance depends on the a and b. As such the magnitude of the impedance obtained through single-frequency EIS can represent, in situ, the surface state of tantalum during ECMP. The friction coefficient was measured using the tribometer during polishing.

4.2.3.4 Processing parameters

All polishing tests followed the same procedure. Before experiments, in order to eliminate the residue on the sample surface, each sample was polished by an alumina (50 nm) paste for 5 min. In the following step, the sample was polished in the slurry for 15 min without applying a potential under the corresponding load and rotating speed. Now the open circuit potential (OCP) of the sample became relatively stable, showing the equilibrium state of the surface condition. Subsequently, a 20 min ECMP was conducted with an applied potential. Meanwhile, the impedance was measured using the single-frequency EIS during polishing and the friction coefficient was recorded. Table 4.1 lists the text conditions in detail.

In Table 4.1, the variables were loads, rotating speeds, applied potentials, and abrasive. The slurry used in polishing tests was a mixer of H_2O_2, alumina abrasive, KCl, and deionized water. Its pH value was adjusted by adding acetic acid to 2.20. The concentration of both H_2O_2 and KCL was fixed at 1 wt% in all tests and that of the abrasive was 0.3 wt% that was not a variable. In each test group, there was only one variable designated. In Groups 1, 2, and 3, we aimed to study effects of load, rotating speed, and applied potential on the removal process during ECMP. In Group 4, the role of the abrasive during ECMP was studied. The applied potentials in Groups 1 to 4 were absolute potential between the tantalum sample and the reference electrode, while the potential in Group 5 was set at 0 V vs. OCP. Basically, no applied potential existed in the condition of 0 V vs. OCP. The potential was measured after 20 min polishing as the OCP. Due to the kinetic nature of CMP, the OCP fluctuates rather than being constant. In the present research, such fluctuation was less than 1 mV, which was sufficiently low, compared with the applied potential in ECMP. We therefore ignored this fluctuation. We compared CMP and ECMP in Group 5. Each test was repeated for at least twice.

Table 4.1 Polishing parameters for ECMP tests

Group	Fixed parameters	Variables
1	60 rpm/1 V	2 N
		4 N
		6 N
2	4N/1 V	20 rpm
		50 rpm
		60 rpm
3	60 rpm/4 N	2 V
		1 V
		0 V
4	60 rpm/4 N/1 V	0.3 wt% abrasive
		1 wt% abrasive
5	60 rpm/4 N	OCP*

Note: rpm, rotations per minute; V, volts; N, Newton; OCP, open circuit potential. OCP* is the OCP at 0 V.

In impedance measurements, the AC signal was generated and measured at the frequency of 5 Hz with the amplitude of 10 mV per second. If the frequency is too low, the impedance measurement cannot be finished in 1 s; if the frequency is too high, the variation of the impedance is not pronounced. The shorter the repeated time, the consistent the results because of the large amount of data. The minimum sampling time for the potentiostat was set for 1s at the frequency of 5 Hz.

4.2.3.5 Triboelectro-impedance study

The potentiostat and tribometer recorded the impedance and friction coefficient, respectively during polishing. The corresponding results are presented as follows.

4.2.3.5.1 Impedance

Figure 4.14 shows a typical impedance plot when the tantalum was polished with 60 rpm rotating speed, 4 N load, and 1 V potential for 20 min. The *X*-axis is the polishing time in seconds and the *Y* the impedance in ohm. The impedance varied periodically. Impedance curves obtained under different polishing conditions have different periods except two cases.

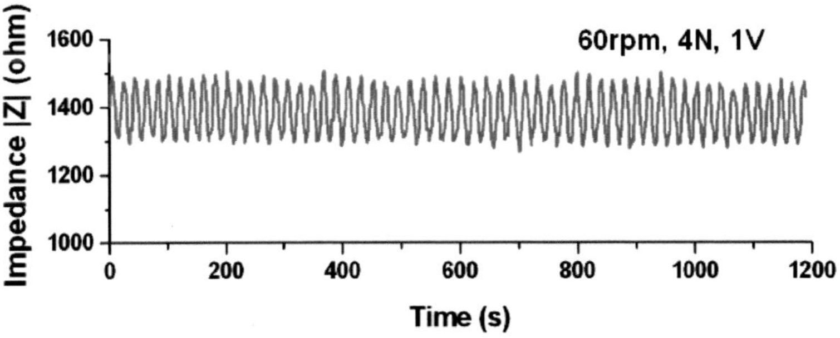

Figure 4.14 Impedance curve obtained at 60 rpm, 4 N, and 1V.

Sine functions were used to calculate the period of each impedance curve. First, the impedance data from 500 to 700 s in ECMP process were extracted in order to observe in detail. The data shown was in the middle of the polishing. Second, a sine function (Eq. (4.3)) was designed to simulate the curve shape. The period of the sine function is the period of the impedance variation by comparison.

$$|Z| = A + B\sin\left(\frac{2\pi}{T}t + \theta\right), \qquad (4.3)$$

where $|Z|$ is the magnitude of the impedance, A is the average of the impedance during the whole ECMP process, B is the amplitude of the sine function, which can be set arbitrarily, T is the period, t is the polishing time, and θ is the phase shift; A and T are the average of repeated experiments.

Figure 4.15 shows the impedance curves and their corresponding sine curves between 500 and 700 s when different loads were employed with the same rotating speed (60 rpm) and potential (1 V). On the right, there are periods T of simulated sine functions, and the average numbers A of the impedance curves obtained during 20 min ECMP. The standard deviations are also listed here. The amplitude of simulated curve was set slightly higher in order to highlight the original curve clearly. It can be seen both period and average impedance decrease with the increase of the applied load.

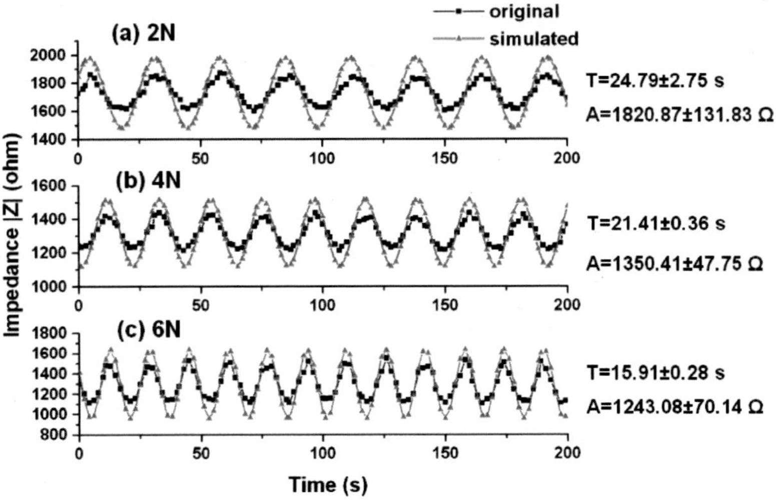

Figure 4.15 The enlarged impedance curves and sinusoidal simulation curves with different loads at 60 rpm and 1 V. (a) 2 N, $T = 24.79 \pm 2.75$ s, $A = 1820.87 \pm 131.83$ Ω; (b) 4 N, $T = 21.41 \pm 0.36$ s, $A = 1350.41 \pm 47.75$ Ω; (c) 6 N, $T = 15.91 \pm 0.28$ s, $A = 1243.08 \pm 70.14$ Ω.

In Fig. 4.16, different rotating speeds were applied in polishing, as the load and potential were kept constant. When the sample was polished at 20 rpm, the data points of the impedance were random, unlike those in other curves and therefore no sine function was found. The average number of the impedance, when 50 rpm was applied, was higher than that at 60 rpm, while the period of the former was shorter than that of the latter.

Figure 4.17 shows the impedance results under different potentials when the load and rotating speed were fixed at 4 N and 60 rpm, respectively. The periods obtained under different potentials are similar, compared with those shown in Fig. 4.15 and 4.16. The average impedance decreases gradually with the decreased potential.

Figure 4.18 shows the influence of the abrasives in the slurry on the period of the impedance variation and the impedance average when the load, rotating speed, and potential were 4 N, 60 rpm, and 1 V, respectively. When the concentration of the abrasive changed from 0.3 to 1 wt%, both period and average impedance decreased.

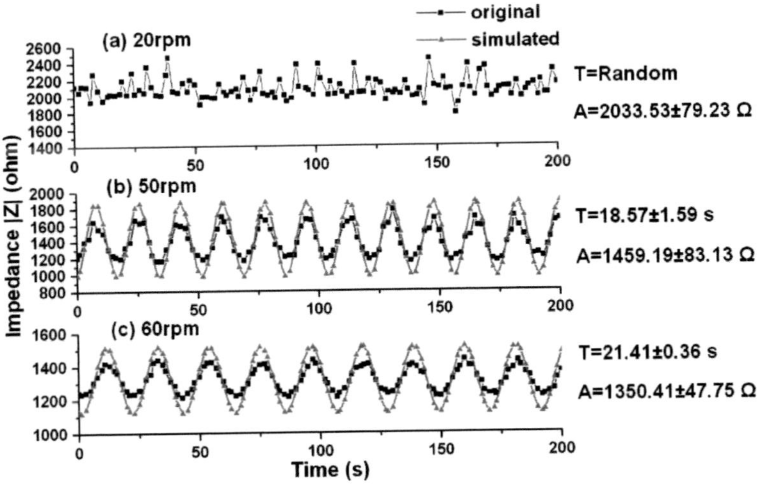

Figure 4.16 The enlarged impedance curves and sinusoidal simulation curves with different rotating speed under 4 N and 1 V. (a) 20 rpm, no obvious period, A = 2033.53 ± 79.23 Ω; (b) 50 rpm, T = 18.57 ± 1.59 s, A = 1459.19 ± 83.13 Ω; (c) 60 rpm, T = 21.41 ± 0.36 s, A = 1350.41 ± 47.75 Ω.

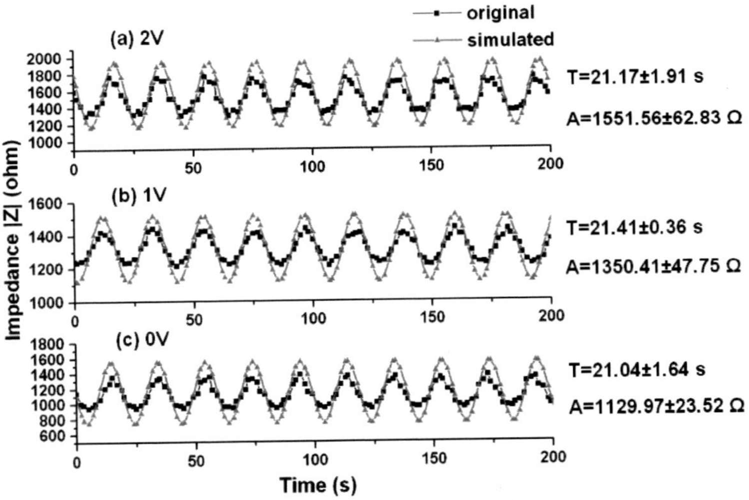

Figure 4.17 The enlarged impedance curves and sinusoidal simulation curves with different potentials when 4 N and 60 rpm were applied. (a) 2 V, T = 21.17 ± 1.91 s, A = 1551.56 ± 62.83 Ω; (b) 1 V, T = 21.41 ± 0.36 s, A = 1350.41 ± 47.75 Ω; (c) 0 V, T = 21.04 ± 1.64 s, A = 1129.97 ± 23.52 Ω.

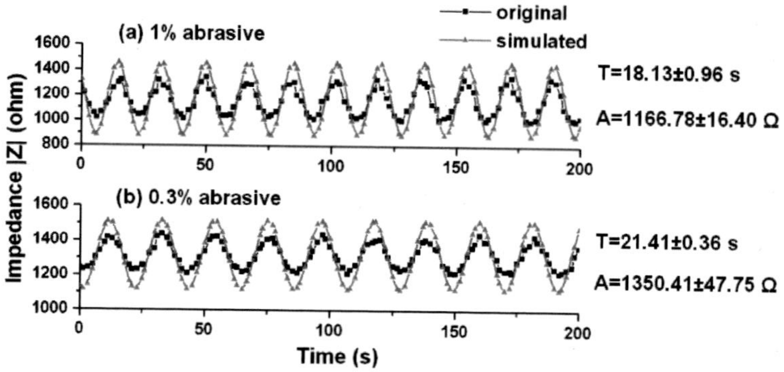

Figure 4.18 The enlarged impedance curves and sinusoidal simulation curves with different concentrations of abrasive when 4 N, 60 rpm and 1 V were applied. (a) 1% abrasive, $T = 18.13 \pm 0.96$ s, $A = 1166.78 \pm 16.40\ \Omega$; (b) 0.3% abrasive, $T = 21.41 \pm 0.36$ s, $A = 1350.41 \pm 47.75\ \Omega$.

Figure 4.19 presents the CMP and ECMP processes. The potential applied in Fig. 4.19a was 0 V vs. OCP, showing a CMP process. Figure 4.19b reveals a typical ECMP process. The average impedance in CMP is lower than that in ECMP. Although the impedance curve of the CMP is not as apparently periodic as that of the ECMP, we still can think it periodic except the period of each circulation is not identical and much shorter than that of the ECMP.

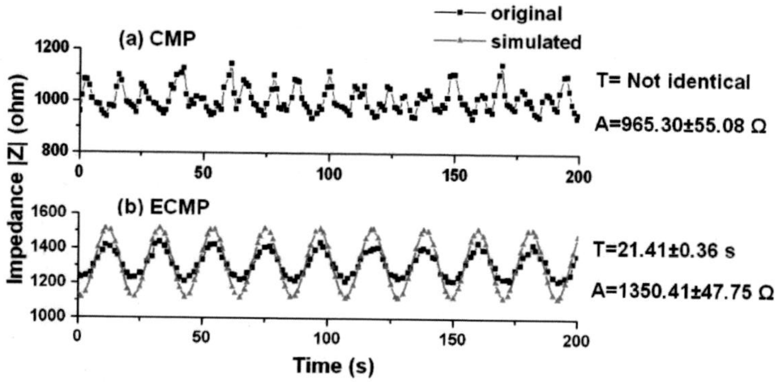

Figure 4.19 The enlarged impedance curves and sinusoidal simulation curves of CMP and ECMP when 4 N and 60 rpm were applied. (a) CMP, 0 V vs. OCP, T is not identical, $A = 965.30 \pm 55.08\ \Omega$; (b) ECMP, 1 V vs. reference, $T = 21.41 \pm 0.36$ s, $A = 1350.41 \pm 47.75\ \Omega$.

4.2.3.5.2 Mechanical power and work

During polishing, the material on the sample surface was removed mechanically by the friction force existing between the sample surface and polishing pad as well as abrasive particles. The mechanical abrasion through friction contributes to the material removal. In order to understand the material removal process, it is necessary to calculate the mechanical abrasion that can be represented by the mechanical power introduced by the friction force.

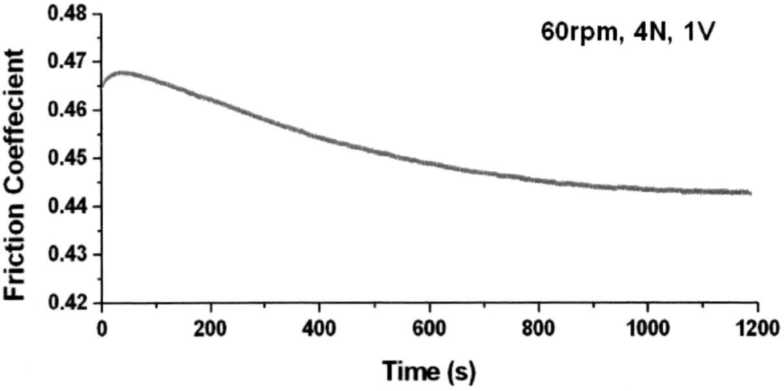

Figure 4.20 A typical friction coefficient curve obtained under the polishing conditions of 60 rpm, 4 N, 1 V.

First, we calculate the friction force. The friction force is obtained by multiplying the friction coefficient and the vertical load. The friction coefficient was recorded during CMP or ECMP. Figure 4.20 shows a typical friction coefficient curve when 60 rpm, 4 N, and 1 V were applied during polishing. The friction coefficient deceases with time. This is consistent with our previous results [24]. Since the friction coefficient was not constant during the polishing process, the average friction coefficient was obtained by using Eq. (4.4).

$$\bar{\mu} = \frac{\int_0^t \mu dt}{\tau}, \tag{4.4}$$

where $\bar{\mu}$ is the average friction coefficient, μ is the friction coefficient, and τ is the total polishing time in second. Second, calculate the average friction force with Eq. (4.5).

$$F = \bar{\mu} F_{\text{n}}, \tag{4.5}$$

where F is the average friction force in Newton, F_{n} is the vertical load. Finally, the pure mechanical power resulted from the friction force can be calculated by Eq. (4.6).

$$P = VF = 2\pi\omega RF, \tag{4.6}$$

where P is the mechanical power in watt, V is linear speed of the polished sample, ω is the rotating speed, and R is the distance (20 mm) between the sample and the spinning center of the platen.

Table 4.2 lists the calculated mechanical power induced by the friction force. Additionally, the periods and average impedance are listed for comparison. In the first group, the mechanical power increased with the increased load. The power was the highest in all polishing tests when 6 N was applied. In Group 2, the lowest mechanical power appeared at 20 rpm. The mechanical power inclined when a higher rotating speed was used. Different potentials were applied in the third group of experiments. The higher the potential, the lower the mechanical power. However, compared with data in Groups 1 and 2, the difference of the mechanical power obtained under different potentials was relatively small, i.e., less than 0.008 W. It is thus reasonable to believe that the power gained under different potentials was similar. In the fourth group, a higher concentration of the abrasives in the slurry resulted in high mechanical power. When the same load and rotating speed were applied, a similar mechanical power was obtained in CMP and ECMP processes.

In order to examine how the mechanical abrasion induced by the load and rotating speed affected the oxide layer in ECMP process, Fig. 4.21 was plotted to show their relationships. The potential was 1 V and the concentration of abrasives was 0.3 wt%. The average impedance decreased with the increase of the calculated mechanical power.

Using the technique of single-frequency EIS, the material removal during tantalum CMP and ECMP was measured carefully. Furthermore, the factors affecting the material removal were pinpointed using different polishing parameters. We will discuss this technique and the results in the following.

Table 4.2 The mechanical power and work introduced by the friction force

Group	Fixed parameters	Variables	Period (s)	Average impedance (Ω)	Power (W)
1	60 rpm/1 V	2 N	24.79 ± 2.75	1820.87 ± 131.83	0.1336 ± 0.0035
		4 N	21.41 ± 0.36	1316.65 ± 47.75	0.2237 ± 0.0048
		6 N	15.91 ± 0.28	1193.48 ± 70.14	0.3297 ± 0.0143
2	4N/1 V	20 rpm		2033.53 ± 79.23	0.0903 ± 0.0056
		50 rpm	18.57 ± 1.59	1459.19 ± 83.13	0.1962 ± 0.0043
		60 rpm	21.41 ± 0.36	1316.65 ± 47.75	0.2237 ± 0.0048
3	60 rpm/4N	2 V	21.17 ± 1.91	1551.56 ± 62.83	0.2173 ± 0.0182
		1 V	21.41 ± 0.36	1316.65 ± 47.75	0.2237 ± 0.0048
		0 V	21.08 ± 1.70	1129.97 ± 23.52	0.2315 ± 0.0002
4	60 rpm/4 N/1V	0.3 wt% abrasive	21.41 ± 0.36	1316.65 ± 47.75	0.2237 ± 0.0048
		1 wt% abrasive	18.13 ± 0.96	1166.78 ± 16.40	0.2490 ± 0.0013
5	60 rpm/4 N	CMP	—	965.30 ± 55.08	0.2355 ± 0.0137
		ECMP	21.41 ± 0.36	1316.65 ± 47.75	0.2237 ± 0.0048

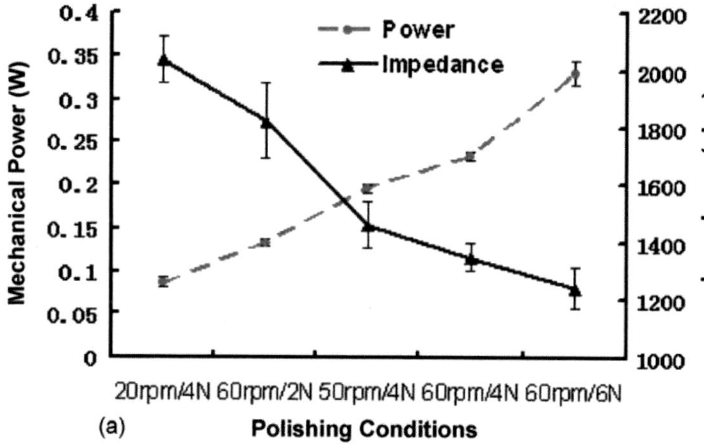

Figure 4.21a The relationship between the mechanical power and average impedance. The load and rotating speed are the only variables. The potential is 1 V and the concentration of abrasives is 0.3 wt%.

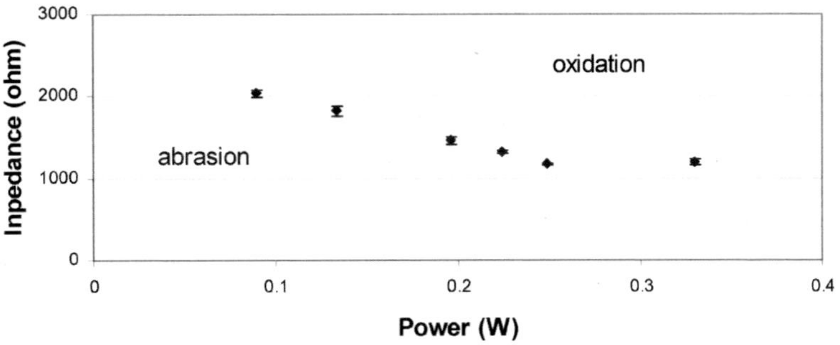

Figure 4.21b Impedance as a function of mechanical power. The curve divides the removal mechanisms into two regions. The upper is dominated by oxidation and lower by mechanical abrasion.

In single-frequency EIS, the magnitude of the impedance of tantalum oxide film can be expressed by Eq. (4.2), where a is loosely referred to the resistance of the oxide layer, and c is the nominal capacitance which can refer to the capacitance of the oxide film [15, 23, 24]. The resistance increases as the oxide layer becomes thicker [25]. The capacitance of an oxide layer can be defined as

$$C = \varepsilon_0 \varepsilon_r \frac{A}{d}, \tag{4.7}$$

where C is the capacitance, ε_r is the dielectric of the oxide layer, ε_0 is the dielectric of free space, A is the surface area of the layer, and d is its thickness. As a result, the thicker the oxide film, the smaller the capacitance, and the greater the reciprocal of the capacitance, and vice versa. The influence of the thickness of an oxide layer on the capacitance has been accepted [26]. Based on previous report and above discussion, according to Eq. (4.2), the nominal resistance a and the reciprocal of the nominal capacitance b increase or decrease simultaneously when the oxide layer thickness increases or decreases correspondingly. Consequently, the change of the impedance magnitude reflects the variation of the thickness of the oxide layer.

As shown in Fig. 4.14, the magnitude of the impedance $|Z|$ of the tantalum sample varies periodically in ECMP process under certain polishing conditions, indicating the periodical change of the thickness of the oxide layer. In ECMP, an anodic potential is applied to help the formation of the oxide layer through electrochemical reactions, making the oxide layer thicker. Meanwhile, the mechanical abrasion resulting from load, rotating speed, abrasives, and polishing pad removes the formed oxides from the metal surface, reducing the thickness of the oxide layer. The electrochemical reactions and mechanical abrasion compete, i.e., the growth and removal of the oxide layer depend on the dominating factor. The mechanical abrasion during ECMP is relatively consistent under the fixed polishing parameters, like load and rotating speed. It has been accepted that the growth rate of tantalum oxide layer was inversely proportional to the thickness [27]. At a constant potential, the growth rate of the oxide layer declines with the increase of its thickness, being shown by the exponentially decreasing current density [26]. Further electrochemical reactions are suppressed by the passive layer. On the contrary, once the oxide film is thinned, its growth rate increases. In the ECMP process, when the oxide is removed by the mechanical abrasion to a certain thickness, the electrochemical reactions are dominant to grow the oxide layer at a higher rate, resulting in the increase of the impedance. The electrochemical reactions forming a thick oxide layer are weakened by the increase of the thickness. Then, the mechanical abrasion is more pronounced than the electrochemical reactions to remove the oxide, causing the declination of the impedance. Additionally, the oxide layer cannot form or be removed instantaneously. In other words, it takes time to form and remove the oxide. As such, the increase and decrease of the oxide layer thickness are periodic, reflected by the periodic variation of the impedance. Furthermore, the periods are different due to the different thickness of the oxide layer and abrasion conditions when the polishing parameters are changed. In summary, the synergy of the electrochemical reactions and mechanical abrasion affecting the thickness of the oxide layer accomplishes the metal removal in ECMP process. This is the mechanism of the material removal in ECMP.

The mechanical abrasion is a function of the rotating speed and the friction force existing at the interface between the metal surface and polishing pad. Hence, the mechanical power calculated previously with the friction force and rotating speed can characterize the mechanical abrasion. There are two questions regarding the mechanical abrasion need to be answered. One is the thickness and the other the oxidation/removal rate. The former corresponds to the average impedance, which shows the thickness of the oxide layer, while the latter the period of the impedance curve. Let us take a bulldozer as an example to explain those two questions. In one single plough, a bulldozer with higher power can remove soil more and faster than one with lower power. Correspondingly, the mechanical abrasion with higher power can remove thicker oxide material.

Figure 4.15 shows the impedance curves obtained under different loads while the rotating speed and potential are 60 rpm and 1 V, respectively. All curves are periodic, showing the mechanism of the material removal. Comparing the data listed in Table 4.2, it is found that the higher the mechanical power, the lower the average impedance and the shorter the period. The mechanical abrasion with a high power removes more oxide layer resulting in a low impedance. As mentioned previously, the growth rate of the oxide layer is inversely proportional to its thickness at a constant potential. That means the thinner oxide film takes less time to grow to the thickness at which the electrochemical reactions are no longer dominant, resulting in a shorter period. Consequently, the tantalum sample abraded with a higher load has shorter period and thinner oxide layer shown by lower average impedance.

The impedance curves (Fig. 4.16) were obtained at different rotating speeds, as the load and applied potential were constant. They are periodic except the curve obtained at 20 rpm. The data in Table 4.2 shows that at 20 rpm the average impedance is the highest, indicating the thickest oxide layer and the mechanical power is the lowest. Only little oxide material is removed from the surface with such low mechanical power and hence the oxide growth is relatively stable. As a result, the impedance curve does not show the thickness change nor is periodic. From the Table 4.2, a higher mechanical power appears when polished at 60 rpm, compared with that at 50 rpm, inducing more oxide removal at higher rotating speed. Thus, the oxide layer is thinner at 60 rpm, reflected by the lower average impedance. The results of periods of these two impedance curves are opposite to the corresponding results that the higher the mechanical power, the shorter the period. There should be some extra power to help remove the surface material. Under a constant load, the smaller distance between the sample surface and a polishing pad is attained at lower rotating speed. Subsequently, the abrasive particles collide with the sample surface frequently due to the small space available. That may be the source of extra power. Although at 50 rpm the growth rate of the oxide layer is lower due to a thick film, the oxide thickness is less than that at 60 rpm because of the existence of the extra power. Accordingly, the period of the impedance curve obtained at 50 rpm is shorter than that at 60 rpm.

In Fig. 4.17, different potentials were applied on the tantalum sample during ECMP while the load and rotating speed were fixed at 4 N and 60 rpm. The average impedance increases with the elevated potential, showing the thickness growth of the oxide layer. Macagno and Schultze reported that the thickness of the tantalum oxide film grew linearly with the electrode potential in acidic solution [23]. It can be seen that the mechanical power obtained under different potentials in Table 4.2 is similar. That implies the mechanical removal is similar in those three cases. The growth rate of the oxide film is not only a function of its thickness, but also the applied potential. The higher the potential, the higher the growth rate due to extra energy. Although the oxide layer formed under 0 V is thinner than that under a higher potential, the growth rate of the former is lower than that of the latter in terms of lower potential. Therefore, the growth rate under different potentials is close, resulting in similar periods.

As shown in Fig. 4.18, when the concentration of abrasives increased from 0.3 wt% to 1 wt%, both average impedance and period decreased with rest polishing parameters fixed. A high concentration of the abrasive particles results in a large contact area between particles and the sample surface, causing a high friction force [28]. Hence, the mechanical power induced through friction is high. Subsequently, more oxide material is removed by mechanical abrasion with a high power. As shown earlier, the average impedance and period decreases with the increase in the amount of the abrasives.

Figure 4.19 shows the difference between CMP and ECMP as the load and rotating speed at 4 N and 60 rpm. In ECMP, both potential and H_2O_2 contribute to the formation of the oxide layer, while in CMP process tantalum is oxidized only by H_2O_2. Thus, without the contribution of the potential, the thickness of the oxide layer in CMP is much smaller than that in ECMP, reflected by the low impedance for the former. In Fig. 4.19a, it can be seen that the impedance oscillates with nonidentical period. The period obtained in the CMP process is much shorter than that in the ECMP, because the thin oxide layer of the former results in a high growth rate of the oxide layer corresponding to a stronger mechanical stimulation. The oscillation of the impedance curve in CMP reveals its material removal mechanism, the formation, removal, and reformation of the oxide material.

We have plotted the mechanical power and average impedance against the samples tested (or conditions used), as shown in Fig. 4.21a. It is interesting to see the reverse relationship between power and impedance. The average impedance reflects the thickness of the oxide layer while the mechanical power is to remove it. Here, we only consider the scenarios in which the load and rotating speed are variables because they are readily adjusted in practice. The composition of the slurry and applied potential (1 V) remained unchanged. In order to highlight the relationship, the impedance is plotted against the input power, as shown in Fig. 4.21b. The impedance reflects the thickness of the oxide layer and it is found to be a function of the mechanical power through load or rotating speed. This plot divides the synergetic process into two regions: oxidation and removal. The oxidation and mechanical removal compete with each other. The upper region is dominated by the oxidation and the lower region by the mechanical removal.

Material removal mechanisms of tantalum CMP and ECMP were investigated using a technique of single-frequency EIS on an electrochemical-combined tribometer. Single-frequency EIS was found to be able to fulfill in situ observation in tantalum CMP and ECMP by measuring the impedance of the sample. The impedance results revealed that there were two competing mechanisms in ECMP, the formation and removal of Ta oxide during polishing. The mechanical abrasion and electrochemical passivation dominate underwent synergetically reflecting the thickness/impendence change of the oxide. Our results indicated that the period of growing and removing the oxide film was a function of the mechanical power.

This research provided a new methodology in CMP and ECMP exploration. The technique of single-frequency EIS can be used to investigate other metals' polishing processes. This method can be used to optimize CMP and ECMP processes.

Acknowledgement

The authors wish to thank the support by the National Science Foundation (0535578) and Dr. Paul Fischer at Intel for his support and discussion.

References

1. D. Landolt, Electrochemical and materials aspects of tribocorrosion systems, *J. Phys. D (Appl. Phys.)*, 39(15) (2006) 3121–3127.
2. S. Mischler, A. Speigel, D. Landolt, The role of passive oxide films on the degradation of steel in tribocorrosion systems, *Wear*, 225–229(2) (1999) 1078–1087.
3. S. W. Watson, F. J. Friedersdorf, B. W. Madsen, S. D. Cramer, Methods of measuring wear-corrosion synergism, *Wear*, 181 (1995) 476–484.
4. H. Abd-El-Kader, S. M. El-Raghy, Wear-corrosion mechanism of stainless steel in chloride media, *Corrosion Sci.*, 26(8) (1986) 647–653.
5. M.-H. Hong, S.-I. Pyun, Corrosive wear behaviour of 304-L stainless steel in 1 N H_2SO_4 solution part 2. Effect of chloride ion concentration, *Wear*, 147(1) (1991) 69–78.
6. M.-H. Hong, S.-I. Pyun, Corrosive wear behaviour of 304-L stainless steel in 1 N H_2SO_4 solution Part 1. Effect of applied potential, *Wear*, 147 (1991) 59–67.
7. Y. Yahagi, Y. Mizutani, Corrosive wear of cast iron in sulphuric acid, *J. Tribol.*, 109(2) (1987) 238–242, doi:10.1115/1.3261345.
8. V. Mitrovic-Scepanovic, B. MacDougall, M. J. Graham, Nature of passive films on $Fe_{26}Cr$ alloy, *Corrosion Sci.*, 24 (1984) 479–490.
9. G. Okamoto, Passive film of 18-8 stainless steel structure and its function, *Corrosion Sci.*, 13(6) (1973) 471–489.
10. S. Sato, Z. Yasuda, M. Ishihara, N. Komai, H. Ohtorii, A. Yoshio, Y. Segawa, H. Horikoshi, Y. Ohoka, K. Tai, S. Takahashi, T. Nogami, Newly developed electro-chemical polishing process of copper as replacement of CMP suitable for damascene copper inlaid in fragile

low-k dielectrics, *Proceedings of the IEEE International Electron Devices Meeting IEDM 2001* (2001) 4.4.1–4.4.4.

11. L. Economikos, X. Wang, A. Sakamoto, P. Ong, M. Naujok, R. Knarr, L. Chen, Y. Moon, S. Neo, J. Salfelder, A. Duboust, A. Manens, W. Lu, S. Shrauti, F. Liu, S. Tsai, W. Swart, Integrated electro-chemical mechanical planarization (ECMP) for future generation device technology, *Proceedings of the IEEE, Interconnect Technology Conference 2004* (2004) 233–235.
12. Q. L. Feng, D. Tianbao, A. Duboust, S. Tsai, H. Wei-Yung, Cu Planarization in Electrochemical Mechanical Planarization, *J. Electrochem. Soc.*, 153(6) (2006) C377–C381.
13. I. I. Suni, D. Bing, Cu Planarization for ULSI Processing by Electrochemical Methods: A Review, *IEEE Trans. Semicond. Manuf.*, 18(3) (2005) 341–349.
14. F. Gao, H. Liang, Material removal mechanisms in electrochemical–mechanical polishing of tantalum, *Electrochim. Acta*, 54 (2009) 6808–6815.
15. F. Gao, H. Liang, In situ observation of friction-induced electrochemical reactions and impedance in tantalum ECMP semiconductor devices, materials, and processing, *J. Electrochem. Soc.*, 156 (2009) H80–H86.
16. F. Gao, H. Liang, Transformable oxidation of tantalum in electrochemical mechanical polishing (ECMP), *J. Electron. Mater.*, 40 (2011) 134–140.
17. C. M. Pettit, D. Roy, Role of iodate ions in chemical mechanical and electrochemical mechanical planarization of Ta investigated using time-resolved impedance spectroscopy, *Mater. Lett.*, 59 (2005) 3885–3889.
18. D. Ng, T. Sen, F. Gao, H. Liang, Friction and wear-mode comparison in copper electrochemical mechanical polishing, *J. Electrochem. Soc.*, 155(7) (2008) H520–H524.
19. F. Guanghui, A. Chandra, S. Guha, G. Subhash, A plasticity-based model of material removal in chemical-mechanical polishing (CMP), *IEEE Trans. Semicond. Manuf.*, 14 (2001) 406–417.
20. K. Qin, B. Moudgil, C.-W. Park, A chemical mechanical polishing model incorporating both the chemical and mechanical effects, *Thin Solid Films*, 446(2) (2004) 277–286.
21. D.-H. Eom, J.-S. Ryu, Y.-K. Hong, J.-J. Myung, K.-S. Kim, and J.-G. Park, Electrochemical and frictional behaviors of Cu surface in organic acid based slurries during copper CMP, *Electrochemical Society Proceedings,* 2003–21 (2003) 44–51.

22. H. J. Kim, H. Y. Kim, H. D. Jeong, E. S. Lee, Y. J. Shin, Friction and thermal phenomena in chemical mechanical polishing, *J. Mater. Proc. Technol.*, 130–131 (2002) 334–338.

23. G. W. Walter, A comparison of single frequency and wide frequency range impedance tests for painted metals, *Corrosion Sci.*, 30(6–7) (1990) 617–629.

24. A. Foyet, T. H. Wu, L. van der Ven, A. Kodentsov, G. de With, R. van Benthem, Influence of mixing ratio on the permeability of water and the corrosion performance of epoxy/amine coated un-pretreated Al-2024 evaluated by impedance spectroscopy, *Prog. Org. Coat.*, 64 (2009) 138–141.

25. S. N. Wosu, Anodic oxidation of tantalum in water and biological solutions using current limiting constant voltage method, *J. Mater. Sci.*, 42(11) (2007) 4087–4097.

26. Y. M. Li, L. Young, Effect of incorporated electrolyte species on the ac response of the high field ionic conduction process in anodic oxide films on tantalum, *Electrochim. Acta*, 44 (1998) 605–611.

27. K. Lehovec, J. D'Amico, Kinetics of Tantalum Corrosion in Aqueous Electrolytes, I, *J. Electrochem. Soc.*, 114 (1967) 363–367.

28. W. Choi, J. Abiade, S.-M. Lee, R. K. Singh, Effects of slurry particles on silicon dioxide CMP, *J. Electrochem. Soc.*, 151 (2004) G512–G522.

Chapter 5

Concluding Remarks and Prospects: Toward the Unified Understanding of Tribocatalysis, Tribochemistry, and Tribocorrosion

Ken'ichi Hiratsuka[a] and Czesław Kajdas[b]

[a]*Department of Innovative Mechanical and Electronic Engineering, Chiba Institute of Technology, Japan*
[b]*Automotive Industry Institute in Warsaw, Poland*

hiratsuka@sea.it-chiba.ac.jp

Tribology is an interdisciplinary subject in science and engineering. It covers vast areas from atomic level to cosmic one, because friction, wear, and lubrication exist wherever two materials are in sliding/rolling contact. In spite of wide areas concerned, research themes often focus only on the limited areas in a particular tribological system and conditions. As a consequence, research is categorized and further subdivided into smaller fields. One can notice this tendency from long research paper titles in the recent years. Unfortunately, the division of research leaves a number of "gaps" between the studies. Many papers, particularly review ones, consider only mechanochemistry, totally neglecting tribochemistry. For instance, very extensive overview paper by Boldyrev [1] considers and discusses only mechanochemistry and mechanical activation issues.

Tribocatalysis, Tribochemistry, and Tribocorrosion
Edited by Czesław Kajdas and Ken'ichi Hiratsuka

ISBN 978-981-4316-95-8 (Hardcover), 978-1-315-36498-8 (eBook)
www.panstanford.com

Although reactions of polymerization of various classes of organic compounds have been recently discovered, and the dramatic acceleration of the polymerization with respect to analogous reactions in the liquid phase was achieved, tribochemical reactions have not been mentioned in the above-discussed overview work by Boldyrev. On the other hand, in 2006, Z. V. Todres [2] published a book on organic mechanochemistry and its practical applications, which considers both mechanochemistry and tribochemistry.

Interestingly to note here, the book mentioned in Chapter 4 in ref. [2], *Mechanically Induced Organic Reactions*, presents and discusses in detail mechanisms of mechanochemically initiated polymerization, depolymerization, and mechanolysis. Very special interdisciplinary research on mechanisms of tribopolymerization is presented in paper [3]. Gutman and Bobovitch [4] presented the polymerization process of crystalline pentabromobenzyl acrylate with simultaneous grafting of the polymer onto magnesium hydroxide filler.

Negative-ion-radical action mechanism (NIRAM) [5] is in line with the NIRAM-HSAB action mechanism [6]. Chapter 4 [2] stresses that polymers can also be made by vibromilling monomers with steel balls, and such polymerization is initiated under the action of electronic stream, generated by mechanoemission under the vibratory milling. Simionescu et al. [7] have demonstrated that applying the vibratory milling technique, acryl and methacryl amides produce anion radicals, being key species in the following reaction type:

$CH_2{=}CRCONH_2$ + electron (e) $\rightarrow$ $(CH_2{=}CRCONH_2)^*$ – anion-radical

and the further growth of the polymer goes in the regular way. Chapter 7 in the book on surface modification and mechanisms [8] overviews all details relating to tribopolymerization and tribopolycondensation processes. Other papers relate to mechanochemistry of molecular organic crystals [9] and the nature of mechanochemical reactions [10].

On the other hand, mechanochemistry of catalysts has been taken into account by Molchanov and Buyanov [11]. They emphasized that the use of mechanochemical activation for the preparation of catalysts is important but usually generalized. It is clearly noted that the effect of mechanochemical activation on the catalytic properties is important for various systems. The hypotheses concerning the nature of the effect of mechanochemical activation on the reactivity (including catalytic properties) are considered and well supported by many references [12–20].

In spite of a number of gaps among studies, the underlying principles of physics and chemistry are the same in all fields in tribology. Thus, it is much better to find the similarities in other fields in order to bridge the gap.

For instance, wear without any lubricant is usually called dry wear. However, this does not mean that materials are worn out in real dry air. In almost all the "dry" conditions, the atmosphere contains some amount of water vapor. It means that wear tests are still conducted in "wet" conditions even when the humidity is low. The relative humidity in laboratory air atmosphere is generally adjusted from 40% to 50%. In these situations, friction surface is adequately covered by water layers, when the surface is not in contact with the mating surface but open to the atmosphere. Although the water layer is not visible, it would affect wear. From this point of view, dry wear in air environment is to be discussed in terms of tribocorrosion, i.e., effect of water on wear should be incorporated. The terms of active surface and passive surface in corrosion can indicate severe wear and mild wear in adhesive wear, respectively.

In this book, the editors selected three subjects, tribocatalysis, tribochemistry, and tribocorrosion, and tried to understand these tribological fields in terms of a common denominator, which is electron emission/transfer at the interface.

In corrosion science, electrochemistry is a powerful tool to analyze phenomena and is applied to tribocorrosion. Although the present state of its understanding is by far incomplete, the electrochemical approach, as represented by the redox process, will be useful. A new approach to such processes has been recently presented in work [6]. It is focused on the novel and better understanding of tribochemistry by combining redox process with the NIRAM concept. The redox process is very important because lubricant used in tribocorrosion is conductive and controllable electrochemically from the outside.

On the contrary, in tribochemistry, the electrochemical approach has not been fully developed. This is due to the fact that lubricants are often used as electrical insulants. However, most chemical reactions are governed by the movement of electrons, although it is complicated to measure from the outside. Similar to the tribocorrosion governed by electron transfer, the tribochemical phenomena should be attributed to the transfer/emission of electrons. In fact, electron emission is measured in a number of friction experiments. Unfortunately, it is rather difficult to measure electron emission simultaneously with chemical reactions. This is because electrons are just consumed in the course of chemical reactions, as in the case of the redox process in corrosion where electrons are transferred from one side to another. Tribochemical reactions are now expected to be applied in chemical reactions more extensively. Its understanding in terms of electrochemistry will be a key to enhancing their activity.

A catalyst is widely used to save energy for chemical reactions as well as to alter the final product. A number of reactions that are enhanced by friction may be related to tribocatalysis, because any substance can be a catalyst when appropriate energy is input. The acceptance of emitted electron can activate a catalyst to increase reaction rates as discussed in Chapter 2.

New discovery in other fields of science can be applied to one's own science. This would also be true for the three fields in tribology chosen in this book, because underlying mechanisms are common.

References

1. V. V Boldyrev, Mechanochemistry and mechanical activation of solids, *Russ. Chem. Rev.,* 75(3) (2006) 177–189 doi:10.1070/RC2006v075n03ABEH001205.
2. Z. V. Todres, *Organic Mechanochemistry and Its Practical Applications,* CRS Taylor & Francis Group, Boca Raton, Florida, USA (2006).
3. J. Zheng, L. Zhang, Z. Du, C. Zhang, H. Li, Tribopolymerization of n-butyl acrylate on the steel-steel rubbing surface, *Tribol. Int.,* 41 (2008) 769–777.
4. E. M. Gutman, A. Bobovitch, Mechanopolymerization of pentabromobenzyl (mono) acrylate, *Int. J. Mechanochem. Mech. Alloying,* 1(3) (1994) 153–158.

5. C. Kajdas, Importance of anionic reactive intermediates for lubricant component reactions with friction surfaces, *Lubrication Sci.,* 6(3) (1994) 203–228.

6. C. Kajdas, A novel approach to tribochemical reactions: Generalized NIRAM-HSAB action mechanism, *Proceedings of the ITC Yokohama, Satellite Forum on Tribochemistry,* Tokyo Institute of Technology (1998) 31–35.

7. C. Simionescu, C. V. Oprea, J. Nicoleanu, Mechanochemically initiated polymerization-5. polymerization by vibratory milling of acrylamide and methacrylamide, *Eur. Polym. J.,* 19, 525–528. http://dx.doi.org/10.1016/0014-3057 (83) 90204-5.

8. M. J. Furey, C. Kajdas, Tribopolymerization as a mechanism of boundary lubrication, in *Surface Modification and Mechanisms* (G. E. Totten, H. Liang, eds.), Marcel Dekker Inc., New York, Basel 2004. Chapter 7 (2004) pp. 165–201.

9. E. A. Varentsov, Y. A. Khrustalev, Mechano-emission and mechanochemistry of molecular organic crystals (in Russian), *Uspehi Khimii (Adv. Chem.)* 64 (1995) 834–839.

10. P. Y. Butiagin, Kinetics and nature of mechano-chemical reactions (in Russian), *Chem. Rev.,* 40 (1971) 1935–194.

11. V. V. Molchanov, R. A. Buyanov, Mechanochemistry of catalysts, *Russ. Chem. Rev.,* 69(5) (2000) 435–450. DOI 10.1070/RC2000v069n05ABEH000555.

12. D. B. Knorr, Jr., T. O. Gray, R. M. Overney, Cooperative and submolecular dissipation mechanisms of sliding friction in complex organic systems, *J. Chem. Phys.,* 129 (2008) 074504. DOI: 10.1063/1.2968548.

13. S. Ikeda, T. Takata, M. Komada, M. Hara, J. Kondo, K. Domen, A. Tanaka, H. Hosono, H. Kawazoe, Mechano-catalysis: A novel method for overall water splitting, *Phys. Chem. Chem., Phys.,* 1 (1999) 4485–4491.

14. K. Hiratsuka, M. Kuzuya, T. Sasada, Friction catalysis in the synthesis of H_2O during adhesive wear, *Proceeding of the 33rd Japan Congress on Materials Research,* March, (1990) pp. 191–195.

15. C. Kajdas, K. Hiratsuka, Tribochemistry, tribocatalysis, and the negative-ion-radical action mechanism, *Proc. IMechE Part J: J. Eng. Tribol.,* 223 (2009) 827–848.

16. K. Hiratsuka, C. Kajdas, Mechanochemistry as a key to understand the mechanisms of boundary lubrication, mechanolysis and gas evolution during friction, *Proc. IMechE Part J: J Eng. Tribol.,* 227(11) (2013) 1191–1203 DOI: 10.1177/1350650113483222pij.sagepub.com.

17. A. Kulczycki, C. Kajdas, A new attempt to better understand Arrhenius equation and its activation energy, in: *Tribology in Engineering* (H. Pihtili, ed.), Chapter 4 (2013). ISBN 978-953-51-1126-9. http://dx.doi.org/10.5772/54503.
18. R. C. Dante, C. Kajdas, A. Kulczycki, Theoretical advances in the kinetics of tribochemical reactions, *React. Kinetics Mech. Catal.,* 99 (2010) 37–46.
19. C. Kajdas, Physical and chemical phenomena related to tribochemistry, in: *Advances in Contact Mechanics: Implications for Materials Science, Engineering and Biology* (R. Buzio, U. Valbusa, eds.), Transworld Research Network, Kerala, India, 2006, Chapter 12 (2006) pp. 383–412.
20. I. C. Engin, Theories on rock cutting, grinding and polishing mechanisms, in: *Tribology in Engineering* (H. Pihtili, ed.), Chapter 10 (2013). ISBN 978-953-51-1126-9. http://dx.doi.org/10.5772/56046.

Index